AF352902

Assessment and Remediation of Soils Contaminated by Potentially Toxic Elements (PTE)

Assessment and Remediation of Soils Contaminated by Potentially Toxic Elements (PTE)

Editors

Matteo Spagnuolo
Paola Adamo
Giovanni Garau

MDPI • Basel • Beijing • Wuhan • Barcelona • Belgrade • Manchester • Tokyo • Cluj • Tianjin

Editors
Matteo Spagnuolo
Department of Soil, Plant and
Food Sciences, University of
Bari "A. Moro",
Bari, Italy

Paola Adamo
Department of Agricultural
Sciences, University of Napoli
"Federico II",
Portici (Napoli), Italy

Giovanni Garau
Department of
Agricultural Sciences,
University of Sassari,
Sassari, Italy

Editorial Office
MDPI
St. Alban-Anlage 66
4052 Basel, Switzerland

This is a reprint of articles from the Special Issue published online in the open access journal *Soil Systems* (ISSN 2571-8789) (available at: https://www.mdpi.com/journal/soilsystems/special_issues/AssessmentRemediation_SoilsContaminated_PTE).

For citation purposes, cite each article independently as indicated on the article page online and as indicated below:

LastName, A.A.; LastName, B.B.; LastName, C.C. Article Title. *Journal Name* **Year**, *Volume Number*, Page Range.

ISBN 978-3-0365-5505-8 (Hbk)
ISBN 978-3-0365-5506-5 (PDF)

© 2022 by the authors. Articles in this book are Open Access and distributed under the Creative Commons Attribution (CC BY) license, which allows users to download, copy and build upon published articles, as long as the author and publisher are properly credited, which ensures maximum dissemination and a wider impact of our publications.
The book as a whole is distributed by MDPI under the terms and conditions of the Creative Commons license CC BY-NC-ND.

Contents

About the Editors

Matteo Spagnuolo

Prof. Matteo Spagnuolo is Associate Professor in Agricultural Chemistry at the Department of Soil, Plant and Food Science of the University of Bari. He received a degree (equivalent of M.Sc. degree) in Agricultural Chemistry in 1994 from the University of Bari with 110/110 and honours and earned his PhD in 1998 in Agricultural Chemistry. In the academic year 1998/1999, he spent one year at Cornell University as Post Doc working on a project of soil remediation. An important part of the research activity concerned environmental chemistry. The transformation processes of different organic xenobiotics and pesticides has been studied in the presence of inorganic soil components. He deeply investigated the processes influencing the bioavailability of organic and inorganic pollutants in soil systems with EPR and X-ray spectroscopic speciation studies, resins and earthworms. Most of the studies are being conducted at our settled "micro X-ray LAB" (www.microxraylab.com).

Paola Adamo

Prof. Paola Adamo is a Full Professor of Agricultural Chemistry at the Department of Agricultural Sciences of the Naples University Federico II. Her main research topics are: (i) monitoring and mitigation of soils and sediments pollution in agricultural, urban and industrial areas; (ii) biomonitoring of air quality by lichens and mosses; (iii) soil–plant interactions and effects on nutrient biogeochemical cycling; (iv) food authenticity and traceability by NIR spectroscopy, multielement analysis and isotopic profiling combined with chemometrics; (v) bioweathering of rocky substrates and biomineralization. President-Elect (2019–2020) and Past-President (2021–2022) of the Italian Society of Soil Science. Director of the Museum of Mineralogy 'Antonio Parascandola' and member of the Scientific and Technical Committee of the Museum Centre "Agricultural Science Museums"—MUSA of Naples Federico II University. Peer-reviewed publications (ISI-JCR) with Impact Factor [132]. H-index: 41; Citations: 4920 (Scopus September 2022).

Giovanni Garau

Giovanni Garau is Associate Professor of Agricultural Chemistry at the Department of Agricultural Sciences of the University of Sassari (UNISS, Italy). He has got a PhD in Microbial Biotechnologies (UNISS, 2004) and, as a visiting PhD student, he undertook research at the *Centre for Rhizobium Studies* of the Murdoch University of Perth (Western Australia, Oct 2002–July 2003). In 2007, he spent almost one year as Post Doc at Murdoch University (Perth, WA) investigating *quorum sensing* in rhizobia and characterizing new symbiotic N-fixing *Burkholderia* adapted to acidic soils. Since 2008, he obtained several Post Doc research fellowships at UNISS mainly addressing soil chemistry, biochemistry and microbiology aspects. In particular, his research activity mainly focuses on the mobility, bioavailability and bioaccessibility of potentially toxic elements in polluted soils, their impact on soil microbial communities and soil functioning and sustainable remediation approaches.

Editorial

Assessment and Remediation of Soils Contaminated by Potentially Toxic Elements (PTE)

Matteo Spagnuolo [1,*], Paola Adamo [2] and Giovanni Garau [3]

[1] Department of Soil, Plant and Food Sciences, University of Bari A. Moro, 70126 Bari, Italy
[2] Department of Agricultural Sciences, University of Napoli Federico II, 80055 Portici, Italy; paola.adamo@unina.it
[3] Department of Agricultural Sciences, University of Sassari, 07100 Sassari, Italy; ggarau@uniss.it
* Correspondence: matteo.spagnuolo@uniba.it

Citation: Spagnuolo, M.; Adamo, P.; Garau, G. Assessment and Remediation of Soils Contaminated by Potentially Toxic Elements (PTE). *Soil Syst.* **2022**, *6*, 55. https://doi.org/10.3390/soilsystems6020055

Received: 6 June 2022
Accepted: 9 June 2022
Published: 15 June 2022

Publisher's Note: MDPI stays neutral with regard to jurisdictional claims in published maps and institutional affiliations.

Copyright: © 2022 by the authors. Licensee MDPI, Basel, Switzerland. This article is an open access article distributed under the terms and conditions of the Creative Commons Attribution (CC BY) license (https://creativecommons.org/licenses/by/4.0/).

Potentially toxic elements (PTE) can cause significant damage to the environment and human health in the functions of mobility and bioavailability [1]. Given the urgency to remediate polluted soils all over the world, appropriate innovative and sustainable remediation strategies need to be developed, assessed, and promoted [2–4].

Before that, a detailed knowledge of PTE bioavailability and bioaccessibility as well as of soil processes affecting contaminant dynamics, in terms of lixiviation, colloidal transport, redox conditions, or microbial activity, is essential in order to assess the actual danger/risk posed by contamination [5]. It is widely recognized that the bioavailability of toxic elements in soils depends on their solubility and geochemical forms, rather than on their origin and total concentration. Therefore, the knowledge of their spatial distribution and chemical speciation in soil is of paramount importance to perform an accurate risk assessment. Investigating these aspects requires the use of analytical techniques able to solve the high complexity of the soil matrix with a spatial resolution down to the micrometer—or even nanometer—scale [6].

In addition, a correct evaluation of remediation intervention requires detailed knowledge of the geochemical forms into which PTE have been converted following the soil treatment. This information is crucial to predict any possible transformation PTEs might naturally undergo over time or as consequence of physical–chemical perturbations that might impact the soil system.

In this Special Issue we invited the submission of articles to address the assessment of PTE contamination in soil systems using innovative approaches, the study of soil processes affecting pollutant dynamics, and the application of new sustainable remediation techniques for the long-term reduction in the threat posed by PTE towards the health of the human being and the environment. This volume contains ten original research articles. Four articles deal with the assessment of bioavailability of PTEs in contaminated soils [7–10], three articles report results on the application of phytoremediation to PTEs contaminated soils [11–13], one paper is related to the source–sink relationships of PTEs at basin scale [14], and two manuscripts address the issue of PTEs contamination in urban soils [15,16].

The assessment of the risk posed by the presence of PTEs in soil has been studied by Porfido et al. [7] investigating the Pb availability in a former polluted shooting range. Micro-XRF and SEM-EDX analyses showed that most of the Pb underwent stabilization processes: a weathering crust (mixture of orthophosphates) around Pb-containing bullet slivers dispersed within the soil. Moreover, no toxicity effects and low bioavailability were measured in earthworm tissues. Kaur et al. [8] assessed the risk of the presence of several PTEs in industrial effluents and soils through *Allium cepa* root chromosomal aberration assay and the potential ecological and human health risks and bioaccumulation in plants, respectively. The study of Diquattro et al. [9] assessed the mobility, phytotoxicity, and bioavailability of antimony (Sb) in soils after the addition of municipal solid waste compost (MSWC). The Sb mobility decreased in amended soils as well as phytotoxicity in triticale plants, whereas soil metabolic activity and catabolic diversity increased. Ahmad et al. [10] assessed the phytoextraction of PTEs by different vegetable crops in soil irrigated with city wastewater, evidencing the possibility of using some species for phytoremediation as well as the significant risk to human health and the environment due to PTE content in their tissues.

The adoption of a sustainable remediation strategy was proposed by Gorelova et al. [11] in a study of the bioaccumulation of PTEs in *Echinochloa frumentacea* grown in different contaminated soils. Results obtained by chemical, biochemical, microbiological, and metagenomic (16S rRNA) methods of analysis recommend *E. frumentacea* for phytoremediation of PTEs contaminated soils. Pietrini et al. [12] confirmed the crucial role of plant–microbe interaction in the phytoremediation of PTEs polluted soil by investigating the inoculation of microcosms of *Brassica juncea* and *Helianthus annuus* with a selected microbial consortium. Adopting a phytoextraction strategy, Fedje et al. [13] used sunflowers and rapeseed to extract Zn from the mineral fraction of the incinerator bottom ash in order to meet the increasing worldwide demand of the metal.

The acquisition of soil and sediment geochemical data in a basin located in the eastern Amazon enabled the source distribution of PTEs content and evidenced that local anomalies were mostly influenced by the predominant lithology rather than any anthropogenic impact [14].

Finally, two articles studied the source and distribution of PTEs in soils of two important cities. Rate [15] performed a spatial statistics analysis to define geochemical zones characterized by the presence of PTEs because of historical waste disposal in public recreation areas in Perth, Western Australia. Silva et al. [16] determined the soil PTEs content in six locations (traffic zone, residential area, urban park, and mixed areas) of the city of Lisbon (Portugal), evidencing the low levels of pollution.

We would like to thank all contributing authors in this Special Issue on "Assessment and remediation of soils contaminated by potentially toxic elements (PTEs)" and all reviewers who dedicated their time and constructive efforts to improve the quality of science during the review process.

Funding: This research received no external funding.

Conflicts of Interest: The authors declare no conflict of interest.

References

1. Roberts, D.; Nachtegaal, M.; Sparks, D.L. Speciation of metals in soils. In *Chemical Processes in Soil*; SSSA Book Series; Tabatabai, M.A., Sparks, D.L., Eds.; Soil Science Society of America: Madison, WI, USA, 2005; Volume 8.
2. Bolan, N.; Kunhikrishnan, A.; Thangarajan, R.; Kumpiene, J.; Park, J.; Makino, T.; Kirkham, M.B.; Scheckel, K. Remediation of heavy metal(loid)s contaminated soils-To mobilize or to immobilize? *J. Hazard. Mater.* **2014**, *266*, 141–166. [CrossRef] [PubMed]
3. Fiorentino, N.; Ventorino, V.; Rocco, C.; Cenvinzo, V.; Agrelli, D.; Gioia, L.; Di Mola, I.; Adamo, P.; Pepe, O.; Fagnano, M. Giant reed growth and effects on soil biological fertility in assisted phytoremediation of an industrial polluted soil. *Sci. Total Environ.* **2017**, *575*, 1375–1383. [CrossRef] [PubMed]
4. Garau, G.; Silvetti, M.; Vasileiadis, S.; Donner, E.; Diquattro, S.; Deiana, S.; Lombi, E.; Castaldi, P. Use of municipal solid wastes for chemical and microbiological recovery of soils contaminated with metal(loid)s. *Soil Biol. Biochem.* **2017**, *111*, 25–35. [CrossRef]
5. Kim, R.Y.; Yoon, J.K.; Kim, T.S.; Yang, J.E.; Owens, G.; Kim, K.R. Bioavailability of heavy metals in soils: Definitions and practical implementation—A critical review. *Environ. Geochem. Health* **2015**, *37*, 1041–1061. [CrossRef] [PubMed]
6. Terzano, R.; Santoro, A.; Spagnuolo, M.; Vekemans, B.; Medici, L.; Janssens, K.; Göttlicher, J.; Denecke, M.A.; Mangold, S.; Ruggiero, P. Solving Mercury (Hg) Speciation in Soil Samples by Synchrotron X-ray Microspectroscopic Techniques. *Environ. Pollut.* **2010**, *158*, 2702–2709. [CrossRef] [PubMed]
7. Porfido, C.; Gattullo, C.E.; Allegretta, I.; Fiorentino, N.; Terzano, R.; Fagnano, M.; Spagnuolo, M. Investigating Lead Bioavailability in a Former Shooting Range by Soil Microanalyses and Earthworms Tests. *Soil Syst.* **2022**, *6*, 25. [CrossRef]
8. Kaur, J.; Bhatti, S.S.; Bhat, S.A.; Nagpal, A.K.; Kaur, V.; Katnoria, J.K. Evaluating potential ecological risks of heavy metals of textile effluents and soil samples in vicinity of textile industries. *Soil Syst.* **2021**, *5*, 63. [CrossRef]
9. Diquattro, S.; Garau, G.; Garau, M.; Lauro, G.P.; Pinna, M.V.; Castaldi, P. Effect of municipal solid waste compost on antimony mobility, phytotoxicity and bioavailability in polluted soils. *Soil Syst.* **2021**, *5*, 60. [CrossRef]

10. Ahmad, I.; Malik, S.A.; Saeed, S.; Rehman, A.U.; Munir, T.M. Phytoextraction of Heavy Metals by Various Vegetable Crops Cultivated on Different Textured Soils Irrigated with City Wastewater. *Soil Syst.* **2021**, *5*, 35. [CrossRef]
11. Gorelova, S.V.; Muratova, A.Y.; Zinicovscaia, I.; Okina, O.I.; Kolbas, A. Prospects for the Use of *Echinochloa frumentacea* for Phytoremediation of Soils with Multielement Anomalies. *Soil Syst.* **2022**, *6*, 27. [CrossRef]
12. Pietrini, I.; Grifoni, M.; Franchi, E.; Cardaci, A.; Pedron, F.; Barbafieri, M.; Petruzzelli, G.; Vocciante, M. Enhanced Lead Phytoextraction by Endophytes from Indigenous Plants. *Soil Syst.* **2021**, *5*, 55. [CrossRef]
13. Fedje, K.K.; Edvardsson, V.; Dalek, D. Initial Study on Phytoextraction for Recovery of Metals from Sorted and Aged Waste-to-Energy Bottom Ash. *Soil Syst.* **2021**, *5*, 53. [CrossRef]
14. Salomão, G.N.; de Lima Farias, D.; Sahoo, P.K.; Dall'Agnol, R.; Sarkar, D. Integrated Geochemical Assessment of Soils and Stream Sediments to Evaluate Source-Sink Relationships and Background Variations in the Parauapebas River Basin, Eastern Amazon. *Soil Syst.* **2021**, *5*, 21. [CrossRef]
15. Rate, A.W. Spatial Analysis of Soil Trace Element Contaminants in Urban Public Open Space, Perth, Western Australia. *Soil Syst.* **2021**, *5*, 46. [CrossRef]
16. Silva, H.F.; Silva, N.F.; Oliveira, C.M.; Matos, M.J. Heavy Metals Contamination of Urban Soils—A Decade Study in the City of Lisbon, Portugal. *Soil Syst.* **2021**, *5*, 27. [CrossRef]

soil systems

Article

Integrated Geochemical Assessment of Soils and Stream Sediments to Evaluate Source-Sink Relationships and Background Variations in the Parauapebas River Basin, Eastern Amazon

Gabriel Negreiros Salomão [1,*], Danielle de Lima Farias [2], Prafulla Kumar Sahoo [3,4], Roberto Dall'Agnol [1,4] and Dibyendu Sarkar [5,*]

[1] Programa de Pós-graduação em Geologia e Geoquímica (PPGG), Instituto de Geociências (IG), Universidade Federal do Pará (UFPA), Belém 66075-110, Pará, Brazil; roberto.dallagnol@itv.org
[2] Secretaria de Estado de Meio Ambiente e Sustentabilidade (SEMAS), Belém 66093-677, Pará, Brazil; danielle.ambiental@hotmail.com
[3] Department of Environmental Science and Technology, Central University of Punjab, Ghudda, Bathinda 151401, Punjab, India; prafulla.iitkgp@gmail.com
[4] Instituto Tecnológico Vale (ITV), Belém 66055-090, Pará, Brazil
[5] Department of Civil, Environmental and Ocean Engineering, Stevens Institute of Technology, Hoboken, NJ 07030, USA
*   Correspondence: salomao.gn@gmail.com (G.N.S.); dsarkar@stevens.edu (D.S.)

Citation: Salomão, G.N.; Farias, D.d.L.; Sahoo, P.K.; Dall'Agnol, R.; Sarkar, D. Integrated Geochemical Assessment of Soils and Stream Sediments to Evaluate Source-Sink Relationships and Background Variations in the Parauapebas River Basin, Eastern Amazon. *Soil Syst.* **2021**, *5*, 21. https://doi.org/10.3390/soilsystems5010021

Academic Editors: Matteo Spagnuolo, Paola Adamo and Giovanni Garau

Received: 15 February 2021
Accepted: 17 March 2021
Published: 22 March 2021

Publisher's Note: MDPI stays neutral with regard to jurisdictional claims in published maps and institutional affiliations.

Copyright: © 2021 by the authors. Licensee MDPI, Basel, Switzerland. This article is an open access article distributed under the terms and conditions of the Creative Commons Attribution (CC BY) license (https://creativecommons.org/licenses/by/4.0/).

Abstract: This study aims to handle an integrated evaluation of soil and stream sediment geochemical data to evaluate source apportionment and to establish geochemical threshold variations for Fe, Al, and 20 selected Potentially Toxic Elements (PTE) in the Parauapebas River Basin (PB), Eastern Amazon. The data set used in this study is from the Itacaiúnas Geochemical Mapping and Background Project (ItacGMBP), which collected 364 surface soil (0–10 cm) samples and 189 stream sediments samples in the entire PB. The <0.177 mm fraction of these samples were analyzed for 51 elements by ICP-MS and ICP-AES, following an aqua regia digestion. The geochemical maps of many elements revealed substantial differences between the north (NPB) and the south (SPB) of PB, mainly due to the geological setting. The new statistically derived threshold values of the NPB and SPB regions were compared to the threshold of the whole PB, reported in previous studies, and to quality guidelines proposed by Brazilian environmental agencies. The natural variation of geochemical background in soils and stream sediments of PB should be considered prior to defining new guideline values. At the regional scale, the local anomalies are mostly influenced by the predominant lithology rather than any anthropogenic impact.

Keywords: surface geochemistry; geoprocessing; geogenic enrichment; Itacaiúnas river watershed; Carajás mineral province

1. Introduction

Multi-element geochemical mapping is useful for both, mineral exploration surveys and environmental studies as previously demonstrated [1–4]. For both applications, the unusually high concentrations (also known as anomalies) are strong pieces of evidence of a potential mineral deposit or a possible anthropogenic impact. Evaluating the spatial distribution of an element, in a given sampling medium (e.g., soils, sediments, water, and rocks), along with different geographic information (land cover and land use, hydrology, geomorphology, geology, and mineralized zones) may assist in accurately identifying the source of enrichment (for some authors, contamination), whether anthropic or not [5–7].

Traditional studies in environmental geochemistry usually carry out impact assessments separately for each sampling medium [8–10]. Taking this into consideration, an integrated assessment, using geochemical data from different sampling mediums, might be a powerful strategy for identifying source apportionment and evaluating the geochemical variability in terrestrial environments. In relation to the large geochemical data set, this strategy became more affordable due to recent advances in data analysis using a robust programming language in addition to geoprocessing techniques. Similar approaches have been conducted elsewhere [11–13].

Prior to conducting an integrated assessment methodology, it is indispensable to understand the limitations in terms of physicochemical properties and representativeness, regarding the medium of sampling [14]. For instance, soil samples have local representativeness and medium (A horizon) to low (B horizon) susceptibility in terms of changes in local conditions. In general, soils have a strong correlation to the parent geological material. In contrast, stream sediment samples have a wider representativeness, restricted to the whole upstream catchment area of the sampling point, and high susceptibility, not only due to the different lithologies occurring in the catchment area, but also due to changes in the hydrosedimentological dynamics, influenced by local seasonal variations, and changes in land cover and land use [15,16].

In addition, the determination of geochemical background is a mandatory approach for this assessment. The background concept has not been clearly established [17–19], but the term is often used as a naturally occurring concentration range of an element or chemical compound [20], determined by a given analytical method [21], which can be established by direct, indirect and integrated methods [6]. Some authors consider that the upper limit of the geochemical background, also known as the threshold value, can be used to distinguish between natural and anthropogenic influences [17]. The threshold concentration value is regularly used as a reference to define action levels in environmental legislation [22,23], whereas, the lower background limit has not been widely discussed, perhaps due to its low relevance for exploration and environmental purposes.

The Parauapebas River Basin (PB) is particularly relevant for geochemical studies because: (i) The basin is located in the Carajás Mineral Province, the largest mineral province of Brazil, and large open pit Fe (N4, N5 and S11D), Cu (Sossego) and Cu-Au (Antas North) mines, besides other mineral deposits, are situated on it (Figure 1d; (ii) The PB has protected areas that are covered by preserved tropical forest but most of it was deforested. Hence, the study area has strong contrasts of land use and cover; (iii) The possible impact in the environment related to mining activities and the effects of deforestation and increase of human occupation, with agricultural development and livestock production, should be evaluated. All these aspects can be evaluated by studying the spatial distribution of chemical elements particularly those of iron, the most voluminous mineral resource produced in the region, and Potentially Toxic Elements (PTE).

This study is associated with the Itacaiúnas Geochemical Mapping and Background Project (ItacGMBP), a large geochemical mapping project carried out by the Instituto Tecnológico Vale (ITV) in the entire Itacaiúnas River Watershed (IRW; Figure 1b). In the PB area (Figure 1c), a previous study about the soil [24] geochemistry has been conducted. However, an integrated and refined interpretation using soil and stream sediment geochemical data sets of the PB is presented here for the first time. To address the source apportionment and establishing geochemical threshold variations, a series of advanced data analyses were implemented, mainly by using geoprocessing techniques and traditional statistical methods. This integrated assessment of geochemical data is of great importance to fill the knowledge gaps identified in previous studies.

This study aims to handle an integrated evaluation of soil and stream sediment geochemical data to evaluate source apportionment and to establish geochemical threshold variations for Fe, Al, and 20 selected PTE (Ag, B, As, Ba, Bi, Cd, Co, Cr, Cu, Hg, Mn, Mo, Ni, Pb, Sb, Se, Sn, U, V, and Zn) in the PB. The study was designed to answer several questions:

- How is the spatial distribution of the PTE in soils and stream sediments of the PB? Are they similar or not?
- Are the soil and stream sediment samples correlated? Do they share the same source apportionment?

- How do the geological domains affect the variations of the geochemical background and distribution of anomalies in the PB and what is their relevance for the definition of environmental guidelines?
- Is there any evidence of anthropogenic impact in the area?

2. Study Area and Geological Setting

The study area of the present research is the Parauapebas River Basin (PB), situated in the eastern Amazon Rainforest (Figure 1a). The PB is classified as one of the eight sub-basins of the IRW (Figure 1b), which covers an area of approximately 9600 km^2, encompassing two main urban areas (Parauapebas and Canaã dos Carajás; Figure 1c), parts of conservation areas, represented mainly by the remaining preserved forest in the Center-West of the PB (Figure 1b), and active open-pit Fe (N4, N5, and S11D), Cu (Sossego) and Cu-Au (Antas North) mines (Figure 1d).

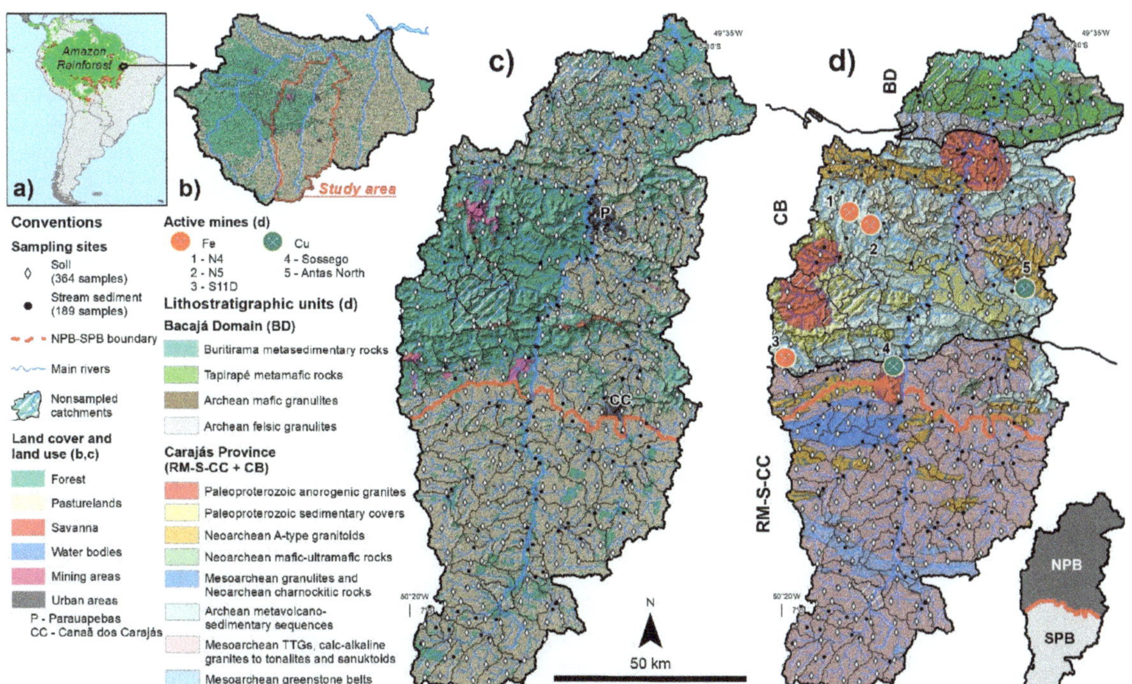

Figure 1. Location of the Parauapebas River Basin (PB) in the eastern Amazon region (**a**), situated in the Itacaiúnas River watershed (**b**). the land cover and land use map (**c**) and the geological map (**d**) are presented along with the sampling sites for soil and stream sediments. These two maps are overlaid on a shaded-relief digital elevation model (DEM). Inset shows the Northern (NPB) and Southern (SPB) PB. Map coordinates are in degrees (WGS84). Source: modified from [24].

For the purpose of the present study, the geology of the PB can be simplified into three different domains (Figure 1d):

- Rio Maria—Sapucaia—Canaã dos Carajás domains (RM-S-CC): It is restricted to the south of PB (Figure 1d), and is composed essentially of Mesoarchean tonalite—trondhjemite—granodiorite (TTG) series associated with greenstone belt sequences and calc-alkaline granites to tonalites and sanukitoids [25–28]. Neoarchean A-type like granitoids, charnockitic rocks, mafic-ultramafic bodies, and Paleoproterozoic anorogenic granites crosscut the Mesoarchean units [25,29–33].

- Carajás Basin (CB): It is restricted to the center of PB (Figure 1d), and is composed dominantly of Neoarchean mafic to felsic metavolcanic rocks and banded-iron formations (BIF), which hold the world-class Fe deposits of Carajás [34,35]. Neoarchean A-type like granite plutons [29,36] and mafic-ultramafic stratified bodies [31,37], as well as Paleoproterozoic anorogenic granites [38–40] also occur.
- Bacajá Domain (BD): It is restricted to the north of PB (Figure 1d), and is mainly composed of high-grade charnockitic rocks, with subordinate mafic orthogranulitesand para-derived granulites [41], supracrustal mafic metamorphic rocks of the Tapirapé Formation, and metasedimentary lithologies of the Buritirama Formation [42].

3. Materials and Methods

3.1. Sampling Procedure

The sampling strategies, data storage, screening, and validation of the ItacGMBP were entirely assisted by a computer-based framework associated with a geographic information system environment. Firstly, a sampling protocol was structured to guide the field teams in storing all field data using tablets in a geochemical database. Further information can be accessed elsewhere [43].

The soil survey was designed on a nominal scale of 1:250,000, with a sampling density of 1 sample/25 km^2, using grids of 5 km $\times$ 5 km, homogeneously distributed along the whole IRW (Figure 1b). The surface soil samples were collected at a constant depth of 10 cm. Each surface soil sample was collected and then bulked to provide one composite sample (approximately 5 kg).

The stream sediment survey was designed on a nominal scale of 1:1,000,000, with a sampling density of 1 sample/100 km^2, using 2nd–3rd order drainage microcatchments along the entire IRW (Figure 1b). The collection of stream sediment samples (approximately 3 kg) was carried out in active sedimentation areas at the catchment outlet, preferably in the middle of the channel to minimize contamination occurring at the margins.

For the quality control procedure, a duplicate sample was collected for every 20 samples collected. Samples were collected as close as possible to the location defined in the planning field map. Most of the unsampled sites are located in remote areas with severe access limitations. The detailed sampling procedure and description of different sampling media used in the ItacGMBP can be found elsewhere [24,43–46].

In the PB, 364 samples of surface soil and 189 samples of stream sediment were collected. The sampling sites are plotted over the land use and land cover (Figure 1c) and geological layers (Figure 1d) for further inspections.

3.2. Sample Preparation and Chemical Analyses

The ItacGMBP's analytical program was extensive, but only the relevant analyses for the present study are described. For more information, refer to previous studies [24,43–46]. Both soil and stream sediment samples were submitted to the same preparation and analytical protocols. The samples were oven-dried at 70 °C, disaggregated, homogenized, and sieved through a <0.177 mm (80 mesh) sieve. Approximately 50 g of the samples were grounded, sieved through a <75 μm (200 mesh) sieve, and submitted to aqua regia (1:3 M HNO$_3$:HCl) digestion. The aqua regia soluble concentrations of 51 elements (the most relevant for the present study are Fe, Al, Ag, B, As, Ba, Bi, Cd, Co, Cr, Cu, Hg, Mn, Mo, Ni, Pb, Sb, Se, Sn, U, V, and Zn) were determined via Inductively Coupled Plasma—Atomic Emission Spectrometry (ICP-AES) and Inductively Coupled Plasma—Mass Spectrometry (ICP-MS). All samples were prepared and analyzed in laboratories of ALS Brasil Ltda (Belo Horizonte, Brazil), a certified/accredited laboratory.

3.3. Data Processing and Statistical Analysis

Firstly, the censored data (values below the lower limit of detection—LLD), commonly observed in the geochemical data set, was replaced by LLD/2) indicated for each element, according to the analytical method. This procedure is widely used in the literature [22,23,45,46]. It is important to highlight that, eventually, two sets of data were used in this study, one for the NPB (soils = 223 samples; stream sediments = 122 samples) and another one for the SPB (soils = 141 samples; stream sediments = 67 samples) (cf. Figure 1). Standard statistical procedures [22,23,47] were undertaken in this research, which involves calculating several descriptive statistics parameters (amount of data below LLD, arithmetic mean, standard deviation—SD, minimum—Min, median—Med, and maximum—Max), constructing graphs based on the exploratory data analysis (e.g., box plots, scatter plot and quantile-quantile plot) and conducting hypothesis test (e.g., Mann–Whitney–Wilcoxon test). Data analysis was performed using the R programming language in RStudio.

3.4. The Determination of Geochemical Background Values

A variety of statistical methods was used for the calculation of geochemical threshold values, which is highly recommended by experts in the field. These methods are described elsewhere [19,20,22,23,48], and the general steps are highlighted herein.

Median + 2 * Median Absolute Deviation (M_{MAD}): The M_{MAD} is considered one of the most prestigious methods for deriving geochemical threshold values [19]. Firstly, a requirement for using the M_{MAD} method is that the raw data should be prior transformed to a common logarithm ($\log_{10}$) scale to get close to a normal distribution, unusually seen in geochemical data set [23]. In this approach, the median absolute deviation (MAD; Equation (1)), leads to a robust estimation of the variability of univariate geochemical data [23].

$$MAD_{(y)} = 1.48 * median_i \, | \, y_i - median_j(y_j) \, | \tag{1}$$

Afterward, the *MAD* value should be multiplied by two and added to the median. At this stage, the result should be back-transformed, from logarithmic scale to natural scale, to derive the M_{MAD} threshold value (Equation (2)):

$$M_{MAD(y)} = 10^{(median(y) \, + \, 2 \, * \, MAD(y))} \tag{2}$$

Tukey's Inner Fence (TIF): The TIF method depends solely on the data distribution. In addition, background values can be calculated even if no outliers are present in the data set (threshold value greater than the maximum value). Moreover, the threshold value obtained from this approach is most informative if the true number of outliers is below 10% of the data [19,23]. Firstly, the log10 transformation is required prior to using this approach [19]. The TIF method consists of determining the upper (Q3 = 75th percentile) and lower (Q1 = 25th percentile) quartiles of the boxplot. Then, the inner fence is determined as the interquartile range (IQR = Q3 − Q1) extended by 1.5 times. The threshold value based on the TIF method is calculated following Equation (3):

$$TIF = 10^{(Q3 \, + \, 1.5 \, * \, IQR(y))} \tag{3}$$

Percentile-based approaches: These approaches are considered as the most straightforward methods for deriving threshold values [48,49], widely used by many environmental agencies from different countries (e.g., Brazil [50], Australia [51], and Finland [52]). The 98th percentile (P98) has become widespread as a 2% outliers (1 in 50 rate) is deemed acceptable, and it distances the method from the 97.5 percentile, 2.5% 1 in 40 rate, which delivers similar results to the "Mean + 2 * Standard Deviation" (outdated method) when a normal distribution of a given element is satisfied [19]. The 95th percentile (P95) corresponds to a more restricted value, as it considers 5% of all samples as outliers. Other percentile values are also used (e.g., 90th, 85th, and 75th), however, they are not going to be considered as threshold values in the present study.

3.5. Geoprocessing and Spatial Representation of Soil and Stream Sediment Geochemical Data

Spatial analyses were processed using the software ArcGIS 10.4.1, under the World Geodetic System 1984 (WGS84) datum. The construction of integrated stream sediment and soil geochemical maps was based on two techniques: (i) Catchment-based representation (polygon) for stream sediment data [53–55] which, simply involves attributing the element concentrations to their respective catchment watershed, with the sample collected at the outlet; (ii) Geochemical dot representation (point) for surface soil data, attributing the uni-element concentration to the sampling site [55,56]. The class intervals were defined according to the quantile method, usually applied to right-skewed distribution, commonly seen in geochemical data [57]. For the correct construction of integrated geochemical maps, the same class interval was used for the representation of the concentration (usually in mg kg^{-1}, except for Fe and Al in wt.%) of a given element in soil and in stream sediment. Finally, spatial analyses using geoprocessing techniques (e.g., buffer, clip, intersect and merge) were eventually used for further interpretation.

4. Results

In this study, integrated soil and stream sediments geochemical maps are presented for selected elements (Fe, Al, Cu, Mo, As, and Hg in Figure 2; Ni, Cr, Co, V, Mn, and Zn in Figure 3). Additional geochemical maps for the remaining PTE (Ag, B, Ba, Bi, Cd, Pb, Sb, Se, Sn, and U) are provided in the supplementary material (Figure S1). In these maps, the catchment-based representation (polygons) was used to display the distribution of elements in stream sediment samples, whereas the point representation was used for soil samples.

The spatial distribution of each element (cf. Figures 2 and 3, Figures S1 and S2) revealed a clear difference between the Northern PB (NPB) and Southern PB (SPB). The NPB region corresponds almost exactly to the BD and CB and the SPB to the RM-S-CC. The chemical compositions of soils and stream sediments from those regions are strongly influenced by the dominant lithologies occurring in their respective geological domains (Figure 1d). Boxplots for Fe, Al, and selected PTE (Figure 4) were constructed by taking into account the compositions of soil and stream sediment samples from the three different geological domains distinguished in the PB. Additional boxplots are provided in the supplementary material (Figure S3). It is apparent that the distributions of some elements (e.g., Al, As, Bi, and Cr; Figure 4) are similar in both BD and CB, representing the NPB, and contrast with that observed in the RM-S-CC domains that occupy the SPB. On the contrary, Cu spatial distribution exhibit remarkable differences among all geological domains (Figure 4).

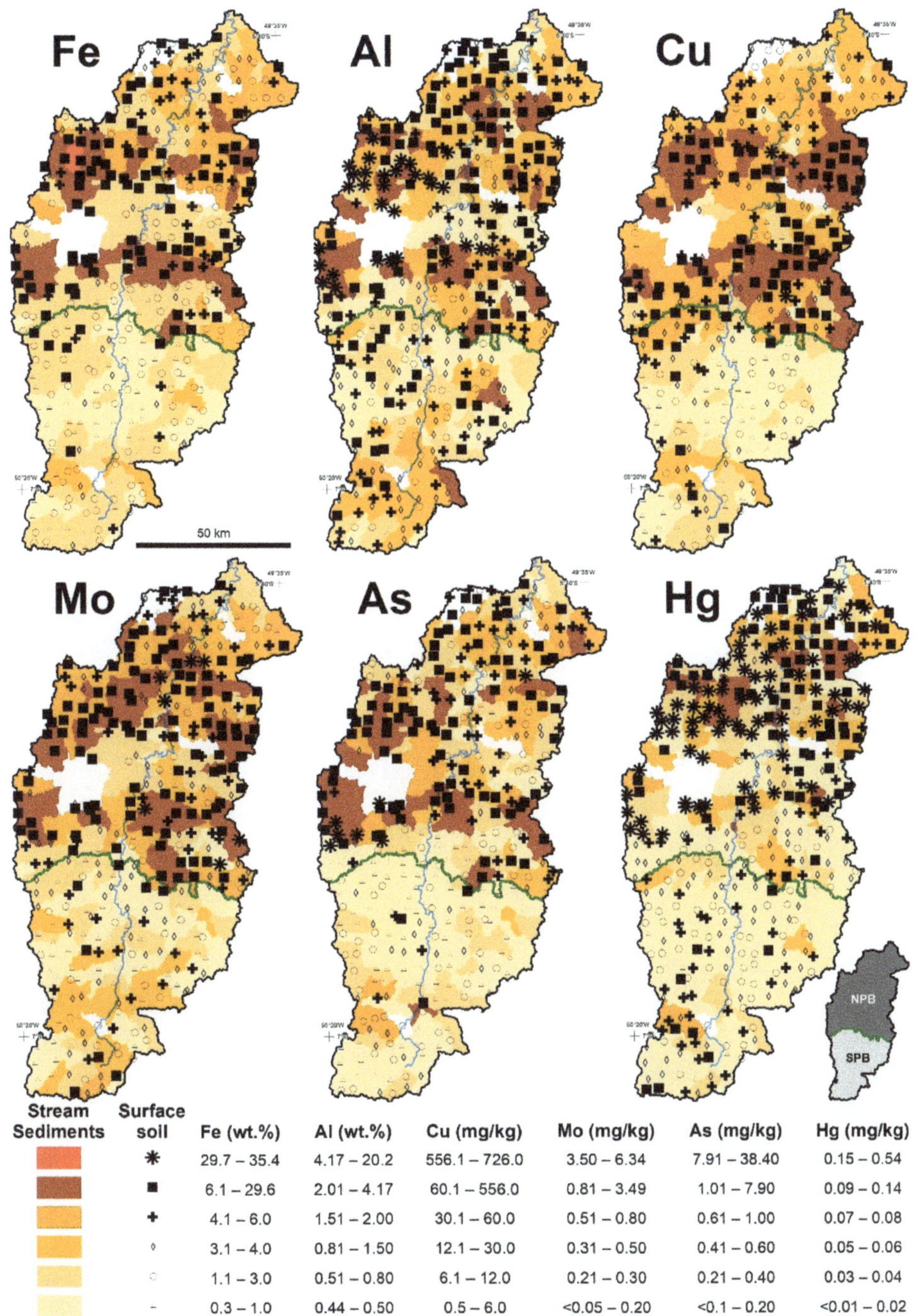

Figure 2. Geochemical maps for Fe, Al, Cu, Mo, As, and Hg in stream sediments (catchment area—polygon) and surface soil (in-situ sampling site—point) in the Parauapebas River Basin (PB). Note that the same concentration range for each element is used for the map representation on both sampling media. Inset shows the Northern (NPB) and Southern (SPB) PB. Refer to Figures S1 and S2 for the geochemical maps of the remaining PTE.

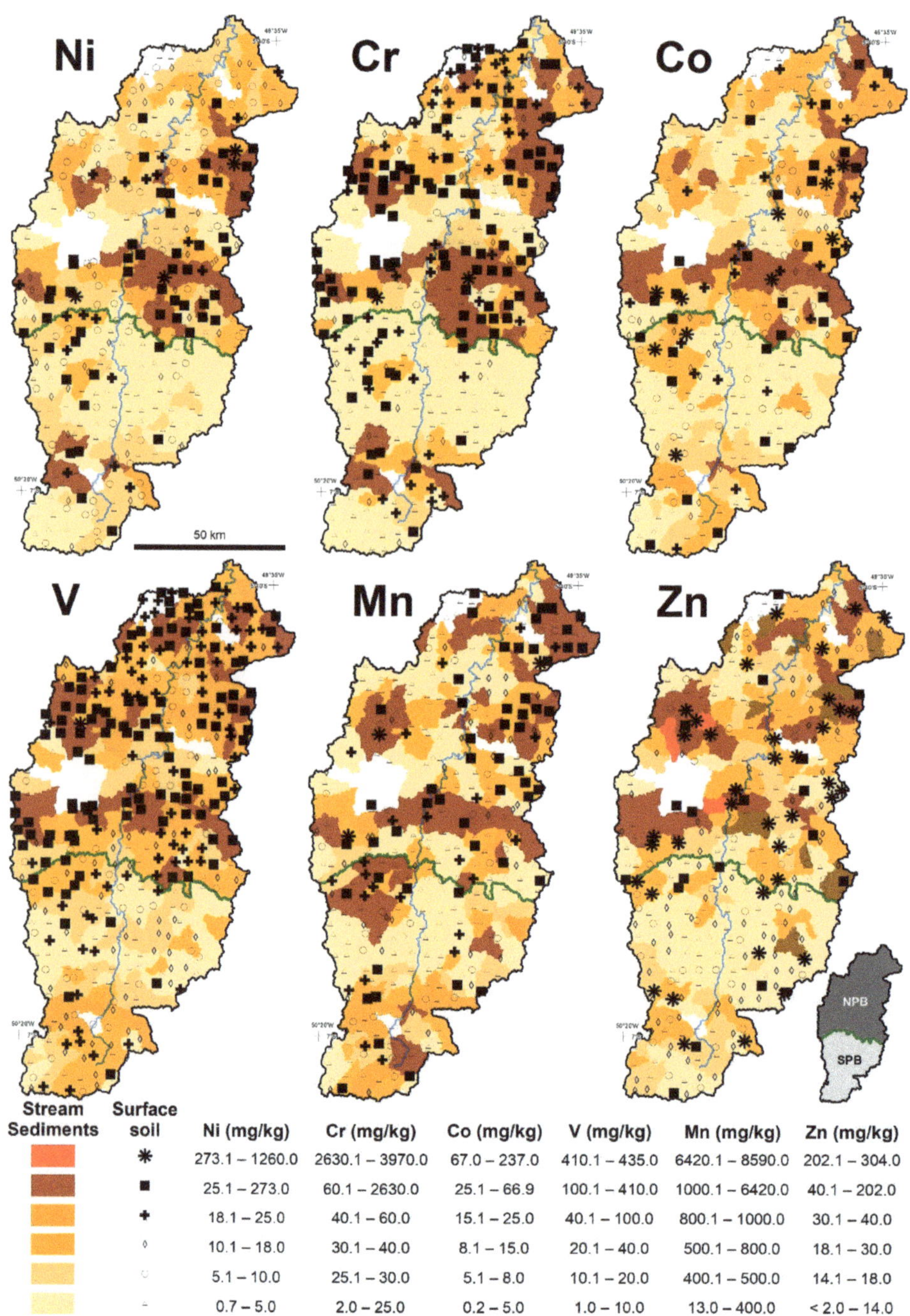

Stream Sediments	Surface soil	Ni (mg/kg)	Cr (mg/kg)	Co (mg/kg)	V (mg/kg)	Mn (mg/kg)	Zn (mg/kg)
	*	273.1 – 1260.0	2630.1 – 3970.0	67.0 – 237.0	410.1 – 435.0	6420.1 – 8590.0	202.1 – 304.0
	■	25.1 – 273.0	60.1 – 2630.0	25.1 – 66.9	100.1 – 410.0	1000.1 – 6420.0	40.1 – 202.0
	+	18.1 – 25.0	40.1 – 60.0	15.1 – 25.0	40.1 – 100.0	800.1 – 1000.0	30.1 – 40.0
	◊	10.1 – 18.0	30.1 – 40.0	8.1 – 15.0	20.1 – 40.0	500.1 – 800.0	18.1 – 30.0
	○	5.1 – 10.0	25.1 – 30.0	5.1 – 8.0	10.1 – 20.0	400.1 – 500.0	14.1 – 18.0
	–	0.7 – 5.0	2.0 – 25.0	0.2 – 5.0	1.0 – 10.0	13.0 – 400.0	< 2.0 – 14.0

Figure 3. Geochemical maps for Ni, Cr, Co, V, Mn, and Zn in stream sediments (catchment area—polygon) and surface soil (in-situ sampling site—point) in the Parauapebas River Basin (PB). Note that the same concentration range for each element is used for the map representation on both sampling media. Inset shows the Northern (NPB) and Southern (SPB) PB. Refer to Figures S1 and S2 for the geochemical maps of the remaining PTE.

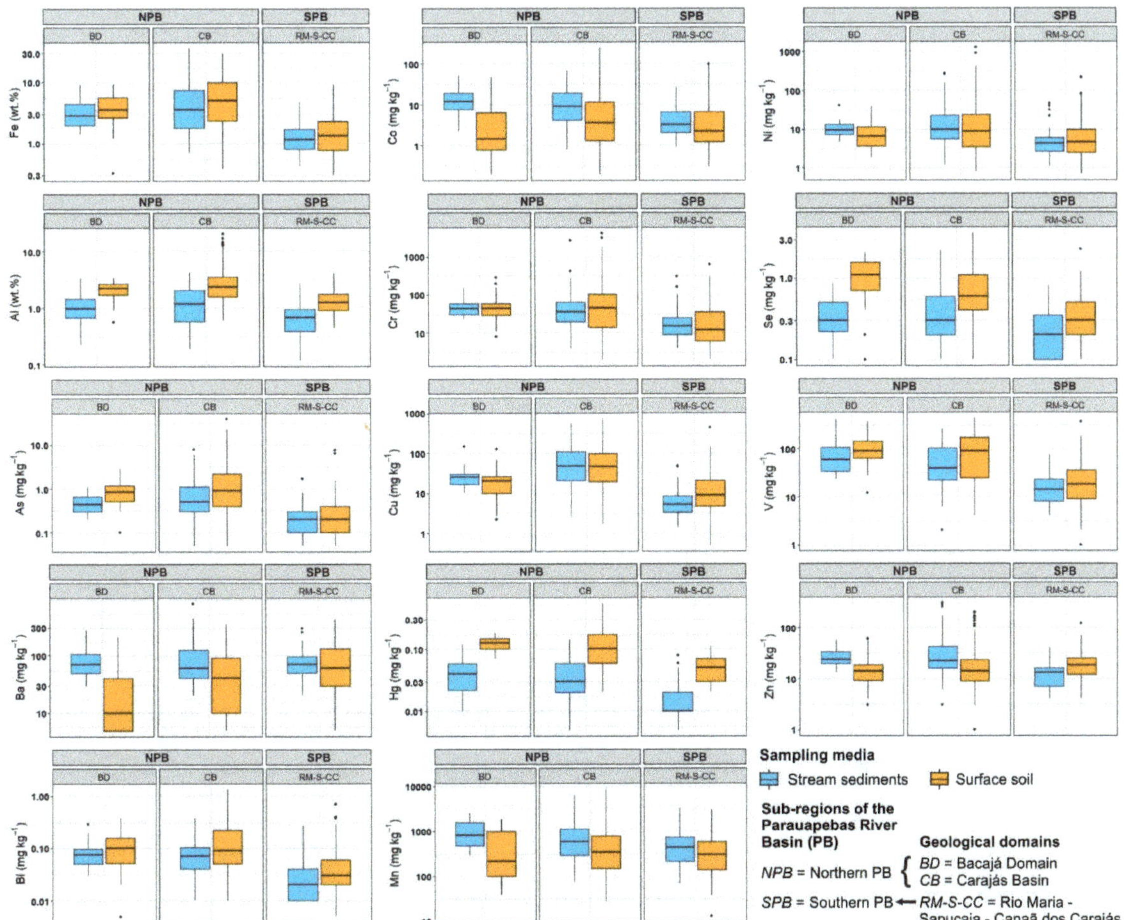

Figure 4. Boxplots for Fe, Al, and 12 potentially toxic elements (PTE; As, Ba, Bi, Co, Cr, Cu, Hg, Mn, Ni, Se, V, and Zn) in surface soils and stream sediments samples of the Parauapebas River Basin (PB), according to the major geological domains of the study area: BD = Bacajá Domain and CB = Carajás Basin, both domains comprised at the Northern PB (NPB); and RM-S-CC = Rio Maria—Sapucaia—Canaã dos Carajás domains, comprised at the Southern PB. Refer to Figure 1 for the geological setting of the study area. Refer to Figure S3 for the boxplot of the remaining PTE.

In addition, the MWW test was carried to devaluate if the composition of soil and stream sediment samples from BD and CB are statistically similar. Table 1 summarizes the *p*-values of the MWW calculated for each element in surface soils and stream sediments separately. In these results, the null hypothesis states that there is no difference between the distribution of a given element in a given sampling media in BD and CB. By assuming a significance level (σ) of 0.05, the *p*-values below σ imply rejecting the null hypothesis and inform that the distributions of the respective element in BD and CB are statistically different. In summary, Al, Ag, As, Bi, Cr, Hg, Ni, Sn, V, and Zn presented similar distribution in both sampling medium; Fe, B, Ba, Cd, Co, Mo, Pb, Sb, Se, and U presented similar distribution only in stream sediments; Mn presented similar distribution only in soils, and Cu presented different distribution in both sampling medium.

The integrated evaluation of the data (geochemical maps, Figures 2 and 3, Figures S1 and S2; boxplots, Figure 4 and Figure S3; and, MWW results, Table 1) highly indicates that the study area can be divided into two separate geochemical regions, the NPB and SPB.

For this reason, the original data set was split into two subsets for further investigation. Table 2 summarizes the descriptive statistics (mean, SD, Min, Med, and Max) of the studied elements in surface soils and stream sediments in the entire PB (soils = 364 samples; stream sediments = 189 samples), which was retrieved from previous studies [24,58] using the same data set, and for each subregion, NPB (soils = 223 samples; stream sediments = 122 samples) and SPB (soils = 141 samples; stream sediments = 67 samples).

Table 1. Summary of the *p*-values obtained from Mann–Whitney–Wilcoxon (MWW) Test used to verify the similarities in the distribution of the studied elements in a sampling medium (Surface soils and stream sediments) from two different geological domains (BD = Bacajá Domain and CB = Carajás Basin). Refer to Figure 1d for the location of the geological domains.

Element	Do the Samples from BD and CB Have Similar Distribution? (*p*-Values of the MWW Test)		Element	Do the Samples from BD and CB Have Similar Distribution? (*p*-Values of the MWW Test)	
	Surface Soils	**Stream Sediments**		**Surface Soils**	**Stream Sediments**
Fe		0.40	Hg	0.22	0.21
Al	0.13	0.46	Mn	0.12	
Ag	0.18	0.37	Mo		0.29
As	0.37	0.43	Ni	0.09	0.90
B		0.12	Pb		0.09
Ba		0.93	Sb		0.50
Bi	0.75	0.50	Se		0.92
Cd		0.59	Sn	0.43	0.74
Co		0.20	U		0.29
Cr	0.88	0.13	V	0.30	0.07
Cu			Zn	0.80	0.86

Note: Blank entries indicate *p*-value < 0.05.

Table 2. Descriptive statistics for Fe, Al, and 10 potentially toxic elements (PTE; As, Co, Cr, Cu, Hg, Mn, Mo, Ni, V, and Zn) in surface soils and stream sediments. Values were calculated for the entire Parauapebas River Basin (PB; soils = 364 samples; stream sediments = 189 samples) and for the northern (NPB; soils = 223 samples; stream sediments = 122 samples) and southern (SPB; soils = 141 samples; stream sediments = 67 samples) regions separately. Refer to Table S1 for the descriptive statistics of the remaining elements.

Element	Region	Surface Soil						Stream Sediments					
		LLD (% < LLD)	Mean	SD	Min	Med	Max	LLD (% < LLD)	Mean	SD	Min	Med	Max
Fe	PB *	0.01 (0)	4.96	5.82	0.3	2.71	29.6	0.01 (0)	4.01	5.17	0.42	1.98	35.4
	NPB	0.01 (0)	6.95	6.60	0.32	4.75	29.6	0.01 (0)	5.48	5.92	0.68	3.5	35.4
	SPB	0.01 (0)	1.84	1.63	0.3	1.34	9.04	0.01 (0)	1.34	0.81	0.42	1.16	4.79
Al	PB *	0.01 (0)	2.48	2.51	0.44	1.80	20.2	0.01 (0)	1.14	0.84	0.12	0.87	4.17
	NPB	0.01 (0)	3.17	2.96	0.57	2.29	20.2	0.01 (0)	1.37	0.91	0.19	1.11	4.17
	SPB	0.01 (0)	1.40	0.66	0.44	1.26	4.06	0.01 (0)	0.74	0.48	0.12	0.67	2.69
As	PB *	0.1 (5.2)	1.24	2.9	<0.1	0.5	38.4	0.1 (10.1)	0.66	1.00	<0.1	0.3	7.9
	NPB	0.1 (1.8)	1.79	3.50	<0.1	0.9	38.4	0.1 (3.3)	0.90	1.17	<0.1	0.5	7.9
	SPB	0.1 (10.6)	0.37	0.82	<0.1	0.2	7.5	0.1 (22.4)	0.22	0.24	<0.1	0.2	1.7
Co	PB *	0.01 (0)	10.71	22.6	0.2	2.9	237	0.01 (0)	11.60	12.88	0.8	6.8	66.9
	NPB	0.01 (0)	12.24	25.98	0.2	3.3	237	0.01 (0)	14.96	14.52	0.8	9.7	66.9
	SPB	0.01 (0)	8.27	15.69	0.3	2.3	97.5	0.01 (0)	5.49	5.21	0.9	3.3	26.8
Cr	PB *	1 (0)	82.34	296.3	2	31	3970	1 (0)	58.38	196.57	4	30	2630
	NPB	1 (0)	114.35	371.76	2	44	3970	1 (0)	75.83	241.32	4	37.5	2630
	SPB	1 (0)	32.63	65.82	2	12	612	1 (0)	26.61	41.41	4	15	294
Cu	PB *	0.2 (0)	48.88	79	0.5	21.8	726	0.2 (0)	48.92	73.69	1.4	20	556
	NPB	0.2 (0)	66.36	90.89	1.7	34.9	726	0.2 (0)	71.55	83.32	2.6	41.65	556
	SPB	0.2 (0)	21.39	42.75	0.5	8.7	438	0.2 (0)	7.73	8.49	1.4	5.2	47.4
Hg	PB *	0.01 (0)	0.1	0.1	0.02	0.07	0.54	0.01 (9.0)	0.03	0.03	<0.01	0.02	0.14
	NPB	0.01 (0)	0.13	0.08	0.02	0.11	0.54	0.01 (5.7)	0.04	0.03	<0.01	0.03	0.14
	SPB	0.01 (0)	0.05	0.02	0.02	0.05	0.11	0.01 (14.9)	0.02	0.01	<0.01	0.01	0.08
Mn	PB *	5 (0)	606.2	941.5	13	308	8590	5 (0)	826.32	933.84	69	547	6420
	NPB	5 (0)	710.33	1136.67	27	318	8590	5 (0)	950.10	1050.42	74	609.5	6420
	SPB	5 (0)	440.04	453.61	13	300	2970	5 (0)	600.94	617.48	69	428	3270

Table 2. *Cont.*

Element	Region	Surface Soil						Stream Sediments					
		LLD (% < LLD)	Mean	SD	Min	Med	Max	LLD (% < LLD)	Mean	SD	Min	Med	Max
Mo	PB *	0.05 (0.5)	0.85	0.9	<0.05	0.48	6.34	0.05 (0)	0.57	0.60	0.06	0.36	3.49
	NPB	0.05 (0)	1.16	1.02	0.06	0.84	6.34	0.05 (0)	0.75	0.68	0.11	0.545	3.49
	SPB	0.05 (1.4)	0.36	0.35	0.025	0.26	2.17	0.05 (0)	0.24	0.13	0.06	0.23	0.76
Ni	PB *	0.2 (0)	21.47	87.8	0.7	6.3	1260	0.2 (0)	15.73	32.24	1.1	6.9	273
	NPB	0.2 (0)	28.20	110.36	0.8	8.3	1260	0.2 (0)	20.95	38.76	1.2	9.7	273
	SPB	0.2 (0)	10.89	22.43	0.7	4.6	220	0.2 (0)	6.21	7.97	1.1	4.2	47.1
V	PB *	1 (0)	81.09	87.3	1	43	435	1 (0)	50.37	57.18	2	29	410
	NPB	1 (0)	111.46	93.79	4	89	435	1 (0)	68.57	63.63	2	43	410
	SPB	1 (0)	33.90	45.80	1	18	369	1 (0)	17.22	12.81	4	14	75
Zn	PB *	2 (0.5)	22.35	25.7	<2	16	202	2 (0)	30.04	40.99	3	19	304
	NPB	2 (0.9)	23.52	30.92	<2	14	202	2 (0)	39.28	48.40	3	24	304
	SPB	2 (0)	20.53	13.95	4	18	120	2 (0)	13.22	6.80	4	13	40

Note: The concentration unit is expressed in mg kg^{-1}, except for Fe and Al in wt.%; DL = Lower limit of detection; % < DL = % of data below LLD; Mean = Arithmetic mean; SD = Standard deviation; Min = Minimum; Med = Median; Max = Maximum; '*' = Surface soil data from [24].

The use of intersection function under geoprocessing environment allowed plotting and merging of soil samples with stream sediment catchment areas. It is important to highlight that a single microcatchment, which is represented by only one stream sediment sample, can be merged to multiple soil samples, or even none in the rare cases of no collected samples in the catchment area. Scatter plots comparing the concentrations of selected elements in surface soil (y-axis) and stream sediment (x-axis) samples were constructed to verify a possible relationship between them by using linear regression analysis (Figure 5 and Figure S4). Spearman's correlation coefficient (ρ) and *p*-value of the linear regression are presented. In general, the majority of the selected elements presented a significant correlation ($p < 0.05$), excepting for Cd, Ba, B, and Zn ($p > 0.05$; Figure S4). Three groups were observed: (i) Strongly ($\rho \geq 0.5$) correlated materials: Fe, As, Cu, and U in Figure 5; Bi, Sn, Mo, and V in Figure S4; (ii) Moderately ($0.49 < \rho \leq 0.3$) correlated materials: Hg, Ni, Cr, and Sb in Figure 5; (iii) Weakly ($\rho \leq 0.29$) to non-significant ($p > 0.05$) correlated materials: Al, Mn, and Co in Figure 5; Se, Ag, Pb, Cd, Ba, B, and Zn in Figure S4.

In this study, geochemical threshold concentration values were determined for all 22 studied elements in soils and stream sediments, but contrarily to the approach adopted in previous studies [24,58], distinct threshold values for the NPB and SPB were calculated, by using a variety of statistical methods (M_{MAD}, TIF, and percentile-based techniques, e.g., P98 and P95; Table 3). Threshold values of each individual subregion were then compared to threshold values proposed for the whole PB, determined in previous studies of soil [24] and stream sediments [58].

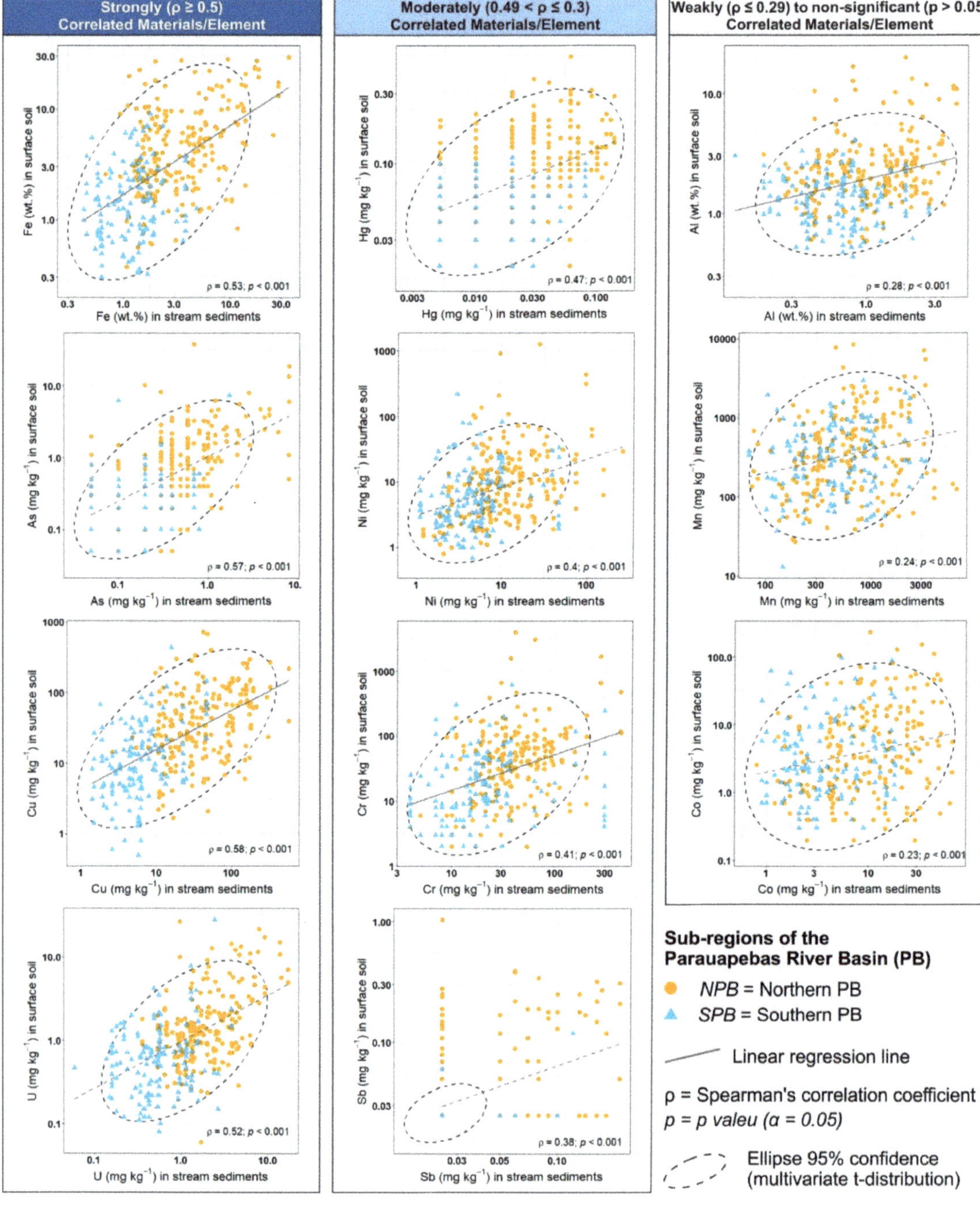

Figure 5. Scatter plot comparing the concentrations of selected elements in surface soil samples against stream sediment samples. Three groups of elements were observed: (i) Strongly correlated materials (e.g., Fe, As, Cu, and U); (ii) Moderately correlated materials (e.g., Hg, Ni, Cr, and Sb); (iii) Weakly to non-significant correlated materials (e.g., Al, Mn, and Co); Additional scatter plots for the remaining elements are presented in Figure S4.

Table 3. Geochemical threshold values for Fe, Al, and 10 potentially toxic elements (PTE; As, Co, Cr, Cu, Hg, Mn, Mo, Ni, V, and Zn) in surface soils and stream sediments, estimated by a variety of methods [Median + 2 Median Absolute Deviation (M_{MAD}), Tukey's inner fences (TIF), the 98th (P98) and 95th (P95) percentiles]. Values are provided for the entire Parauapebas River Basin (PB, retrieved from previous studies), for the northern (NPB) and the southern (SPB) regions separately, determined in the present study. Guideline values reposted by Brazilian environmental agencies are presented as references. Refer to Table S2 for the results of the remaining elements.

Element	Regions	Geochemical Threshold in Surface Soil				Soil Guidelines		Geochemical Threshold in Stream Sediments				Stream Sediment Guidelines	
		M_{MAD}	TIF	P98	P95	QRV [b]	PGV [c]	M_{MAD}	TIF	P98	P95	L1 [d]	L2 [d]
Fe	PB [a]	25.7	*58.21*	*24.55*	18.2			12.02	29.51	17.76	12.94		
	NPB	*34.24*	*68.63*	*27.51*	22.65			22.93	42.89	26.74	16.73		
	SPB	6.93	*12.24*	6.92	5.99			3.54	5.20	3.74	2.88		
Al	PB [a]	5.62	8.51	11.22	8.32			▲	▲	3.32	2.76		
	NPB	6.51	8.89	12.45	10.57			*6.25*	*10.72*	3.44	3.16		
	SPB	3.34	*4.74*	3.05	2.65			*2.87*	*3.63*	2.01	1.51		
As	PB [a]	7.2	14.1	7.2	4.5			2.4	4.6	3.9	2.4		
	NPB	7.0	14.9	9.4	5.2	3.5	15	2.3	6.1	4.5	2.9	5.9	17
	SPB	1.6	3.2	1.1	0.6			1.6	1.6	0.8	0.5		
Co	PB [a]	69.2	201.8	79.4	47.9			▲	▲	51.4	42.5		
	NPB	96.4	*284.9*	96.5	48.9	13	25	*93.2*	*190.1*	52.2	48.3		
	SPB	35.7	91.7	64.5	37.0			16.7	*39.6*	18.3	16.3		
Cr	PB [a]	562	1413	468	191			191	331	275	138		
	NPB	564	1401	661	254	40	75	160	277	348	139	37.3	90
	SPB	161	529	204	117			68	121	134	89		
Cu	PB [a]	398.1	*1216.2*	275.4	162.2			512.9	▲	288.5	177.0		
	NPB	365.9	*926.3*	302.7	223.7	35	60	*482.4*	*1223.9*	306.0	190.5	35.7	197
	SPB	72.7	203.1	101.9	70.7			20.0	35.1	37.8	22.8		
Hg	PB [a]							0.16	▲	0.13	0.10		
	NPB	0.40	*0.55*	0.32	0.27	0.05	0.5	*0.23*	*0.31*	0.13	0.11	0.17	0.486
	SPB	*0.14*	0.25	0.10	0.09			0.08	0.06	0.06	0.04		
Mn	PB [a]	3890	*8810*	2884	1995			3548	6383	3387	2564		
	NPB	4835	*14,783*	3908	2359			4503	*8660*	3732	3117		
	SPB	2538	*5083*	1793	1250			3079	4643	2766	1839		
Mo	PB [a]	4.79	*10.59*	3.31	2.88			1.95	▲	2.92	1.69		
	NPB	6.11	*11.43*	3.60	3.14	<4	30	2.82	*4.63*	3.02	2.36		
	SPB	1.10	1.58	1.70	0.99			0.73	*1.02*	0.51	0.46		
Ni	PB [a]	60.3	142.9	107.2	52.5			45.7	88.1	100.3	46.8		
	NPB	98.8	173.0	174.2	64.1	13	30	56.4	127.0	137.3	71.0	18	35.9
	SPB	32.6	80.0	78.7	33.5			12.6	19.5	36.4	17.5		
V	PB [a]	*891*	*2512*	331	269			234	▲	188	162		
	NPB	*769*	*1831*	344	284			360	*931*	239	183		
	SPB	141	288	160	121			61	*112*	47	40		
Zn	PB [a]	63	89	115	65			60	88	27	18		
	NPB	53	84	144	74	60	300	83	169	216	114	123	315
	SPB	54	75	55	43			47	*55*	27	24		

Note: The concentration unit is expressed in mg kg^{-1}, except for Fe and Al in wt.%; *Italic* or "▲" (value not available): threshold values greater than the maximum value; [a] Results reported in previous studies of soil [24] and stream sediments [58] of the PB; [b] Quality reference value (QRV) reported by the Sao Paulo Sanitation Technology Company (CETESB) [59]; [c] Prevention guideline value (PGV) reported by the National Council of the Environment (CONAMA) of Brazil [50]; [d] Threshold levels 1 (L1) and 2 (L2) reported by CONAMA [60]. Blank entries indicate data not available.

5. Discussion

5.1. Regional Geochemical Distribution of Fe, Al, and PTE in Soils and Stream Sediments

The geochemical signature of an element in a given sampling medium provides important information about external controlling conditions, either natural or anthropic. The geochemical maps for Fe, Al, and 20 PTE (Figures 2 and 3, Figures S1 and S2) in the two-sampling media, as assessed by visual inspection, are notoriously similar. The composition of soils and stream sediments appears to be strongly controlled by bedrock lithologies in the area (Figure 1c), with evident differences between the north and south of the study area. Higher concentrations for the majority of the elements in both sampling media are observed in the NPB in comparison to SPB (Figure 1b).

The spatial distribution of these elements is well-documented in previous studies of soil [24,44] and stream sediment [45,58]. For instance, the concentrations of Fe, V, As (Figure 2) (±Cd, Sb and Ag; Figure S1) and relative concentrations of Al, Ba, Mn, and Zn (Figures 2 and 3, Figure S2), along with two E-W corridors (tectonic trend of the region) in the NPB are controlled by the occurrences of Fe-Al-rich duricrusts, formed by the intense weathering of metamafic rocks and BIF of the Carajás Formation, which hosts the world-class Fe deposits (N4, N5, and S11D) of Carajás [61–63].

Similarly, the concentrations of Ni, Cr, Co, and ±V (Figure 3) are strongly controlled by the metavolcanic rocks of the Parauapebas Formation, generally intercalated with rocks from the Carajás Formation, and cross-cut by local mafic to ultramafic rocks of the Cateté Intrusive Suite (the Luanga Complex, located in the NE of the NPB; the Vermelho Complex, in the southern area of the NPB) [37,64]. Local anomalies also occur in the south of the SPB, due to the occurrences of Sapucaia greenstone belt [65].

The spatial distribution of Cu, Mo (Figure 2), and to a lesser extent Se (Figure S1) is a response of mineralization zones along two E-W corridors (northern and southern copper belts [27]) in the NPB, similarly observed for Fe and related elements. To some extent, higher concentrations of Se (Figure S1) were observed in stream sediments and, especially, in soils of the north of NPB, along the northern copper belt in comparison to the southern copper belt. This evidence indicates that these two corridors do not share common surface multi-element signatures, perhaps, due to different metallogenic evolution [27].

Uranium, Pb, and Sn tend to concentrate in quartz-feldspar rocks, such as granites and felsic granulites, particularly those with an alkaline tendency. In the NPB, the main sources for the enrichment of these elements are A-type granitoids (e.g., Estrela and Igarapé Gelado granites) and Paleoproterozoic anorogenic granites (e.g., Seringa and Serra dos Carajás granites) [38]. In addition, unusually high concentrations of Mn, Co, and Ba (Figure 3 and Figure S2) in soils and stream sediments and B in soils (Figure S1) were observed in the NW region of the SPB, under the influence of charnockitic rocks (Figure 1d) of the Diopside-Norite Pium unit [28].

Using integrated geochemical maps is important not only for understanding the regional distribution of a given element but also to compare two (or three) different sampling mediums in terms of concentration magnitude and source. Although, as described, the different lithologies of the area are the main source of enrichment for Fe, Al, and the 20 PTE. However, anthropogenic contributions cannot be disregarded and further studies should be conducted.

5.2. The Geochemical Compartmentation of PB as Subsidy to Territorial and Watershed Management

Watershed management is essential for planning the sustainable use of natural resources. It supports many kinds of human needs in terms of water consumption, food production, recreation, and most importantly the maintenance of ecologic function [66]. Each watershed has its own particularities, for instance, climate conditions, land use, human occupation, and, for the purpose of the present study, geochemical background. All these multidisciplinary concerns have increased the need of developing robust watershed-management strategies aiming for realistic environmental policies [67].

As presented in previous studies, the chemical composition of soils [24,44] and stream sediments [45] changes with the geological domain across the IRW, including the study area (Figure 1b). Nevertheless, minor differences were also observed between the BD and CB in terms of spatial distribution for many elements, as discussed previously. The boxplots for the studied elements (Figure 4 and Figure S3) display the difference among the geological domains of the PB area (Figure 1d), suggesting that many elements actually present similar distribution (e.g., Al, As, Bi, and Cr; Figure 4). For this reason, the element concentration in soil and stream sediment samples from these two domains were statistically compared using the MWW test (Table 1). The obtained results suggest that dividing the study area into NPB and SPB is statistically acceptable, at least for surface geochemical studies at a regional scale. However, the geochemical data of Cu in the PB represents a clear exception to this approach. Not only the MWW test revealed different statistical distribution between BD and CB (p-value < 0.05; cf. Table 1), but also the boxplots (Figure 4 and Figure S3) and the integrated geochemical map (Figure 2). The Cu signature in the PB is a clear response from highly mineralized areas that constitute a geochemical and metallogenic province [68], which cannot be generalized in terms of watershed management.

5.3. The Relationship between Surface Soil and Stream Sediment Geochemistry

A visual inspection of the boxplots (Figure 4 and Figure S3) and the comparison of descriptive statistics results (Table 2 and Table S2) revealed that element concentrations in soils are generally greater than in stream sediments of PB. This is particularly true for Fe, Al, Ag, As, B, Bi, Cd, Cr, Cu, Hg, Mo, Sb, V, Se, and Sn, and it is not observed for Ba, Co, Mn, Ni, Zn, Pb, and U. By using geoprocessing techniques, it was possible to identify the soil samples that are geographically located at a given catchment area, which is represented by a stream sediment sample collected at the outlet, and compare them. Scatter plots used to evaluate the relationship between surface soil and stream sediment samples revealed that these sampling media are strongly to moderately correlated for many elements (Fe, As, Bi, Cr, Cu, Hg, Mo, Ni, Sb, Sn, U, and V; Figure 5 and Figure S4), but weakly to other elements (Al, Ag, B, Ba, Cd, Co, Mn, Pb, Se, and Zn; Figure 5 and Figure S4). In terms of source apportionment, in the case of the elements with strong to moderate correlations the composition of soils and stream sediments seems to be both controlled by the bedrock lithologies or, alternatively, the soils themselves are the primary source for the constitution of stream sediment, driven by erosion processes that have been directly influenced by the deforestation in the area over the past decades [69]. The weak correlations may be due to three reasons: (i) Mineralogical sorting during hydrodynamic and sedimentation processes. For instance, Al is a major constituent of kaolinites, which is a naturally occurring mineral in the soils of the Amazon [70], but to a lesser extent in active river sediments; (ii) low sensitivity of the analytical method, which is clearly the case for Ag, B, Ba, Cd, and Se; (iii) Different sources contributing to the enrichment of an element in a given sampling media. For this case, Ba, Co, Mn, Pb, and Zn captured our attention because these elements are well-known for being part of important biogeochemical processes that take place in nature [71].

5.4. Geochemical Threshold Variation in Soils and Stream Sediments vs. Environmental Guidelines

The existence of solid legislations proposing reference-quality guideline values for different regions, demarked by considering a multidisciplinary approach, including the definition of geochemical compartments, instead of simply using political boundaries would be ideal. Under this context, the variation of the background concentrations in soils and stream sediments of the PB and the quality guideline values proposed by Brazilian environmental agencies [50,59,60] should be critically evaluated.

Firstly, the threshold values calculated by using statistical techniques widely applied in geochemical studies [72–74] pointed out different results among each statistical method (Table 3 and Table S3). These differences occur due to the statistical approach and the central criteria of the method, which is widely discussed in the literature [17,22,23,44,45]. In general, the highest background values were obtained by the TIF method, with several overestimated values (cf. Table 3 and Table S3). The P98 and P95 deliver threshold values by considering a fixed percentage of outliers, within the range of values in the data set, which is not entirely appropriate. Among the methods used herein to derive threshold concentration values, the M_{MAD} appears to derive more consistent and realistic results. This conclusion has been also achieved in other studies [17,22,23]. For this reason, the following discussion will be addressed by using the M_{MAD} results (Table 3 and Table S3).

The comparison of the threshold values, in both sampling media, between the NPB and the SPB regions revealed significant discrepancies. For instance, the threshold concentrations of Fe in soils of the NPB and SPB are 34.24 and 6.93 wt.%, respectively, and in stream sediment are 22.93 and 3.54 wt.%, respectively (Table 3). Similar behavior is observed for the majority of the studied elements (cf. Table 3 and Table S3). The differences observed in each region are a direct influence of the natural geological/geochemical variation, already described in this study, confirming the consistency of the results presented here.

The previous results demonstrated the existence of a large natural spatial variation of PTE in the territory of PB and indicated that establishing threshold values for the NPB and SPB regions is more adequate than assuming a uniform value for the entire PB. This conclusion is also relevant in terms of defining quality guideline values for large areas elsewhere. For instance, the comparison of threshold values in NPB and SPB (cf. Table 3 and Table S3) for some PTE [As (NPB = 2.3; SPB = 1.6), Cd (NPB = 0.16; SPB = 0.01), Pb (NPB = 24.4; SPB = 14.2), and Zn (NPB = 83; SPB = 47; all values in mg kg^{-1}] contemplated in the Brazilian environmental resolution of stream sediments [60] shows that they have their threshold concentrations below the Level 1 (L1; also known as the Threshold Effect Level—TEL [75]) proposed by the National Council of the Environment (CONAMA) of Brazil [60] for the mentioned elements (cf. Table 3 and Table S3). Mercury exhibits a threshold value in NPB (0.23 mg kg^{-1}) greater than the L1 (0.17 mg kg^{-1}), but the referenced value in SPB (0.08 mg kg^{-1}) is lower than L1. Nickel shows in NPB threshold value (56.4 mg kg^{-1}) greater than the Level 2 (L2, also known as the Probable Effect Level—PEL [75]; L2 of Ni = 35.9 mg kg^{-1}), but in the SPB the obtained value (12.6 mg kg^{-1}) is lower than L1 (18 mg kg^{-1}). The greatest differences were observed for Cu (482.4 mg kg^{-1}) and Cr (160 mg kg^{-1}), for which threshold values in the NPB are considerably greater than the L2 (Cu = 197; Cr = 90 mg kg^{-1}). On the other hand, the threshold values in SPB for Cr (68 mg kg^{-1}) are between the L2 and L1) (respectively, 90 mg kg^{-1} and 37.3 mg kg^{-1}), and for Cu (20 mg kg^{-1}) below L1 (35.7 mg kg^{-1}).

The quality guidelines of soils in Brazil are based on two resolutions, a Federal resolution [50] applicable for the entire Brazilian territory, which presents Prevention Guideline Values (PGV), and a State resolution for the definition of Quality Reference Value (QRV), which corresponds to the geochemical background. In the absence of QRV for the State of Pará (PA), where the PB is situated, the State Secretariat of the Environment and Sustainability (SEMAS-PA) adopted for the Pará territory the same QRV presented by the São Paulo Sanitation Technology Company (CETESB) [59], derived for the State of São Paulo.

Firstly, it should be emphasized that it is profoundly inadequate to establish geochemical background values for a given area (e.g., the PB situated in the State of Pará) based on values of another region (e.g., the State of São Paulo), with completely different environmental, geological and geochemical characteristics. This is clear when comparing the threshold values of the PTE with the QRV proposed by the State resolution [59]. Among the 14 PTE contemplated in the resolution, ten (As, Ba, Co, Cr, Cu, Hg, Mo, Ni, Pb, and Se) presented generally, in both NPB and SPB, higher threshold values in comparison to the QRV. From this group, when compared to the PGV, which values are somewhat higher than those of QRV (Table 3 and Table S3), the discrepancies with assumed background values of PB are reduced. In contrast, Ag, Cd, and Zn presented lower threshold values than the proposed QRV and PGV, whereas Se showed in NPB and SPB background values higher than QRV and lower than PGV (Table 3 and Table S3).

Secondly, instead of defining realistic threshold values for the entire State and some relevant areas, it appears that environmental agencies tend to define even more conservative QRV values, which is clearly seen by comparing the QRV and PGV values (Table 3 and Table S3). It is highly recommended that new QRV values should be established for the different large geological domains of the State of Pará, in particular for the Carajás region, by using soil samples from the area of investigation and considering the complexity of the geological setting, as similarly conducted in previous studies in the State of Pará [76,77].

The issues addressed above show how understanding the variation of the geochemical background is important for territorial and watershed management. Therefore, the source of the anomalies of the PTE needs to be carefully investigated on a case-by-case basis, considering the local scenario (geology, land use, possible anthropic interventions). It is not demonstrated that high concentrations of some PTE, above the threshold values, are indeed influenced by anthropogenic activities. In the case of PB, the majority of the soil and stream sediment samples with values above the threshold is actually due to bedrock lithologies and hydrothermal mineralized zones (e.g., northern and southern copper belts) that naturally occur in the area. This reinforces the need for new environmental resolutions, which consider the regional characteristics and can thus provide more realistic guideline values.

6. Conclusions

Soils and stream sediments of the PB are strongly influenced by the geological environment of the region. The geochemical maps revealed substantial differences between the NPB and the SPB regions. At the regional scale, the local anomalies are mostly influenced by the bedrock lithologies rather than by any anthropogenic impact. Significant evidence has shown that the metavolcanosedimentary rocks of the CB and, particularly, the mineralizations along the northern and southern Cu belts in the NPB are the main source for several anomalous concentrations of PTE in both sampling media. Considering the elements under investigation in the PB area, the geochemical compositions of soils and, especially, stream sediments of the BD and CB are statistically similar. The soils and stream sediments are strongly to moderately correlated for many elements (Fe, As, Bi, Cr, Cu, Hg, Mo, Ni, Sb, Sn, U, and V), which suggest the same rock source or, in some cases, the soils themselves as the primary source for the constitution of stream sediments, driven by erosion processes and intensified by the deforestation in the area. The comparison between new statistically derived threshold values of the NPB and SPB regions with the threshold values of the whole PB and to quality guidelines proposed by Brazilian environmental agencies demonstrates that a uniform value of quality guideline is not adequate, because it does not consider the natural geological/geochemical variation of the area. For this reason, geochemical compartments, instead of political boundaries, should be considered prior to defining new guideline values. The integrated assessment used here is easily replicated and remarkably useful for territorial and watershed management, important for the Sustainable Development Goals.

Supplementary Materials: The following are available online at https://www.mdpi.com/2571-878 9/5/1/21/s1, Figure S1 Geochemical maps for Ag, B, Cd, Sb, and Se; Figure S2: Geochemical maps for Ba, Bi, Pb, Sn, and U; Figure S3: Boxplots for 8 potentially toxic elements (PTE; Ag, B, Cd, Mo, Sb, Sn, Pb, and U); Figure S4: Scatter plot comparing the concentrations of the studied elements in surface soil and stream sediment samples; Table S1: Descriptive statistics for 10 potentially toxic elements (PTE; Ag, B, Ba, Bi, Cd, Pb, Sb, Se, Sn, and U) in surface soils and stream sediments; Table S2. Geochemical threshold values for 10 potentially toxic elements (PTE; Ag, B, Ba, Bi, Cd, Pb, Sb, Se, Sn, and U) in surface soils and stream sediments.

Author Contributions: Conceptualization, D.S.; Data curation, G.N.S.; Formal analysis, G.N.S.; Funding acquisition, R.D.; Investigation, D.d.L.F.; Methodology, G.N.S.; Supervision, R.D.; Validation, P.K.S.; Writing—original draft, G.N.S. and D.d.L.F.; Writing—review & editing, P.K.S., R.D. and D.S. All authors have read and agreed to the published version of the manuscript.

Funding: This study was financed mostly with resources of the ITV (Instituto Tecnológico Vale) and Gerência Ambiental de Ferrosos Norte (Vale S.A). This study was financed in part by the Coordenação de Aperfeiçoamento de Pessoal de Nível Superior—Brasil (CAPES)—Finance Code 001. This work was partially supported by Conselho Nacional de Desenvolvimento Científico e Tecnológico—CNPq (306108/2014-3 and 304648/2019-1 to RD; 443247/2015-3 and 402727/2018-5 projects coordinated by RD).

Institutional Review Board Statement: Not applicable.

Informed Consent Statement: Not applicable.

Data Availability Statement: The data presented in this study are available on request from the corresponding author. The data are not publicly available due to privacy restrictions.

Acknowledgments: The Instituto Tecnológico Vale provided the geochemical dataset of the Itacaiúnas Geochemical Mapping and Background Project. The authors acknowledge Pedro Walfir Martins e Souza Filho, Wilson da Rocha Nascimento Junior and Jair da Silva Ferreira Júnior for the project conceptualization and sampling design; Marcio Sousa da Silva and Carlos Augusto de Medeiros Filho for the management of the field work and sampling. Marlene F. da Costa (Vale S.A.) for the financial support to the project development.

Conflicts of Interest: The authors declare no conflict of interest.

References

1. Caritat, P.; Cooper, M. A continental-scale geochemical atlas for resource exploration and environmental management: The National Geochemical Survey of Australia. *Geochem. Explor. Environ. Anal.* **2016**, *16*, 3–13. [CrossRef]
2. Cheng, Z.; Xie, X.; Yao, W.; Feng, J.; Zhang, Q.; Fang, J. Multi-element geochemical mapping in Southern China. *J. Geochem. Explor.* **2014**, *139*, 183–192. [CrossRef]
3. Plant, J.; Smith, D.; Smith, B.; Williams, L. Environmental geochemistry at the global scale. *Appl. Geochem.* **2001**, *16*, 1291–1308. [CrossRef]
4. Darnley, A.G.; Bjiirklund, A.; Belviken, B.; Gustavsson, N.; Koval, P.V.; Plant, J.A.; Steenfelt, A.; Tauchid, M.; Xuejing, X.; Xie, X. *A Global Geochemical Database for Environmental and Resource Management. Recommendations for International Geochemical Mapping—Final Report of IGCP Project 259*; UNESCO Publishing: Paris, France, 1995; ISBN 923103085X.
5. Albanese, S.; De Vivo, B.; Lima, A.; Cicchella, D. Geochemical background and baseline values of toxic elements in stream sediments of Campania region (Italy). *J. Geochem. Explor.* **2007**, *93*, 21–34. [CrossRef]
6. Gałuszka, A. Different approaches in using and understanding the term "Geochemical background"—Practical implications for environmental studies. *Pol. J. Environ. Stud.* **2007**, *16*, 389–395.
7. Gałuszka, A.; Migaszewski, Z.M.; Zalasiewicz, J. Assessing the Anthropocene with geochemical methods. *Geol. Soc. London Spec. Publ.* **2014**, *395*, 221–238. [CrossRef]
8. Yuan, G.L.; Sun, T.H.; Han, P.; Li, J. Environmental geochemical mapping and multivariate geostatistical analysis of heavy metals in topsoils of a closed steel smelter: Capital Iron & Steel Factory, Beijing, China. *J. Geochem. Explor.* **2013**, *130*, 15–21. [CrossRef]
9. Smith, S.L.; MacDonald, D.D.; Keenleyside, K.A.; Ingersoll, C.G.; Field, L.J. A preliminary evaluation of sediment quality assessment values for freshwater ecosystems. *J. Great Lakes Res.* **1996**, *22*, 624–638. [CrossRef]
10. Zhuang, W.; Gao, X. Integrated Assessment of Heavy Metal Pollution in the Surface Sediments of the Laizhou Bay and the Coastal Waters of the Zhangzi Island, China: Comparison among Typical Marine Sediment Quality Indices. *PLoS ONE* **2014**, *9*, e94145. [CrossRef] [PubMed]
11. Pereira, V.; Inácio, M.; Ferreira, A.; Pinto, M. Geochemical regional surveys: Comparative analysis of data from soils and stream sediments. In *Proceedings of the 19th World Congress of Soil Science, Soil Solutions for a Changing World*; International Union of Soil Sciences: Brisbane, Australia, 2010; p. 4.
12. Reimann, C. Comparison of stream sediment and soil sampling for regional exploration in the eastern Alps, Austria. *J. Geochem. Explor.* **1988**, *31*, 75–85. [CrossRef]
13. Deschamps, E.; Ciminelli, V.S.T.; Lange, F.T.; Matschullat, J.; Raue, B.; Schmidt, H. Soil and sediment geochemistry of the iron quadrangle, Brazil the case of arsenic. *J. Soils Sediments* **2002**, *2*, 216–222. [CrossRef]
14. Demetriades, A.; Smith, D.; Wang, X. General concepts of geochemical mapping at global, regional, and local scales for mineral exploration and environmental purposes. *Geochim. Bras.* **2018**, *32*, 136–179. [CrossRef]
15. Licht, O. Geochemical background—What a complex meaning has such a simple expression! *Geochim. Bras.* **2020**, *34*, 161–175. [CrossRef]
16. Burenkov, E.; Golovin, A.; Morozova, I.; Filatov, E. Multi-purpose geochemical mapping (1:1,000,000) as a basis for the integrated assessment of natural resources and ecological problems. *J. Geochem. Explor.* **1999**, *66*, 159–172. [CrossRef]
17. Matschullat, J.; Ottenstein, R.; Reimann, C. Geochemical background—Can we calculate it? *Environ. Geol.* **2000**, *39*, 990–1000. [CrossRef]
18. Reimann, C.; Garrett, R.G. Geochemical background—Concept and reality. *Sci. Total Environ.* **2005**, *350*, 12–27. [CrossRef]
19. Reimann, C.; Filzmoser, P.; Garrett, R.G. Background and threshold: Critical comparison of methods of determination. *Sci. Total Environ.* **2005**, *346*, 1–16. [CrossRef]
20. Reimann, C.; Caritat, P. Distinguishing between natural and anthropogenic sources for elements in the environment: Regional geochemical surveys versus enrichment factors. *Sci. Total Environ.* **2005**, *337*, 91–107. [CrossRef]
21. Johnson, C.C.; Demetriades, A. Urban Geochemical Mapping: A Review of Case Studies in this Volume. In *Mapping the Chemical Environment of Urban Areas*; John Wiley & Sons, Ltd.: Hoboken, NJ, USA, 2011; pp. 7–27. [CrossRef]
22. Reimann, C.; Caritat, P. Establishing geochemical background variation and threshold values for 59 elements in Australian surface soil. *Sci. Total Environ.* **2017**, *578*, 633–648. [CrossRef]

23. Reimann, C.; Fabian, K.; Birke, M.; Filzmoser, P.; Demetriades, A.; Négrel, P.; Oorts, K.; Matschullat, J.; de Caritat, P.; Albanese, S.; et al. GEMAS: Establishing geochemical background and threshold for 53 chemical elements in European agricultural soil. *Appl. Geochem.* **2018**, *88*, 302–318. [CrossRef]
24. Sahoo, P.K.; Dall'Agnol, R.; Salomão, G.N.; Ferreira Junior, J.S.; Silva, M.S.; Martins, G.C.; Souza Filho, P.W.M.; Powell, M.A.; Maurity, C.W.; Angelica, R.S.; et al. Source and background threshold values of potentially toxic elements in soils by multivariate statistics and GIS-based mapping: A high density sampling survey in the Parauapebas basin, Brazilian Amazon. *Environ. Geochem. Health* **2020**, *42*, 255–282. [CrossRef] [PubMed]
25. Feio, G.R.L.; Dall'Agnol, R.; Dantas, E.L.; Macambira, M.J.B.; Santos, J.O.S.; Althoff, F.J.; Soares, J.E.B. Archean granitoid magmatism in the Canaã dos Carajás area: Implications for crustal evolution of the Carajás Province, Amazonian Craton, Brazil. *Precambrian Res.* **2013**, *227*, 157–185. [CrossRef]
26. Marangoanha, B.; Oliveira, D.C.; Oliveira, V.E.S.; Galarza, M.A.; Lamarão, C.N. Neoarchean A-type granitoids from Carajás province (Brazil): New insights from geochemistry, geochronology and microstructural analysis. *Precambrian Res.* **2019**, *324*, 86–108. [CrossRef]
27. Moreto, C.P.N.; Monteiro, L.V.S.; Xavier, R.P.; Creaser, R.A.; DuFrane, S.A.; Tassinari, C.C.G.; Sato, K.; Kemp, A.I.S.; Amaral, W.S. Neoarchean and Paleoproterozoic iron oxide-copper-gold events at the Sossego deposit, Carajás Province, Brazil: Re-Os and U-Pb geochronological evidence. *Econ. Geol.* **2015**, *110*, 809–835. [CrossRef]
28. Vasquez, M.L.; Sousa, C.S.; Carvalho, K.M.A. *Mapa Geológico e de Recursos Minerais do Estado do Pará, escala 1:1.000.000. Programa Geologia do Brasil (PGB), Integração, Atualização e Difusão de Dados da Geologia do Brasil, Mapas Geológicos Estaduais*; Serviço Geológico do Brasil (CPRM): Belém, Brasil, 2008.
29. Barros, C.E.M.; Sardinha, A.S.; Barbosa, J.P.O.; Macambira, M.J.B.; Barley, P.; Boullier, A. Structure, petrology, geochemistry and zircon U/Pb and Pb/Pb geochronology of the synkinematic Archean (2.7 Ga) A-type Granites from the Carajás Metallogenic Province, Northern Brazil. *Can. Mineral.* **2009**, *47*, 1423–1440. [CrossRef]
30. Dall'Agnol, R.; Cunha, I.R.V.; Guimarães, F.V.; Oliveira, D.C.; Teixeira, M.F.B.; Feio, G.R.L.; Lamarão, C.N. Mineralogy, geochemistry, and petrology of Neoarchean ferroan to magnesian granites of Carajás Province, Amazonian Craton: The origin of hydrated granites associated with charnockites. *Lithos* **2017**, *277*, 3–32. [CrossRef]
31. Mansur, E.T.; Ferreira Filho, C.F. Chromitites from the Luanga Complex, Carajás, Brazil: Stratigraphic distribution and clues to processes leading to post-magmatic alteration. *Ore Geol. Rev.* **2017**, *90*, 110–130. [CrossRef]
32. Marangoanha, B.; Oliveira, D.C.; Dall'Agnol, R. The Archean granulite-enderbite complex of the northern Carajás province, Amazonian craton (Brazil): Origin and implications for crustal growth and cratonization. *Lithos* **2019**, *350–351*, 105275. [CrossRef]
33. Rosa, W.D. Complexos Acamadados da Serra da Onça e Serra do Puma: Geologia e Petrologia de Duas Intrusões Máfico-ultramáficas com Sequência de Cristalização Distinta na Província Arqueana de Carajás, Brasil. Master's Thesis, Universidade de Brasília, Brasília, Brazil, 2014.
34. Gibbs, A.K.; Wirth, K.R.; Hirata, K.H.; Olszewski Junior, W.J. Age and composition of the Grao Para Group volcanics, Serra dos Carajas, Brazil. *Rev. Bras. Geocienc.* **1986**, *16*, 201–211. [CrossRef]
35. Martins, P.L.G.; Toledo, C.L.B.; Silva, A.M.; Chemale, F.; Santos, J.O.S.; Assis, L.M. Neoarchean magmatism in the southeastern Amazonian Craton, Brazil: Petrography, geochemistry and tectonic significance of basalts from the Carajás Basin. *Precambrian Res.* **2017**, *302*, 340–357. [CrossRef]
36. Sardinha, A.S.; Barros, C.E.M.; Krymsky, R. Geology, geochemistry, and U-Pb geochronology of the Archean (2.74 Ga) Serra do Rabo granite stocks, Carajás Metallogenetic Province, northern Brazil. *J. S. Am. Earth Sci.* **2006**, *20*, 327–339. [CrossRef]
37. Mansur, E.T.; Ferreira Filho, C.F.; Oliveira, D.P.L. The Luanga deposit, Carajás Mineral Province, Brazil: Different styles of PGE mineralization hosted in a medium-size layered intrusion. *Ore Geol. Rev.* **2020**, 103340. [CrossRef]
38. Dall'Agnol, R.; Teixeira, N.P.; Rämö, O.T.; Moura, C.A.V.; Macambira, M.J.B.; Oliveira, D.C. Petrogenesis of the Paleoproterozoic rapakivi A-type granites of the Archean Carajás metallogenic province, Brazil. *Lithos* **2005**, *80*, 101–129. [CrossRef]
39. Machado, N.; Lindenmayer, Z.G.; Krogh, T.E.; Lindenmayer, D. U-Pb geochronology of Archean magmatism and basement reactivation in the Carajás area, Amazon shield, Brazil. *Precambrian Res.* **1991**, *49*, 329–354. [CrossRef]
40. Teixeira, M.F.B.; Dall'Agnol, R.; Santos, J.O.S.; Kemp, A.; Evans, N. Petrogenesis of the Paleoproterozoic (Orosirian) A-type granites of Carajás Province, Amazon Craton, Brazil: Combined in situ Hf O isotopes of zircon. *Lithos* **2019**, *332–333*, 1–22. [CrossRef]
41. Oliveira, F.A.; Almeida, J.A.C.; Figueira, R.L. Magnetic petrology of rocks of the Novolândia Granulite, Cruzeiro do Sul Village, Bacajá Domain. *Rev. Sumaúma* **2018**, *10*, 61–74.
42. Macambira, M.J.B.; Vasquez, M.L.; Silva, D.C.C.; Galarza, M.A.; Barros, C.E.M.; Camelo, J.F. Crustal growth of the central-eastern Paleoproterozoic domain, SW Amazonian Craton: Juvenile accretion vs. reworking. *J. S. Am. Earth Sci.* **2009**, *27*, 235–246. [CrossRef]
43. Souza-Filho, P.W.M.; Sahoo, P.K.; Silva, M.S.; Medeiros Filho, C.A.; Dall'Agnol, R.; Ferreira Junior, J.S.; Nascimento-Junior, W.; Silva, G.S.; Salomão, G.N.; Sarracini, F.; et al. General guidelines for baselines mapping and geochemical background in the Itacaiúnas River watershed, Carajás Mineral Province, Brazilian Amazon. *J. S. Am. Earth Sci.* **2020**. submitted.
44. Sahoo, P.K.; Dall'Agnol, R.; Salomão, G.N.; Ferreira Junior, J.S.F.; Silva, M.S.; Souza Filho, P.W.M.; Costa, M.L.; Angélica, R.S.; Medeiros Filho, C.A.; Costa, M.F.; et al. Regional-scale mapping for determining geochemical background values in soils of the Itacaiúnas River Basin, Brazil: The use of compositional data analysis (CoDA). *Geoderma* **2020**, *376*, 114504. [CrossRef]

45. Salomão, G.N.; Dall'Agnol, R.; Sahoo, P.K.; Angélica, R.S.; Medeiros Filho, C.A.; Ferreira Júnior, J.S.S.; Silva, M.S.S.; e Souza Filho, P.W.M.; Nascimento Junior, W.R.R.; Costa, M.F.; et al. Geochemical mapping in stream sediments of the Carajás Mineral Province: Background values for the Itacaiúnas River watershed, Brazil. *Appl. Geochem.* **2020**, *118*, 104608. [CrossRef]
46. Sahoo, P.K.; Dall'Agnol, R.; Salomão, G.N.; Ferreira Junior, J.S.; Silva, M.S.; Souza Filho, P.W.M.; Powell, M.A.; Angélica, R.S.; Pontes, P.R.; Costa, M.F.; et al. High resolution hydrogeochemical survey and estimation of baseline concentrations of trace elements in surface water of the Itacaiúnas River Basin, southeastern Amazonia: Implication for environmental studies. *J. Geochem. Explor.* **2019**, *205*, 106321. [CrossRef]
47. Grunsky, E.C.; Caritat, P. Advances in the Use of Geochemical Data for Mineral Exploration. In Proceedings of the Exploration 17: Sixth Decennial International Conference on Mineral Exploration; Tschirhart, V., Thomas, M.D., Eds.; Lakehead University: Orillia, ON, USA, 2017; pp. 441–456.
48. Ander, E.L.; Johnson, C.C.; Cave, M.R.; Palumbo-Roe, B.; Nathanail, C.P.; Lark, R.M. Methodology for the determination of normal background concentrations of contaminants in English soil. *Sci. Total Environ.* **2013**, *454–455*, 604–618. [CrossRef]
49. Cave, M.R.; Johnson, C.C.; Ander, E.; Palumbo-Roe, B. Methodology for the determination of normal background contaminant concentrations in English soils. *Br. Geol. Surv. Comm. Rep.* **2012**, *CR/12/003*, 42.
50. Conselho Nacional do Meio Ambiente—CONAMA. *Resolução CONAMA N° 420, de 28 de Dezembro de 2009. Dispõe sobre critérios e valores orientadores de qualidade do solo quanto à presença de substâncias químicas e estabelece diretrizes para o gerenciamento ambiental de áreas contaminadas por essas substâncias em decorrência de atividades antrópicas*; Diário Oficial da República Federativa do Brasil: Brasilia, Brasil, 2009; pp. 81–84.
51. DEC (Department of Environment and Conservation). *Assessment Levels for Soil, Sediment and Water*; Version 4; DEC: Joondalup, Australia, 2010; ISBN 2009641-0110-100.
52. Ministry of the Environment of Finland (MEF) Government Decree on the Assessment of Soil Contamination and Remediation Needs. 214/2007. Available online: http://www.finlex.fi/en/laki/kaannokset/2007/en20070214 (accessed on 19 February 2021).
53. Bonham-Carter, G.F.; Rogers, P.J.; Ellwood, D.J. Catchment basin analysis applied to surficial geochemical data, Cobequid Highlands, Nova Scotia. *J. Geochem. Explor.* **1987**, *29*, 259–278. [CrossRef]
54. Carranza, E.J.M. Geochemical sampling for geological–environmental studies. *J. Geochem. Explor.* **2011**, *111*, 57–58. [CrossRef]
55. Lancianese, V.; Dinelli, E. Different spatial methods in regional geochemical mapping at high density sampling: An application on stream sediment of Romagna Apennines, Northern Italy. *J. Geochem. Explor.* **2015**, *154*, 143–155. [CrossRef]
56. Reimann, C.; Flem, B.; Gasser, D.; Eggen, O.A.; Birke, M. Background values of gold, potentially toxic elements and emerging high-tech critical elements in surface water collected in a remote northern European environment. *Geochem. Explor. Environ. Anal.* **2018**, *18*, 185–195. [CrossRef]
57. Tukey, J.W. *Exploratory Data Analysis*, 1st ed.; Addison-Wesley Publishing Company: Boston, MA, USA, 1977.
58. Farias, D.L. *Geoquímica de Sedimentos Ativos de Corrente e Estimativa de Background Geoquímico na Bacia do Rio Parauapebas—Pará*; Instituto Tecnológico Vale: Belém, Brazil, 2020.
59. Companhia Ambiental do Estado de São Paulo—CETESB. *Decisão de Diretoria-256/2016/E, de 22-11-2016 Dispõe Sobre a Aprovação dos "Valores Orientadores para Solos e Águas Subterrâneas no Estado de São Paulo—2016" e dá Outras Providências*; CETESB: São Paulo, Brazil, 2016; p. 5.
60. Conselho Nacional do Meio Ambiente—CONAMA. *Resolução CONAMA N° 454, de 01 de Novembro de 2012. Estabelece as Diretrizes Gerais e os Procedimentos Referenciais Para o Gerenciamento do Material a Ser Dragado Em águas Sob Jurisdição Nacional*; Diário Oficial da República Federativa do Brasil: Brasilia, Brasil, 2012; p. 17.
61. Tolbert, G.E.; Tremaine, J.W.; Malcher, G.C.; Gomes, C.B. The Recently Discovered Serra dos Carajás Iron Deposits. *Econ. Geol.* **1971**, *66*, 985–994. [CrossRef]
62. Beisiegel, V.R.; Bernardelli, A.L.; Drummond, N.F.; Ruff, A.W.; Tremaine, J.W. Geologia e recursos minerais da Serra dos Carajás. *Rev. Bras. Geociências* **1973**, *3*, 215–242.
63. Docegeo (Rio Doce Geologia e Mineração). Revisão litoestratigráfica da Província Mineral de Carajás. In *Proceedings of the 35 Congresso Brasileiro de Geologia*; CVRD/SBG: Belém, Brazil, 1988; pp. 11–59.
64. Siepierski, L.; Ferreira Filho, C.F. Magmatic structure and petrology of the Vermelho Complex, Carajás Mineral Province, Brazil: Evidence for magmatic processes at the lower portion of a mafic-ultramafic intrusion. *J. S. Am. Earth Sci.* **2020**, *102*, 102700. [CrossRef]
65. Sousa, S.D.; Monteiro, L.V.S.; Oliveira, D.C.; Silva, M.A.D.; Moreto, C.P.N.; Juliani, C. O Greenstone Belt Sapucaia na região de Água Azul do Norte, Província Carajás: Contexto geológico e caracterização petrográfica e geoquímica. *Contrib. Geol. Amaz.* **2015**, *9*, 317–338.
66. Mander, Ü. Watershed Management. In *Encyclopedia of Ecology*; Elsevier: Amsterdam, The Netherlands, 2008; pp. 3737–3748.
67. Tomer, M.D. Watershed Management. In *Encyclopedia of Soils in the Environment*; Elsevier: Amsterdam, The Netherlands, 2005; pp. 306–315.
68. Wang, X.; Chi, Q.; Liu, H.; Nie, L.; Zhang, B. Wide-spaced sampling for delineation of geochemical provinces in desert terrains, northwestern China. *Geochem. Explor. Environ. Anal.* **2007**, *8*, 153–161. [CrossRef]
69. Souza-Filho, P.W.M.; Souza, E.B.; Silva Júnior, R.O.; Nascimento, W.R., Jr.; Mendonça, B.R.V.; Guimarães, J.T.F.; Dall'Agnol, R.; Siqueira, J.O. Four decades of land-cover, land-use and hydroclimatology changes in the Itacaiúnas River watershed, southeastern Amazon. *J. Environ. Manag.* **2016**, *167*, 175–184. [CrossRef]

70. Souza, E.S.; Fernandes, A.R.; Braz, A.M.S.; Oliveira, F.J.; Alleoni, L.R.F.; Campos, M.C.C. Physical, chemical, and mineralogical attributes of a representative group of soils from the eastern Amazon region in Brazil. *Soil* **2018**, *4*, 195–212. [CrossRef]
71. Kabata-pendias, A.; Pendias, H. *Trace Elements in Soils and Plants*, 3rd ed.; CRC Press: Boca Raton, FL, USA, 2001; ISBN 0849315751.
72. Gałuszka, A.; Migaszewski, Z.; Duczmal-Czernikiewicz, A.; Dołęgowska, S. Geochemical background of potentially toxic trace elements in reclaimed soils of the abandoned pyrite–uranium mine (south-central Poland). *Int. J. Environ. Sci. Technol.* **2016**, *13*, 2649–2662. [CrossRef]
73. Yotova, G.; Padareva, M.; Hristova, M.; Astel, A.; Georgieva, M.; Dinev, N.; Tsakovski, S. Establishment of geochemical background and threshold values for 8 potential toxic elements in the Bulgarian soil quality monitoring network. *Sci. Total Environ.* **2018**, *643*, 1297–1303. [CrossRef]
74. Urresti-Estala, B.; Carrasco-Cantos, F.; Vadillo-Pérez, I.; Jiménez-Gavilán, P. Determination of background levels on water quality of groundwater bodies: A methodological proposal applied to a Mediterranean River basin (Guadalhorce River, Málaga, southern Spain). *J. Environ. Manag.* **2013**, *117*, 121–130. [CrossRef]
75. MacDonald, D.D.; Ingersoll, C.G.; Berger, T.A. Development and evaluation of consensus-based sediment quality guidelines for freshwater ecosystems. *Arch. Environ. Contam. Toxicol.* **2000**, *39*, 20–31. [CrossRef]
76. Lima, M.W.; Hamid, S.S.; Souza, E.S.; Teixeira, R.A.; Conceição Palheta, D.; Faial, K.C.F.; Fernandes, A.R. Geochemical background concentrations of potentially toxic elements in soils of the Carajás Mineral Province, southeast of the Amazonian Craton. *Environ. Monit. Assess.* **2020**, *192*, 649. [CrossRef]
77. Fernandes, A.R.; Souza, E.S.; Souza Braz, A.M.; Birani, S.M.; Alleoni, L.R.F. Quality reference values and background concentrations of potentially toxic elements in soils from the Eastern Amazon, Brazil. *J. Geochem. Explor.* **2018**, *190*, 453–463. [CrossRef]

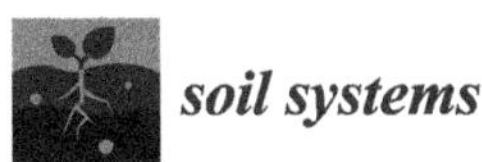

soil systems

MDPI

Article

Heavy Metals Contamination of Urban Soils—A Decade Study in the City of Lisbon, Portugal

Hugo Félix Silva [1,2,3], Nelson Frade Silva [1,2,3], Cristina Maria Oliveira [2] and Manuel José Matos [1,3,4,*]

[1] ISEL-ADEQ, Área Departamental de Engenharia Química do Instituto Superior de Engenharia de Lisboa, Rua Conselheiro Emídio Navarro, 1, 1059-007 Lisboa, Portugal; hugo.felix.silva@isel.pt (H.F.S.); nelson.silva@isel.pt (N.F.S.)

[2] Centro de Química Estrutural, Faculdade de Ciências, Universidade de Lisboa, Campo Grande, 1749-016 Lisboa, Portugal; cmoliveira@fc.ul.pt

[3] CIEQB, Instituto Superior de Engenharia de Lisboa, Rua Conselheiro Emídio Navarro, 1, 1059-007 Lisboa, Portugal

[4] IT, Instituto de Telecomunicações, IST, Avenida Rovisco Pais, 1049-001 Lisboa, Portugal

* Correspondence: manuel.matos@isel.pt

Abstract: There is an intense and continuous growth of the world population living in cities. This increase in population means an increase in car traffic, an increase in new constructions and an increase in the production of waste that translates into an intensive use of land, particularly in terms of soil contaminants. Among other environmental contaminants, toxic metals, such as lead (Pb), cadmium (Cd), nickel (Ni) and chromium (Cr) represent a public health problem. In this study the content of toxic metals in Lisbon's (Portugal) soils was determined. The study was conducted over approximately a decade in six city locations, with a total of about 700 samples. Each site has different urban characteristics: traffic zone, residential area, urban park and mixed areas. The study allowed to verify the heterogeneity of metal content values in the city soils and their dependence on local traffic. Metal contents were determined by graphite furnace atomic absorption spectroscopy (GFAAS). For each site the geo-accumulation index, pollution factor, degree of contamination, pollution load index and ecological risk factor were calculated. The mean concentrations of Cd, Cr, Ni and Pb in soils were 0.463, 44.0, 46.6 and 5.73 mg/kg of dry soil, respectively. In the last year of the study the values were 0.417, 51.5, 62.4 and 8.49 mg/kg of dry soil, respectively. Cd and Ni exceeded the typical content values of these metals in the earth's crust, indicating their anthropogenic origin. The correlation analysis revealed a significant correlation between Cr and Ni, Cd and Ni and Cd and Pb contents in the city soils. Regarding the results obtained in this long monitoring campaign, Lisbon's soils can be considered as having low levels of pollution by these metals.

Keywords: urban soils; heavy metals pollution; soil contamination; pollution indexes; health risk; GFAAS

Citation: Silva, H.F.; Silva, N.F.; Oliveira, C.M.; Matos, M.J. Heavy Metals Contamination of Urban Soils—A Decade Study in the City of Lisbon, Portugal. *Soil Syst.* **2021**, *5*, 27. https://doi.org/10.3390/soilsystems5020027

Academic Editors: Matteo Spagnuolo, Paola Adamo and Giovanni Garau

Received: 10 March 2021
Accepted: 9 April 2021
Published: 13 April 2021

Publisher's Note: MDPI stays neutral with regard to jurisdictional claims in published maps and institutional affiliations.

Copyright: © 2021 by the authors. Licensee MDPI, Basel, Switzerland. This article is an open access article distributed under the terms and conditions of the Creative Commons Attribution (CC BY) license (https://creativecommons.org/licenses/by/4.0/).

1. Introduction

The population living in urban areas has been continuously increasing in the last few decades [1,2] and it is estimated by the United Nations that nowadays the majority of people lives in urban areas, not only in Europe but also in less developed countries [3]. Urban environments have then become extremely important in what human health and wellbeing is concerned. Urban soils are a fundamental component of urban ecosystems contributing directly or indirectly to the general quality of life of cities' inhabitants. They play an important role in many processes such as in the cycling of elements, filtration of water, supporting plants and some built infrastructures. However, soils are very often contaminated by the anthropogenic activities like vehicle traffic or industrial activities [4,5].

Different anthropogenic sources, just like industry, energy production and fossil fuels combustion in vehicle traffic, release to the atmosphere pollutants containing heavy metals that can contribute to the contamination of soils by dry or wet deposition [6,7]. Heavy metals, such as Pb, Zn, Cr, Ni or Cd, are the most frequently studied inorganic contaminants in urban soils and can be linked to adverse health hazard effects [8–10] since they cannot be decomposed by micro-organisms having a long-term toxicity for plants, animals and humans. High concentrations of heavy metals have been identified in many studies of urban soils, especially Pb [11] due to the use of leaded petrol [12,13]. With regulations like the one banning leaded petrol [14], emissions from single vehicles have been reduced, although this effect could be compensated by the worldwide increasing road traffic [15,16]. Emissions resulting from vehicle traffic are mostly caused by wear of vehicular components like break lining, tire wear off and exhaust, as well as from incomplete fuel combustion, fuel additives or oil leaking from vehicles [17].

Due to the importance of the quantitative identification of potential sources of heavy metals in soils, source apportionments for these metals concentrations have been performed by principal component analysis (PCA) and positive matrix factorization (PMF) models. These models assume significant correlations of compounds derived from the same sources [18,19]. Moreover, some indexes are fairly used to evaluate the degree of heavy metal pollution in the soil. The most common one is the enrichment factor (*EF*) [18,20,21] but others, namely the geo-accumulation index (I_{geo}) [20], the contamination factor (C_f) [22–24], the pollution load index (*PLI*) [21,22] or the ecological risk factor [23,24] are also widely used.

In cities, soil pollution is important due to the health risks of its inhabitants. Many particulate pollutants in the air are originated in soil and its composition and size determine its impact on health. In this work, soils from a set of carefully chosen different locations in the city were sampled from 2003 to 2011 and heavy metals contents (Pb, Cr, NI and Cd) were analyzed by GFAAS (graphite furnace atomic absorption spectrometry). The metals under study were considered due to their negative effects on human health and are therefore metals that are subject to surveillance in the European regulation of atmospheric pollution [25] and pollution of soils and groundwater [26]. Results are discussed in terms of the level of contamination of these soils as well as pollution and ecological risks associated with that contamination.

2. Materials and Methods

2.1. Sampling Sites

Soil samples were collected from six different sites in the city of Lisbon. Lisbon is the capital city of Portugal with about half a million inhabitants and 2–2.5 millions of commuters every day. Since industrial activity is almost inexistent in the city, the main anthropogenic source of air and soil pollution relies on vehicle traffic.

The sampling sites (Figure 1) were chosen taking as criteria the level of automotive traffic, the population density, and the topology of the site. The existence of a high population density is important in order to reduce the health risk due to soil pollution, in particular from its use in urban gardens or parks. The topology of the site is critical to interpret whether the dispersion of the pollutants is favored in each location.

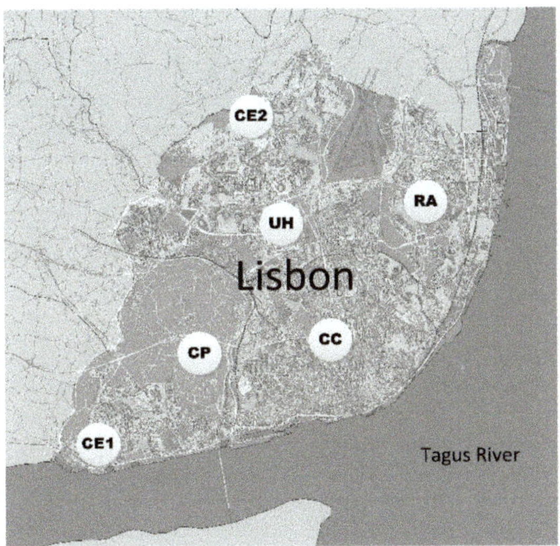

Figure 1. Soil sampling sites in the city of Lisbon.

The characteristics of each location are: CE1—City Entrance in Belém (GPS: 38.69562, −9.19930)—one of the city entrances with an open topology and characterized by a high traffic intensity as well as a high population density; CE2—City Entrance in Calçada de Carriche (38.72653, −9.14992)—another city entrance, with a valley topology and a steep slope of entrance in the city. It is also characterized by a high traffic intensity and a high population density; UH—Urban Highway of Segunda Circular (38.75796, −9.16339)–a city circular road with urban highway characteristics. Moreover, it is characterized by an open topology and a very high traffic intensity; CC—City Center at Marquês de Pombal (38.78163, −9.16551)—it has a valley topology and is a high traffic, as well as a high population density site; RA—Residential Area (38.75741, −9.11784)—in the neighborhood of Olivais with open topology and a low traffic intensity; CP—City Park of Monsanto (38.73387, −9.17786)—a small sloping mountain and the major city green park. It has a low traffic intensity and no residential population or buildings.

2.2. Sampling Procedure

Soil sampling was performed according to ASTM E1727-04/05, Sampling and analysis of soil, part IIIa recommendations [27]. Soil sampling was carried out in November of each year. Samples were collected following a concentric profile around a tree in which leaves were also collected for analysis. The soils were collected according to the four cardinal points (north, south, east and west) at 1, 3 and 5 m radius circle. The total area covered by this systematic sampling procedure was 100 m^2. Soil samples were collected at a depth of 20 cm, using plastic (PVC) tubes with 3.5 cm in diameter and stored in black plastic sampling bags. Sample handling was always performed with plastic instruments in order to ensure non-contamination of the soils by metals from usual sampling tools. Each collected sample was identified with a label encoding information of the sampled site, year, cardinal point and distance to the tree, in order to ensure full traceability of the samples. A total of 648 soil samples were collected in the six sampling sites of the city and for nine years.

2.3. Sample Preparation

In the laboratory, soil samples were dried for 12 h in an oven at about 50 °C, followed by manual cleaning using a plastic tweezer to remove roots, stones, glasses, snails, earthworms and insects. Thereafter samples were sieved for 20 min on a 2 mm nylon sieve using a Retsch AS200 shaker (Retsch, Düsseldorf, Germany). Only the fraction of soil with particles size less than 2 mm was taken. The resulting sample was homogenized and divided by the quartile method into four parts.

Two portions of each quartile were dried at 105 °C, for 24 h in porcelain capsules. After complete drying, soils were stored in desiccants until grinding. Grinding was performed on a Retsch S100 ball mill of agate grinding jars and agate balls for 5 min at 400 rpm. After milling, the diameter of the soil particles was less than 1 μm. About 50 cm^3 of each of the 12 samples from each local and each year were stored and used as the composite sample of that local and that year. Two portions of the composite sample, of about 100 g each, were used to perform the duplicate analysis of the studied metals.

2.4. Soil Characterization

Composite samples of the first (2004) and last (2011) years, from each site were characterized in terms of pH, conductivity, humidity and organic matter content.

The procedure for pH measurement was conducted according to ISO 10390:2015 [28] and involved the preparation of a 10 g of soil in 25 mL of CaCl$_2$, 0.01 M suspension, followed by pH measurement with a calibrated pH meter. For conductivity measurement a 20 g of soil in 100 mL of ultrapure water suspension was used. The humidity of the samples was determined according to ISO 11465:1993 [29] with the samples to be heated at 105 °C until constant weight. The humidity was then calculated by the weight difference between the initial and the dried sample. The determination of organic matter according to European Standard CEN/TC [30] involved heating and calcining the samples at 550 °C for the complete destruction of organic matter. This procedure was performed after the determination of sample humidity and the samples were calcined until constant weight. The organic matter contents were calculated by the weight difference between the dried and the calcined samples.

2.5. Metals Determination

For the determination of the metals under study in soils, 2 g of each sample were subjected to an acidic digestion process as described in ISO 11466:2010 [31]. The extraction of the metal content was conducted with *aqua-regia*, using a conventional heating process (open sand bath system) at (120 ± 10) °C. The residue from the acidic digestion was filtrated using a cellulose filter and diluted to a final volume of 100 mL with ultrapure water (18.2 MΩ cm).

Each metal content in the soil samples was determined by GFAAS, in a Thermo Elemental SOLAAR, Suite M5 Spectrometer, equipped with an automatic sampler FS95 and a graphite furnace GF95, controlled by SOLAAR software, v.11.2.

2.6. Evaluation of Environmental Risks—Pollution and Ecological Index Models

Determining the concentration of toxic metals in soil is not enough to assess the state of soil pollution. It is common to use several indexes to estimate the degree of soil pollution and the ecological risk of the soil for humans and biota. The most important are the enrichment and contamination factors, the pollution and geo accumulation indexes and the ecological risk factor. These indexes are presented and explained below according to their original definition in the references cited for each one.

2.6.1. Geo-Accumulation Index (I_{geo})

The Geo-accumulation index (I_{geo}) is calculated with the formula proposed by Förstner and Müller [32]. This relationship (Equation (1)) is used to assess the degree of contamination of soil and water by metals.

$$I_{geo} = log_2 \frac{C_M}{1.5 \times C_B} \tag{1}$$

In Equation (1) C_M represents the concentration of the metal in the soil and C_B represents the background concentration of that same metal in unpolluted soils.

According to Förstner and Müller [32], I_{geo} values are categorized into seven contamination classes ranging from Class 0 ($I_{geo} < 0$) classified as "uncontaminated" until Class 6 ($I_{geo} > 5$) classified as "extremely contaminated".

It should be noted that I_{geo} value greater than 6 is an indication of soils with 100 times the metal concentration considered as background value, a very high pollution by the metal under study.

2.6.2. Contamination Factor (C_f)

The contamination factor defined almost simultaneously by Tomlinson et al. [22] and Hakanson [23] was used to measure levels of metal pollution in sediments in lakes. However, these factors were widely adopted for the calculation of pollution indexes also in soil. The factor is defined as the ratio between the metal concentration in the soil (C_M) and the reference concentration of that same metal in unpolluted soil (C_b), the background values or preindustrial reference values (Equation (2)).

$$C_f = \frac{C_M}{C_b} \tag{2}$$

Considering the calculated index, soils can be classified according to the contamination levels proposed by Hakanson [23] from "low contamination" ($C_f < 1$) to "very high contamination" ($C_f \geq 6$).

2.6.3. Degree of Contamination (C_d)

The degree of contamination (C_d), also defined by Hakanson [23], represents the sum of the contamination factors. While the contamination factor represents the individual contribution of each metal, the degree of contamination considers all the n polluting metals in a given location, and is defined by Equation (3):

$$C_d = \sum C_f \tag{3}$$

The classification used for this factor was also proposed by Hakanson [23] and goes from "low degree of contamination" for $C_d < 8$ to "very high degree of contamination" for $C_d \geq 32$ indicating serious anthropogenic pollution.

2.6.4. Pollution Load Index (PLI)

The Pollution load index (PLI) proposed by Tomlinson et al. [22] measures the overall degree of contamination in the sample and is often used to calculate the degree of soils pollution and other polluted ecosystems. For a site or zone, it is calculated using Equation (4):

$$PLI_{zone} = \sqrt[n]{C_f(1) \times C_f(2) \times C_f(3) \times \ldots \times C_f(n)} \tag{4}$$

Adapting the concept defined for the calculation of PLI for an estuary, the PLI for a city can be obtained using the PLI of each zone (Equation (5)).

$$PLI_{city} = \sqrt[n]{PLI_{zone\ 1} \times PLI_{zone\ 2} \times PLI_{zone\ 3} \times \ldots \times PLI_{zone\ n}} \tag{5}$$

2.6.5. Ecological Risk Factor (E_r) and Global Potential Ecological Risk (RI)

The ecological risk factor (E_r) proposed by Hakanson [23] is used to assess the ecological risk of an element in the soil and can be obtained using Equation (6):

$$E_r^i = T_r^i \times C_f^i \tag{6}$$

where C_f^i is the contamination factor of element i and T_r^i is the toxic response factor of element i. The toxic response factors for Cd, Cr, Ni and Pb are 30, 2, 5, 5, respectively [23].

E_r values proposed by Hakanson [23] are divided in five ranges from "low potential ecological risk" for $E_r < 40$ to "very high ecological risk" when $E_r > 320$. The sum of the individual potential risks (E_r^i) gives the global potential ecological risk (*RI*) for the soil (Equation (7)).

$$RI = \sum E_r^i = \sum T_r^i \times C_f^i \tag{7}$$

Originally, the global ecological risk was a diagnostic tool for water pollution control but presently it has been used with success for assessing quality of sediments and soils contaminated by heavy metals [33–36]. The classification of the global ecological risk (*RI*), according to Hakanson [23], varies from "low global ecological risk" (*RI* < 150) to "very high global ecological risk" (*RI* > 600).

2.7. Quality Control and Quality Assurance

Analyses were performed carrying out quality assurance and quality control procedures to provide high analytical precision. Special care was taken to minimize cross-contamination, contamination of glass material. Additionally, all efforts were made to minimize contamination from air.

The GFAAS operating conditions were optimized for each metal using reference metal standard solutions, CertiPUR® from Merck (Darmstadt, Germany). Sample concentrations were determined using calibration curves evaluated in terms of linearity and sensitivity, according to the obtained correlation coefficient and slope, respectively. Limits of detection and quantification for each metal were determined. The validity of the calibration curves was performed daily using control standards prepared using CertiPUR® standards from Merck in the same acid matrix used for soil samples. For the control standards the same procedure analysis used for real samples was applied. Two control standards were used to test the precision of the calibration curve of each metal in two points: ~25% and 75% of the linear range. The precision of the calibration curve was maintained between 2% and 5% and never higher than 8%.

The extraction method was validated through a soil certified reference material (BCR CRM-142R Light Sandy Soil) from IRMM (Institute for Reference Materials and Measurements). The reference soil was subjected to the same acid digestion procedure as the soil samples and the recovery percentage values of the CRM were 93% for Pb, 106% for Cd, 93% for Ni and 0% for Cr (since the CRM did not contain chromium).

All samples were digested and analyzed in duplicate, and the final content was always considered as the average value of six measurements made in repeatability conditions.

2.8. Statistical Analysis

The software SPSS® v22.0 from IBM Corp. was used to perform a correlation analysis between the content of each metal in the different sites and the several years studied. Principal component analysis (PCA) was carried out with Varimax rotation with Kaiser normalization, in order to verify the existence of relationships between metal contents in soils analyzed in different city sites over the studied years [37].

3. Results and Discussion

3.1. Soil Characterization

Composite soil samples of the years 2004 and 2011, from each site, were characterized in terms of pH, conductivity, humidity and organic matter content. The values obtained are shown in Table 1.

Table 1. Soil characteristics in two years of sampling.

Year	Rainfall (mm)	Humidity (%)	Organic Mat. (%)	Conductivity (µS/cm)	pH
2004	553	0.79 ± 0.05	7.11 ± 1.74	316 ± 46	7.41 ± 0.16
2011	1045	4.28 ± 0.16	9.58 ± 2.44	328 ± 43	7.34 ± 0.17

When comparing both years, the values of pH, organic matter and electrical conductivity are identical. The main difference is found in the humidity values, as they depend directly on the rain on the days before sampling and not on annual precipitation. The paired *t*-test performed between each of the variables and for the two years, showed that all data pairs are statistically significant ($p = 0.05$) with the exception of humidity data.

3.2. Soil Metals Contents

Metal content values of the sampled soils in the different city sites are presented in Table 2. Overall, and for most metals there is an increase in their content in soils over the years, except for two sites, residential area (RA) and city park (CP). Detailing the analysis of the data for each metal it was found that cadmium levels increased at the entrances (EC1 and EC2) and in the city center (CC), registering however a slight decrease in these places in the last year of sampling. In the urban highway (UH) the levels are stable, showing a decrease over the years in the residential area and in the city park, although being smaller the decreasing trend in these locations. the values for each metal content in each sampled site over the years are shown in Figure 2.

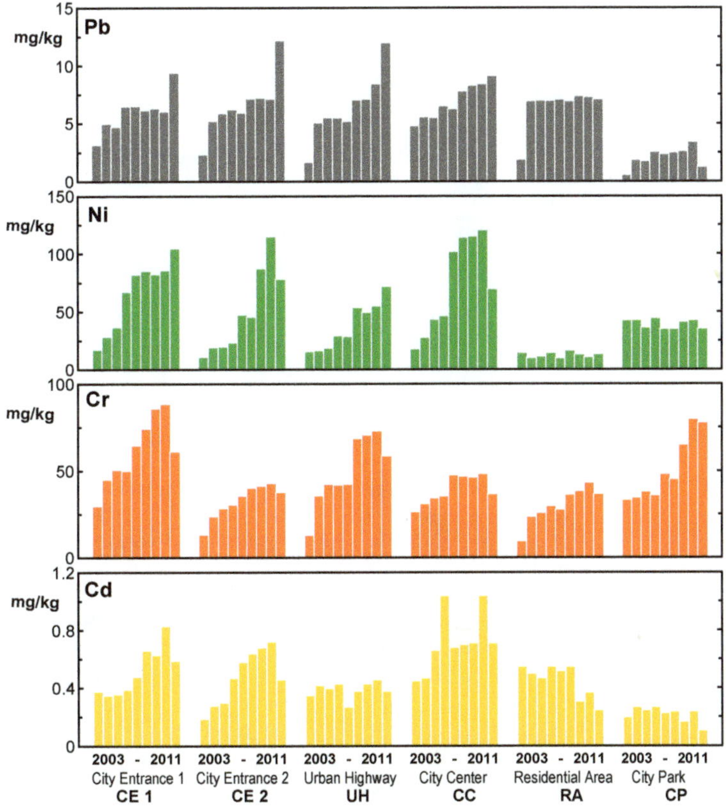

Figure 2. Metal content (Pb, Ni, Cr and Cd) in each sampled site over the years of the campaign (2003–2011).

Table 2. Metal content (mg/kg) values for each sampling site and each year, minimum (Min), maximum (Max), mean (Mean) values and relative standard deviation of the mean (RSD %). Expanded uncertainty values (U), for 95% confidence level, are shown in the last column.

Metal	Site	Year													
		2003	2004	2005	2006	2007	2008	2009	2010	2011	Min	Max	Mean	RSD (%)	U
Cd	CE 1	0.38	0.35	0.36	0.39	0.48	0.66	0.63	0.83	0.59	0.35	0.83	0.52	32	±0.05
	CE 2	0.19	0.28	0.30	0.47	0.58	0.64	0.68	0.72	0.46	0.19	0.72	0.48	40	±0.07
	UH	0.35	0.42	0.40	0.43	0.27	0.38	0.43	0.46	0.38	0.27	0.46	0.39	14	±0.04
	CC	0.45	0.47	0.66	1.04	0.68	0.70	0.71	1.04	0.71	0.45	1.04	0.72	29	±0.07
	RA	0.55	0.50	0.47	0.55	0.52	0.55	0.31	0.37	0.25	0.25	0.55	0.45	25	±0.06
	CP	0.20	0.27	0.25	0.27	0.23	0.24	0.17	0.24	0.11	0.11	0.27	0.22	24	±0.04
Year average		**0.35**	**0.38**	**0.41**	**0.53**	**0.46**	**0.53**	**0.49**	**0.61**	**0.42**	**0.35**	**0.61**	**0.46**	**18**	**±0.14**
Cr	CE 1	29.8	45.3	50.7	50.1	64.7	74.3	85.9	88.5	61.3	29.8	88.5	61.2	32	±1.6
	CE 2	13.3	23.8	28.6	30.4	35.7	40.4	41.5	43.1	37.7	13.3	43.1	32.7	30	±0.9
	UH	12.8	35.8	42.5	42.1	42.4	68.6	70.7	73.1	58.6	12.8	73.1	49.6	40	±1.6
	CC	26.3	31.0	34.4	35.6	47.8	47.0	46.6	48.6	36.9	26.3	48.6	39.4	21	±0.9
	RA	9.6	23.6	25.9	29.7	27.6	36.4	38.5	43.5	36.7	9.6	43.5	30.2	34	±0.9
	CP	33.3	34.4	38.2	35.8	48.4	45.4	65.2	79.9	77.9	33.3	79.9	50.9	37	±1.8
Year average		**20.9**	**32.3**	**36.7**	**37.3**	**44.4**	**52.0**	**58.1**	**62.8**	**51.5**	**20.9**	**62.8**	**44.0**	**31**	**±3.3**
Ni	CE 1	17.8	28.7	37.0	67.5	82.4	85.3	82.7	86.0	104.6	17.8	104.6	65.8	46	±2.5
	CE 2	11.4	19.6	20.2	23.7	47.8	45.9	87.7	114.9	78.5	11.4	114.9	50.0	72	±2.3
	UH	16.0	16.8	19.1	29.6	29.0	53.9	49.6	55.3	72.1	16.0	72.1	37.9	54	±2.1
	CC	18.2	28.2	43.7	46.7	102.0	114.3	115.4	120.4	70.0	18.2	120.4	73.2	55	±2.3
	RA	14.7	10.1	11.6	14.9	9.8	16.9	13.6	10.6	13.7	9.8	16.9	12.9	19	±1.3
	CP	42.8	43.2	36.3	44.9	35.4	35.4	41.5	42.8	35.5	35.4	44.9	39.8	10	±1.2
Year average		**20.2**	**24.4**	**28.0**	**37.9**	**51.1**	**58.6**	**65.1**	**71.7**	**62.4**	**20.2**	**71.7**	**46.6**	**42**	**±4.9**
Pb	CE 1	3.2	5.0	4.7	6.5	6.5	6.1	6.3	6.0	9.4	3.2	9.4	6.0	28	±0.2
	CE 2	2.3	5.2	5.9	6.2	5.9	7.1	7.2	7.1	12.2	2.3	12.2	6.6	39	±0.2
	UH	1.7	5.1	5.5	5.5	5.2	7.0	7.1	8.4	12.0	1.7	12.0	6.4	44	±0.2
	CC	4.8	5.6	5.5	6.5	6.3	7.8	8.3	8.4	9.1	4.8	9.1	6.9	22	±0.2
	RA	1.9	6.9	7.0	7.0	7.1	6.9	7.4	7.3	7.1	1.9	7.4	6.5	27	±0.2
	CP	0.6	1.9	1.7	2.6	2.3	2.5	2.6	3.4	1.3	0.6	3.4	2.1	40	±0.1
Year average		**2.4**	**4.9**	**5.1**	**5.7**	**5.5**	**6.2**	**6.5**	**6.8**	**8.5**	**2.4**	**8.5**	**5.7**	**29**	**±0.4**

Chromium levels exhibited, in all sampled sites, an overall increase between 2003 and 2010 and a slight decrease in the last year (2011). Nickel and lead showed a growing trend of their content in the soils over the years except for the residential area and the city park. In these two places the levels remained stable over time. For nickel, however, two sites (CE2 and CC) presented a decrease in 2011.

The decrease in the levels of some metals in the last year of sampling may be directly linked to the decrease in car traffic in the city of Lisbon. In fact, a serious economic crisis began in 2010 in Portugal, which was reflected in the decline of economic activity and in the increase in unemployment with less people moving around the city, either in private or corporate vehicles, or in public transports.

Table 2 also shows the average annual contents for each metal in the city (values in bold), obtained from the average values of each metal in all locations and for each year.

Figure 3 shows an increasing trend in the levels of the various metals in different places over the years except for the last year of sampling and for Cr, Cd and Ni. The average metal values observed over the years of this campaign are close to 0.46 mg/kg for Cd, 44 mg/kg for Cr, 46 mg/kg for Ni and 5.7 mg/kg for Pb. These values represent moderate metal pollution in the soils of Lisbon as also revealed by the calculated environmental and ecological risk indexes presented below. Other authors [38], present identical results to this study observing low levels of these metals in the soils of Lisbon. The city soils can be used for agriculture, in particular for urban gardens, without significant risk of contamination for the vegetables growing there.

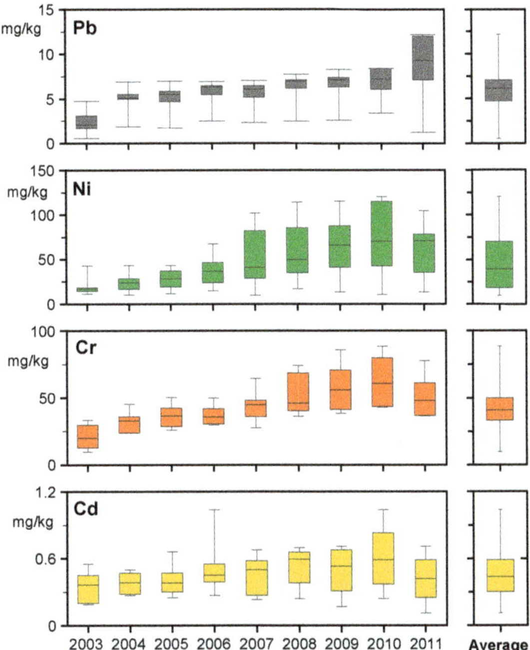

Figure 3. Boxplot of the evolution of metal content in soils over the years of the study.

It is presented in Table 3 the mean, minimum and maximum values of the studied metals content from several studies all around the world, which were used to compare with the values obtained in the present work. Except for lead, the average values for each metal shown in Table 2 are within the limits indicated for the crust. Pb has a higher value in the soils of the studied cities, probably due to emissions from burning leaded gasoline throughout many decades. The values of the metal levels in the soils reveal the reality of each city in terms of anthropogenic emissions, namely the existence of manufacturing facilities and the burning of fossil fuels, both from heating and car traffic. Moreover, the baseline values of metals soil content at each site are important but not available in all or in most cases. Additionally, it must be emphasized that the analytical techniques that allow the determination of metal concentration levels in soils, GFAAS and ICP-MS (inductively coupled plasma mass spectrometry) among others, were only developed and implemented in the decades of 1970s and 1980s. At that time, soils already presented reasonable levels of anthropogenic pollution due to many years of industrialization and low environmental protection by state regulation.

Table 3. Metal content (mg/kg) mean, minimum (min) and maximum (max) values from different places all around the world.

City, Country	Cd		Cr		Ni		Pb		Reference
	Mean	(Min–Max)	Mean	(Min–Max)	Mean	(Min–Max)	Mean	(Min–Max)	
Athens, Greece	0.4	(0.1–3.5)	163	(43–1586)	111	(27–727)	77	(3–2764)	[39]
Aveiro, Portugal	-		59	(15–83)	39	(13–57)	44	(17–64)	[40]
Beijing, China	0.148	(0.01–0.971)	35.6	(7.00–228.2)	27.8	(2.80–168.9)	28.6	(5.00–116.6)	[41]
Berlin, Germany	0.68	(nd–20.3)	30	(nd–168)	10.1	(nd–44.5)	133	(nd–1490)	[42]
Glasgow, U.K.	-		17	(6–37)	16	(6–33)	24	(8–42)	[40]
Huelva, Spain	0.81	(0.02–20.3)	31.7	(3–112.1)	16.1	(0.9–48.7)	135.2	(7.22–5469)	[43]
Kavala, Greece	0.2	(nd–1.2)	232.4	(50–692)	67.9	(25–267)	387	(75–2500)	[44]
Ljubljana, Slovenia	-		41	(24–66)	39	(30–56)	40	(30–57)	[40]
New York, EUA	<0.4	(<0.4–3.1)	12.8	(3.2–366.1)	10	(<2.8–38)	102	(11–2455)	[45]
Oslo, Norway	0.41	(0.06–3.10)	32.5	(2.85–224)	28.4	(2.23–232)	55.6	(<5–1000)	[46]
Palermo, Italy	0.68	(0.27–1.86)	34	(12–100)	17.8	(7.0–38.6)	202	(57–682)	[47]
São Paulo, Brazil	0.1	(0.04–0.22)	37	(15–90)	36	(11–67)	10	(4–23)	[48]
Shanghai, China	0.52	(0.19–3.66)	107.9	(25.5–233)	31.14	(4.95–65-70)	70.69	(13.7–192)	[49]
Stockholm, Sweden	0.16	(0.02–0.25)	35	(6–25)	16	(2.1–27.8)	19	(2.4–32)	[50]
Torino, Italy	-	-	14	(2–25)	20	(4–35)	25	(8–44)	[40]
Lisbon, Portugal	-	-	38	(1.0–172)	43	(2.0–209)	89	(4.8–561)	[51]
Lisbon, Portugal	0.23	(0.01–1.28)	48	(6.8–201)	42	(4.3–194)	114	(4.7–305)	[52]
Lisbon, Portugal	-	-	35.3	(12.2–121)	23.9	(8.0–95)	33.9	(1.0–110)	[38]
Lisbon, Portugal	**0.41**	**(0.11–1.04)**	**51.5**	**(9.61–88.5)**	**62.4**	**(9.77–120.4)**	**8.5**	**(0.55–12.2)**	**This study**
Average	**0.396**	**(nd–20.3)**	**55.6**	**(nd–1586)**	**34.6**	**(nd–727)**	**84.1**	**(nd–5469)**	

3.3. Pollution and Ecological Indexes

In order to estimate the degree of soil pollution and the ecological and environmental risk of Lisbon soils, the most important indexes to assess the level of pollution and ecological risk were calculated: geo-accumulation (I_{geo}), contamination factor (C_f), degree of contamination (C_d), pollution load index (PLI), ecological risk factor (E_r) and global ecological risk (RI). The authors had no knowledge or information concerning any past published values for the levels of metals in Lisbon soils in unpolluted conditions.

In the absence of background values, one of the solutions often used is to consider the estimated values for the earth's crust as a background. The crust values are global values that are intended to represent the average composition of the crust based not on global and systematic measurements but on values resulting from the combination of several studies. There is obviously an over-representation of countries and regions in which studies were carried out, in comparison to less studied areas. In the specific case of Lisbon, the values proposed for the crust are higher than some of the metal contents found. This shows that the values of the crust cannot be used as reference values in this study. Table 4 presents some values from literature taken as reference values for the concentration of metals in soils and earth's continental crust.

It should also be noted that the soils of Lisbon have a reasonable geological diversity [62] including regions with volcanic soils, alluvion soils, medium Cretacic and from Miocenic periods. The sampling sites CC, RA and UH are in a Miocenic formation (sand, clay and limestone). The CP and CE1 are in an antique volcanic zone and the CE2 site is located in a calcaric cambisols zone. However, the reasonable antiquity of Lisbon as a city (more than twenty centuries) and therefore a recipient of anthropogenic pollution for many centuries, as well as the frequent works that take place in cities involving soil movement, prevent the unambiguous geological classification of the soils of each of the studied areas.

Table 4. Typical metal content in earth's continental crust and in soil (mg/kg).

Medium	Reference	Metal			
		Cd	Cr	Ni	Pb
Crust	[53]	0.1–5	200–380	20–8	16–30
Crust		0.13	83.0	58.0	16.0
Soils	[54]	0.50	200.0	40.0	10.0
Urban Soils		0.90	80.0	33.0	54.5
Crust	[55]	0.2	100	75	12.5
Crust	[56]	0.2	100	75	13
Crust	[57]	0.13–0.5	83–370	58–200	12.5–20
Crust	[58]	0.102	35	18.6	17
Crust	[59]	0.02–6	125	2–500	10–50
Soils		0.01–2	40	17–50	2–30
Agriculture Soils	[60]	-	32	10	17
Crust	[61]	-	126–185	20	14–15
Soils		0.02–14	60	29	27
Earth's Crust Range		**0.01–6**	**35–380**	**8–500**	**2–50**

To overcome the limitation of not knowing the background values and being unable to use the crust values method above mentioned, an established statistical method was used to estimate background values for metals in the various locations studied in the city of Lisbon. On this subject, several authors [63,64] emphatically refer to the impossibility of obtaining exact or even reasonably values for the background values of metals (and other pollutants) in soils and other environmental compartments. Among the estimation methods referred by Reimann [65], the iterative 2s-technique was chosen because it is the method that best adapted to the available amount of data. This method is also referred by the authors as the one that presents the most reasonable degree of approximation to real values [66]. The obtained values are shown in Table 5 and were used to calculate the pollution indexes and ecological indexes.

Table 5. Calculated background values (mg/kg) for each sampling site according to the iterative 2s-technique, proposed in [65].

	Cd	Cr	Ni	Pb
CE1	0.260	20.3	11.1	21.3
CE2	0.127	9.21	7.69	15.3
UH	0.240	8.94	10.9	11.4
CC	0.317	18.0	12.4	32.9
RA	0.360	6.57	10.1	12.6
CP	0.133	22.9	31.8	3.77

Results of I_{geo} calculated values are presented in Figure 4. Ni shows the highest values followed by Cr and Pb. Ni exhibits moderate levels of contamination at all sites except in residential area (RA) and the city park (CP), both classified as "uncontaminated". Higher values of I_{geo} for Ni were found in the soils next to the city entrance 2 (CE2). For Pb and Cr the I_{geo} index have classifications between "uncontaminated" and "moderately contaminated". Regarding the values of I_{geo} for Cd, this metal reveals an "uncontaminated" contamination class in UH, RA and CP and an "uncontaminated to moderately" class in the city entrances (CE1, CE2) and at the city center (CC).

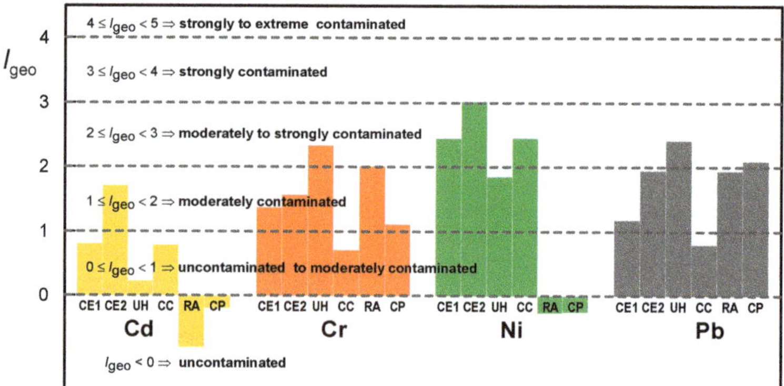

Figure 4. I_{geo} values calculated for each metal (Pb, Ni, Cr and Cd) in each sampled site.

The values for the Contamination factor (C_f) are shown in Figure 5. Nickel presents a "very high contamination" level in three sites, city entrances (CE1, CE2) and city center (CC). The residential area (RA) and the city park (CP) present values in the "moderate contamination" level but very close of the "low contamination" level. Pollution by Cr and Pb also presents some places with a "very high contamination" level, namely in urban highway (UH). It should be noted that contamination by these metals in the city park and residential area is also reasonable. Cd has low values for this index with only one of the city entrances (CE2) showing a "moderate contamination" level.

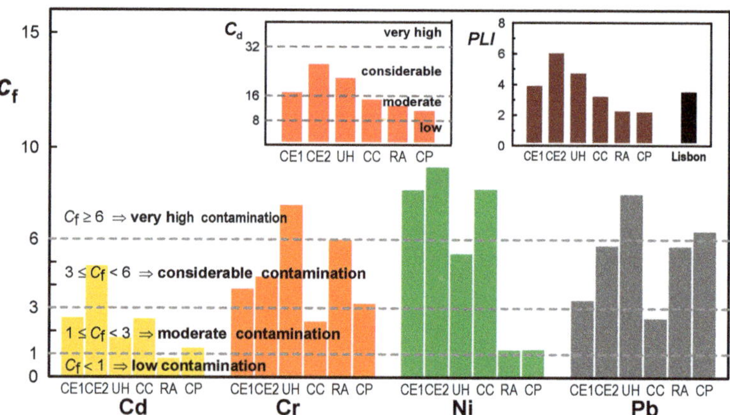

Figure 5. Contamination factor (C_f) values for each metal, in each sampling site. Insert: values of degree of contamination (C_d) and pollution load index (*PLI*) with the values obtained for each sampling site and the global *PLI* calculated for the city of Lisbon.

A global view of pollution in the city of Lisbon can be given by the indexes that include the contribution of different metals: degree of contamination (C_d) and pollution load index (*PLI*). The C_d and *PLI* indexes are important because they considered all the analyzed metals in the soil and not only an individual metal. The insert graphs in Figure 5 show the results for these two indexes. The values of C_d and *PLI* have a quite similar profile for the various studied locals. These global indexes reveal a great accumulation of metals in the city entrances (CE1 and CE2) and in the urban highway (UH) and a little less in the city center (CC). The residential area (RA) and the city park (CP) were the less polluted places. The *PLI* calculated for Lisbon, according to Equation (5), presented a value of 1.6 which represents a "moderate pollution" value for the whole city.

The obtained values for these different pollution indexes (I_geo, C_f, C_d and *PLI*) follow the expected behavior, showing that the city entrances and the urban highway had the highest pollution load, followed by the city center. On the other hand, the residential area and the city park had lower pollution values.

The calculated values for the ecological risk factor (E_r) and global potential ecological risk (*RI*) are shown in Figure 6.

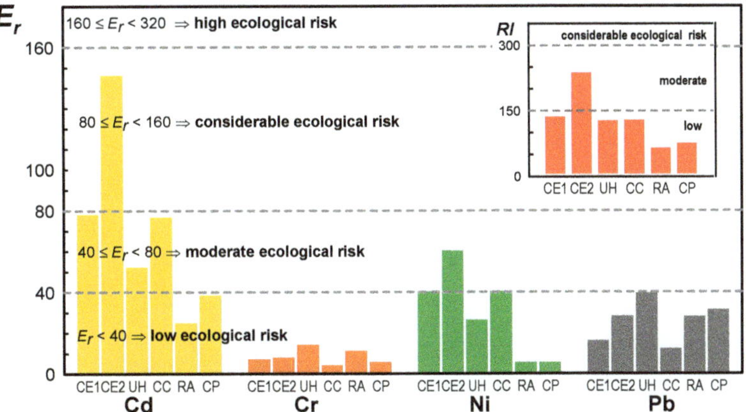

Figure 6. Values of the ecological risk factor (E_r) and the global potential ecological risk (*RI*).

The E_r index presents higher values of risk for cadmium resulting from the greater toxic response factor attributed to this metal. The values of E_r for cadmium are higher for the CE2 site with values classified as "considerable ecological risk". For the remaining locations, the ecological risk is moderate and is even considered as "low ecological risk" for the RA and CP locations. For Ni and Pb, the values of E_r vary between "low" and "moderate ecological risk". For Cr the values of this index are also classified as "low ecological risk" but with globally lower values than Ni and Pb.

The RI values are shown in the insert graph of Figure 6. Moreover, for this risk factor that takes into account the various metals, CE2 is the site with the highest risk but only classified as "moderate ecological risk". The entrance to the city (CE1), the urban highway (UH) and the city center (CC) have a "low" but close to "moderate ecological risk". The city park (CP) and residential area (RA) locations have the lowest values of this ecological risk index.

Recent work [67] shows the usefulness of the various indexes for the classification of soil pollution. The use of indices makes possible to compare the soil pollution from different locations based on their initial state. However, the authors stress the necessary precaution in the use of soil pollution indexes. Other factors than the initial and final metal contents are relevant, including the geological structure of the soil that affects pH, a determining factor in the bioavailability of the elements.

Despite the fact that the present study was focused on these four dangerous metals, there is for sure contamination by other metals in the soils of Lisbon. This contamination certainly increases the pollution and composite ecological indexes, namely C_d, *PLI* and *RI* as these indexes consider the contribution of different metals and reflect the accumulated health risk of all metals in the soils. In fact, the metals considered are the most toxic and those that must be monitored in the soil according to the European Directive [68], except for mercury and arsenic.

3.4. Data Correlation and Principal Component Analysis

Significant correlations with $p = 0.01$ were obtained for the following pairs of metals: (Ni, Cr), (Ni, Cd), (Ni, Pb) and (Pb, Cd). In other words, Ni is positively correlated with all other metals and in addition to that, only the correlation between Pb and Cd is verified. The pairs (Cr, Pb) and (Cr, Cd) do not present a significant correlation.

For the principal component analysis (PCA), the levels of metals in the soil in each location of the city and in each year were considered. It was intended to verify the existence of a relationship between metals, that is, variations in soil metal concentrations that exhibit similar behaviors, in time or in different places in the city. From the data, three main principal components were extracted that explain 78% of the total variance of the results. The first factor (PC1) is responsible for 58% of the variance, PC2 for 11% and PC3 for 8%.

Figure 7 shows the data obtained for PCA analysis, PC1 *vs.* PC2. The grouping of sites with a high correlation in Cr, Ni and Pb concentrations is evident. These sites are located at the entrances to the city, establishing a clear link between the levels of these metals in the soil and vehicle traffic. In addition to those locations, the city center and the urban highway are also part of this set. The third factor (PC3), a factor that explains only a small variance of the global system, may explain the conjoint variation of Pb and Ni in the city entrances and in the urban highway.

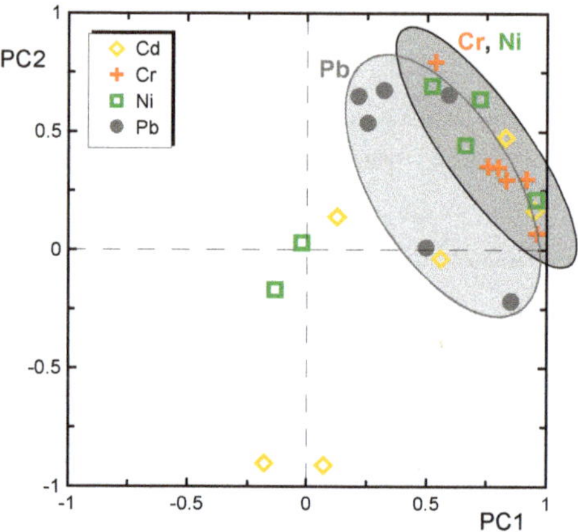

Figure 7. Principal component analysis plot of the soil samples (PC1 and PC2).

Authors such as Imperato et al. [69], Andersson et al. [70], Meza-Montenegro et al. [71], Pant and Harrison [72] and Moaref et al. [73] stressed out that the origin of anthropogenic pollution in cities without large industrial facilities (like Lisbon) is mainly due to car traffic. The principal source of lead was, during the past, the leaded fuels. Nowadays lead is still present in soils despite its drastic reduction due to the abolition of added lead gasoline. The main sources of Cr and Ni are the vehicle braking system and tire wear. However, these two elements also have a strong relationship with parent material. Several studies have detected this trend [40]. So, for Cd and Ni there will be two source contributions, the natural one due to the lithogenic contribution and the anthropogenic one, originated from car traffic pollution.

4. Conclusions

In this work, levels of Cd, Cr, Ni and Pb in the soils of six sites with different vehicle traffic characteristics were monitored in Lisbon for a period of about ten years. To our knowledge, this is the longest soil monitoring program carried out in the city of Lisbon. Time wise, in many of the other studies referenced in this work the monitoring periods for soil contamination by toxic metals are typically only one to two years, making this study one of the longest. For each location, 12 samples were collected each year, bringing up 72 samples per year and a total of 648 soil samples. When compared with values obtained for other cities (Table 3), the data in the monitoring period revealed a reasonable low concentration of metals studied. However, these low values of metals in soils are consistent in the nine years of this study. It should be noted that many of the values listed in Table 3 are single values, obtained generally in a given year and in the weather conditions observed for that year.

The evolution of the levels of metals over the years and in the various places in the city has typically grown, with the exception of the last year. In most locations, the year 2011 showed a decrease in the content of metals in soils. When looking at the average levels for each year, involving all locations, a downward trend in this last year of the campaign is spotted. In 2011 the economic crisis had important impacts on economic activity and this impact was reflected in the circulation of fewer vehicles and, consequently, fewer emissions. That results in a decrease of observed metals in the soil, showing the unequivocal main origin of these pollutants: automotive traffic.

Looking at the levels of metals in each location in the city, the residential area and the city park have the lowest levels of toxic metals in soils, as expected. Exception is noted for Cr in which for those two places the levels are identical to the rest of the city. Still for the same locations, the levels of Cd, Ni and Pb have been maintained or decreased over the years. Cr grew each year, except in 2011. The determination of the pollution indexes (I_{geo}, C_f, C_d, *PLI*) underlined this global conclusion with values indicating that the most serious levels of contamination are connected to locations with great traffic density, namely the city entrances, the city center and the urban highway.

Ecological indexes, E_r and *RI*, also revealed a low or moderate ecological risk except Cd in CE2. When considering the *RI*, in which the joint effect of the metals is considered, moderate ecological risk values are also achieved in CE2. The remaining sites presented a low or moderate level of ecological risk. This index also confirms that the lowest values of potential ecological risk are found in the residential and city park areas.

The low values of metal contents found are also positive for the many urban agriculture projects that are being developed in Lisbon. The municipality has encouraged this practice with the construction of several urban vegetable garden parks throughout the city. This practice should be accompanied by periodic analyzes of soils and plants, as metal levels, even when low, may be transferred from the soil to the plants. It will also be important to follow the pH of the soils and their content in organic matter, which are essential factors in the bioavailability of metals.

Considering that these metals may affect human health by the ingestion of contaminated food (among other forms), the results obtained in this work lead to the conclusion that as the city of Lisbon presents low values of these metals in its soils, it also has a very low potential to affect the health of its inhabitants, workers and visitors.

Author Contributions: Conceptualization M.J.M.; data curation H.F.S., M.J.M.; formal analysis H.F.S.; funding acquisition M.J.M., C.M.O.; investigation H.F.S., N.F.S.; methodology H.F.S., M.J.M.; project administration M.J.M., C.M.O.; resources M.J.M., C.M.O.; software M.J.M.; supervision; validation H.F.S.; visualization M.J.M.; roles/writing—original draft H.F.S., N.F.S.; writing—review and editing C.M.O., M.J.M., N.F.S. All authors have read and agreed to the published version of the manuscript.

Funding: This research was funded by Fundação para a Ciência e a Tecnologia (FCT) through projects UID/QUI/00100/2013 and UIDB/QUI/00100/2020.

Institutional Review Board Statement: Not applicable.

Informed Consent Statement: Not applicable.

Data Availability Statement: The data collected, the results of the chemical and physical analysis and the main results obtained, may be requested to the authors. They will be provided, free of charge or not, depending on its future use.

Conflicts of Interest: The authors declare no conflict of interest.

References

1. Antrop, M. Landscape change and the urbanization process in Europe. *Landsc. Urban Plan.* **2004**, *67*, 9–26. [CrossRef]
2. Pouyat, R.V.; Trammell, T.L.E. Chapter 10—Climate change and urban forest soils. *Dev. Soil Sci.* **2019**, *36*, 189–211. [CrossRef]

3.   UN. 2018. Available online: https://population.un.org/wup/Publications/Files/WUP2018-PopFacts_2018-1.pdf (accessed on 21 November 2020).

4.   Grigoratos, T.; Samara, C.; Voutsa, D.; Manoli, E.; Kouras, A. Chemical composition and mass closure of ambient coarse particles at traffic and urban-background sites in Thessaloniki, Greece. *Environ. Sci. Pollut. Res.* **2014**, *21*, 7708–7722. [CrossRef] [PubMed]

5.   Wang, M.; Liu, R.; Chen, W.; Peng, C.; Markert, B. Effects of urbanization on heavy metal accumulation in surface soils. *J. Environ. Sci.* **2018**, *64*, 328–334. [CrossRef] [PubMed]

6.   Gulan, L.; Milenkovic, B.; Zeremski, T.; Milic, G.; Vuckovic, B. Persistent organic pollutants, heavy metals and radioactivity in the urban soil of Priština city, Kosovo and Metohija. *Chemosphere* **2017**, *171*, 415–426. [CrossRef] [PubMed]

7.   Soleimani, M.; Amini, N.; Sadeghian, B.; Wang, D.; Fang, L. Heavy metals and their source identification in par-ticulate matter (PM2.5) in Isfahan city, Iran. *J. Environ. Sci.* **2018**, *72*, 166–175. [CrossRef]

8.   Jaishankar, M.; Tseten, T.; Anbalagan, N.; Mathew, B.B.; Beeregowda, K.N. Toxicity, mechanism and health effects of some heavy metals. *Interdiscip. Toxicol.* **2014**, *7*, 60–72. [CrossRef]

9.   Yadav, I.C.; Devi, N.L.; Singh, V.K.; Li, J.; Zhang, G. Spatial distribution, source analysis and health risk assess-ment of heavy metals contamination in house dust and surface soil from four major cities of Nepal. *Chemosphere* **2019**, *218*, 1100–1113. [CrossRef]

10.   Yuan, X.; Xue, N.; Han, Z. A meta-analysis of heavy metals pollution in farmland and urban soils in China over the past 20 years. *J. Environ. Sci.* **2021**, *101*, 217–226. [CrossRef]

11.   Guagliardi, I.; Cicchella, D.; de Rosa, R.; Buttafuoco, G. Assessment of lead pollution in topsoils of a southern Italy area: Analysis of urban and peri-urban environment. *J. Environ. Sci.* **2015**, *33*, 179–187. [CrossRef]

12.   Wilkins, D.A. The measurement of tolerance to edaphic factors by means of root growth. *New Phytol.* **1978**, *80*, 623–633. [CrossRef]

13.   European Commission DG Environment. Soil Contamination: Impacts on Human Health, Science for Environment Policy, 2013, Issue 5. Available online: http://ec.europa.eu/science-environment-policy (accessed on 22 January 2021).

14.   Directive 98/70/EC, European Parliament and of the Council. Relating to the quality of petrol and diesel fuels and amending Council Directive 93/12/EEC. *Off. J. L* **1998**, *350*, 58–68.

15.   Migon, C.; Jourdan, E.; Nicolas, E.; Gentili, B. Effects of reduced leaded fuel consumption on atmospheric lead behavior. *Chemosphere* **1994**, *28*, 139–144. [CrossRef]

16.   Tomtom. 2019. Available online: https://www.tomtom.com/en_gb/traffic-index/ (accessed on 15 December 2020).

17.   Werkenthin, M.; Kluge, B.; Wessolek, G. Metals in European roadside soils and soil solution-A review. *Environ. Pollut.* **2014**, *189*, 98–110. [CrossRef] [PubMed]

18.   Jiang, Y.; Chao, S.; Liu, J.; Yang, Y.; Chen, Y.; Zhang, A.; Cao, H. Source apportionment and health risk assessment of heavy metals in soil for a township in Jiangsu Province, China. *Chemosphere* **2017**, *168*, 1658–1668. [CrossRef]

19.   Liu, J.; Liu, Y.J.; Liu, Y.; Liu, Z.; Zhang, A.N. Quantitative contributions of the major sources of heavy metals in soils to ecosystem and human health risks: A case study of Yulin, China. *Ecotoxicol. Environ. Saf.* **2018**, *164*, 261–269. [CrossRef]

20.   Barbieri, M. The importance of Enrichment Factor (EF) and Geoaccumulation Index (Igeo) to evaluate the soil contamination. *J. Geol. Geophys.* **2016**, *5*, 237. [CrossRef]

21.   Wu, J.; Lu, J.; Li, L.; Min, X.; Luo, Y. Pollution, ecological-health risks and sources of heavy metals in soil of the northeastern Qinghai-Tibet Plateau. *Chemosphere* **2018**, *201*, 234242. [CrossRef]

22.   Tomlinson, D.L.; Wilson, J.G.; Harris, C.R.; Jeffrey, D.W. Problems in the assessment of heavy-metal levels in estuaries and the formation of a pollution index. *Helgol. Mar. Res.* **1980**, *33*, 566–575. [CrossRef]

23.   Hakanson, L. An ecological risk index for aquatic pollution control. A sedimentological approach. *Water Res.* **1980**, *14*, 975–1001. [CrossRef]

24.   Varol, M.; Sunbul, M.R.; Aytop, H.; Yılmaz, C.H. Environmental, ecological and health risks of trace elements and their sources in soils of Harran Plain, Turkey. *Chemosphere* **2020**, *245*, 125592. [CrossRef] [PubMed]

25.   Directive 2008/50/EC, European Parliament and of the Council. On ambient air quality and cleaner air for Europe. *Off. J. L* **2008**, *152*, 1–44.

26.   Directive 2010/75/EU, European Parliament and of the Council. On industrial emissions (integrated pollution prevention and control). *Off. J. L* **2010**, *334*, 17–119.

27.   Gallaert, G.; Cools, N.; Delanote, V.; De Vos, B.; Groenemans, R.; Langouche, D.; Roskams, P.; Scheldeman, X.; Mechelen, L.V.; Ranst, E.V. *Sampling and Analysis of Soil–Manual on Methods and Criteria for Harmonized Sampling, Assessment, Monitoring and Analysis of the Effects of Air Pollution on Forests*, 4th ed.; UNECE: Hamburg, Germany, 2003.

28.   ISO 10390. *Soil Quality—Determination of pH*; ISO: Genebra, Switzerland, 2015.

29.   ISO 11465. *Soil quality—Determination of Dry Matter and Water Content on a Mass Basis–Gravimetric Method*; ISO: Genebra, Switzerland, 1993.

30.   CEN/TC. *Chemical Analyses—Determination of Dry Matter and Water Content on a Mass Basis in Sediment, Sludge, Soil, and Waste—Gravimetric Method*; CEN: Brussels, Belgium, 2003.

31.   ISO 11466. *Soil Quality—Extraction of Trace Elements Soluble in Aqua Regia*; ISO: Genebra, Switzerland, 2010.

32.   Förstner, U.; Müller, G. Concentrations of heavy metals and polycyclic aromatic hydrocarbons in river sediments: Geochemical background, man's influence and environmental impact. *GeoJournal* **1981**, *5*, 417. [CrossRef]

33.   Baltas, H.; Sirin, M.; Gökbayrak, E.; Ozcelik, A.E. A case study on pollution and a human health risk assessment of heavy metals in agricultural soils around Sinop province, Turkey. *Chemosphere* **2020**, *241*, 125015. [CrossRef] [PubMed]

34. Mazurek, R.; Kowalska, J.; Gąsiorek, M.; Zadrożny, P.; Jozefowska, A.; Zaleski, T.; Kepka, W.; Tymczuk, M.; Orłowska, K. Assessment of heavy metals contamination in surface layers of Roztocze National Park forest soils (SE Po-land) by indices of pollution. *Chemosphere* **2016**, *168*, 839–850. [CrossRef]
35. Liu, K.; Li, C.; Tang, S.; Shang, G.; Yu, F.; Li, Y. Heavy metal concentration, potential ecological risk assessment and enzyme activity in soils affected by a lead-zinc tailing spill in Guangxi, China. *Chemosphere* **2020**, *251*, 126415. [CrossRef]
36. Kumar, V.; Sharma, A.; Kaur, P.; Singh Sidhu, G.P.; Bali, A.S.; Bhardwaj, R.; Thukral, A.K.; Cerda, A. Pollution assessment of heavy metals in soils of India and ecological risk assessment: A state-of-the-art. *Chemosphere* **2019**, *216*, 449–462. [CrossRef]
37. Massart, D.L.; Vandeginste, B.G.M.; Deming, S.N.; Michotte, Y.; Kaufman, L. *Chemometrics: A Textbook (Data Handling in Science and Technology)*; Fifth Impression 2003; Elsevier Science: Amsterdam, The Netherlands, 1988.
38. Leitão, T.E.; Cameira, M.R.; Costa, H.D.; Pacheco, J.M.; Henriques, M.J.; Martins, L.L.; Mourato, M.P. Environmental quality in urban allotment gardens: Atmospheric deposition, soil, water and vegetable assessment at Lisbon city. *Water Air Soil Pollut.* **2018**, *229*, 31. [CrossRef]
39. Argyraki, A.; Kelepertzis, E. Urban soil geochemistry in Athens, Greece: The importance of local geology in controlling the distribution of potentially harmful trace elements. *Sci. Total Environ.* **2014**, *482*, 366–377. [CrossRef]
40. Ajmone-Marsan, F.; Biasioli, M.; Kralj, T.; Grčman, H.; Davidson, C.M.; Hursthouse, A.S.; Madrid, L.; Rodrigues, S. Metals in particle-size fractions of the soils of five European cities. *Environ. Pollut.* **2008**, *152*, 73–81. [CrossRef]
41. Zheng, Y.; Chen, T.; He, J. Multivariate geostatistical analysis of heavy metals in topsoils from Beijing, China. *J. Soils Sediments* **2008**, *8*, 51–58. [CrossRef]
42. Birke, M.; Rauch, U. Urban Geochemistry: Investigations in the Berlin Metropolitan area. *Environ. Geochem. Health* **2000**, *22*, 233–248. [CrossRef]
43. Guillén, M.T.; Delgado, J.; Albanese, S.; Nieto, J.M.; Lima, A.; de Vivo, B. Heavy metals fractionation and multi-variate statistical techniques to evaluate the environmental risk in soils of Huelva township (SW Iberian Peninsula). *J. Geochem. Explor.* **2012**, *119–120*, 32–43. [CrossRef]
44. Christoforidis, A.; Stamatis, N. Heavy metal contamination in street dust and roadside soil along the major national road in Kavala's region, Greece. *Geoderma* **2009**, *151*, 257–263. [CrossRef]
45. Mitchell, R.; Spliethoff, H.M.; Ribaudo, L.N.; Lopp, D.M.; Shayler, H.A.; Marquez-Bravo, L.G.; Lambert, V.T.; Ferenz, G.S.; Russell-Anelli, J.M.; Stone, E.B.; et al. Lead (Pb) and other metals in New York City community garden soils: Factors influencing contaminant distributions. *Environ. Pollut.* **2014**, *187*, 162–169. [CrossRef] [PubMed]
46. Tijhuis, L.; Brattli, B.; Sæther, O. A geochemical survey of topsoil in the city of Oslo, Norway. *Environ. Geochem. Health* **2002**, *24*, 67–94. [CrossRef]
47. Manta, D.S.; Angelone, M.; Bellanca, A.; Neri, R.; Sprovieri, M. Heavy metals in urban soils: A case study from the city of Palermo (Sicily), Italy. *Sci. Total Environ.* **2002**, *300*, 229–243. [CrossRef]
48. Nogueira, T.A.R.; Abreu-Junior, C.H.; Alleoni, L.R.F.; He, Z.; Soares, M.R.; Vieira, C.S.; Lessa, L.G.F.; Capra, G.F. Background concentrations and quality reference values for some potentially toxic elements in soils of São Paulo state, Brazil. *J. Environ. Manag.* **2018**, *221*, 10–19. [CrossRef] [PubMed]
49. Shi, G.; Chen, Z.; Xu, S.; Zhang, J.; Wang, L.; Bi, C.; Teng, J. Potentially toxic metal contamination of urban soils and roadside dust in Shanghai, China. *Environ. Pollut.* **2008**, *156*, 251–260. [CrossRef]
50. Linde, M.; Bengtsson, H.; Öborn, I. Concentrations and pools of heavy metals in urban soils in Stockholm, Sweden. *Water Air Soil Pollut.* **2001**, *1*, 83–101. [CrossRef]
51. Cachada, A.; Dias, A.C.; Pato, P.; Mieiro, C.; Rocha-Santos, T.; Pereira, M.E.; Ferreira da Silva, E.; Duarte, A.C. Major inputs and mobility of potentially toxic elements contamination in urban areas. *Environ. Monit. Assess.* **2013**, *185*, 279–294. [CrossRef]
52. Costa, C.; Reis, A.P.; Silva, E.F.; Rocha, F.; Patinha, C.; Dias, A.C.; Sequeira, C.; Terroso, D. Assessing the control exerted by soil mineralogy in the fixation of potentially harmful elements in the urban soils of Lisbon, Portugal. *Environ. Earth Sci.* **2012**, *65*, 1133–1145. [CrossRef]
53. Fleischer, M. *Recent Estimates of the Abundances of the Elements in the Earth's Crust*; U.S. Geological Survey: Washington, DC, USA, 1953; Volume 285.
54. Vinogradov, A.P. *The Geochemistry of Rare and Dispersed Chemical Elements in Soils*, 2nd ed.; Consultants Bureau: New York, NY, USA, 1959.
55. Taylor, S.R. Abundance of chemical elements in the continental crust: A new table. *Geochim. Cosmochim. Acta* **1964**, *28*, 1273–1285. [CrossRef]
56. Mason, B.; Moore, C.B. *Principles of Geochemistry*, 4th ed.; John Wiley Sons: New York, NY, USA, 1966.
57. Parker, R.L. *Data of Geochemistry. Chapter D. Composition of the Earth's Crust*, 6th ed.; United States Government Printing Office: Washington, DC, USA, 1967.
58. Wedepohl, K.H. The composition of the continental crust. *Geochim. Cosmochim. Acta* **1995**, *59*, 1217–1232. [CrossRef]
59. Adriano, D.C. *Trace Elements in Terrestrial Environments: Biogeochemistry, Bioavailability and Risks of Metals*, 2nd ed.; Springer: New York, NY, USA, 2001.
60. Reimann, C.; Siewers, U.; Tarvainen, T.; Bityukova, L.; Eriksson, J.; Giucis, A.; Gregorauskiene, V.; Lukashev, V.K.; Matinian, N.N.; Pasieczna, A. *Agricultural Soils in Northern Europe: A Geochemical Atlas*; Schweizerbart Science Publishers: Stuttgart, Germany, 2003.

61. Kabata-Pendias, A. *Trace Elements in Soils and Plants*, 4th ed.; Taylor Francis Group: Boca Raton, FL, USA; London, UK; New York, NY, USA, 2011.
62. Moitinho de Almeida, F. *Carta Geológica do Concelho de Lisboa, Folha 1, 2, 3 e 4, Escala 1:10 000*; Serviços Geológicos de Portugal: Lisbon, Portugal, 1986.
63. Matschullat, J.; Ottenstein, R.; Reimann, C. Geochemical background—Can we calculate it? *Environ. Geol.* **2000**, *39*, 990–1000. [CrossRef]
64. Reimann, C.; Filzmoser, P. Normal and lognormal data distribution in geochemistry:death of a myth. Consequences for the statistical treatment of geochemical and environmental data. *Environ. Geol.* **2000**, *39*, 1001–1014. [CrossRef]
65. Reimann, C.; Filzmoser, P.; Garrett, R.G. Background and threshold: Critical comparison of methods of determination. *Sci. Total Environ.* **2005**, *346*, 1–16. [CrossRef]
66. Reimann, C.; Garrett, R.G. Geochemical background—Concept and reality. *Sci. Total Environ.* **2005**, *350*, 12–27. [CrossRef]
67. Korzeniowska, J.; Krąż, P. Heavy Metals Content in the Soils of the Tatra National Park Near Lake Morskie Oko and Kasprowy Wierch—A Case Study (Tatra Mts, Central Europe). *Minerals* **2020**, *10*, 1120. [CrossRef]
68. Directive 86/278/EEC of the Council. On the protection of the environment, and in particular of the soil, when sewage sludge is used in agriculture. *Off. J. L* **1986**, *181*, 0006–0012.
69. Imperato, M.; Adamo, P.; Arienzo, M.; Naimo, D.; Stanzione, D.; Violante, P. Spatial distribution of heavy metals in urban soils of Naples city (Italy). *Environ. Pollut.* **2003**, *124*, 247–256. [CrossRef]
70. Andersson, M.; Ottesen, R.T.; Langedal, M. Geochemistry of urban surface soils—Monitoring in Trondheim, Norway. *Geoderma* **2010**, *156*, 112–118. [CrossRef]
71. Meza-Montenegro, M.; Gandolfi, A.J.; Santana-Alcántar, M.E.; Klimecki, W.T.; Aguilar-Apodaca, M.G.; Río-Salas, R.D.; De la O-Villanueva, M.; Gómez-Alvarez, A.; Mendivil-Quijada, H.; Valencia, M.; et al. Metals in residential soils and cumulative risk assessment in Yaqui and Mayo agricultural valleys, northern Mexico. *Sci. Total Environ.* **2012**, *433*, 472–481. [CrossRef] [PubMed]
72. Pant, P.; Harrison, R.M. Estimation of the contribution of road traffic emissions to particulate matter concentra-tions from field measurements: A review. *Atmos. Environ.* **2013**, *77*, 78–97. [CrossRef]
73. Moaref, S.; Sekhavatjou, M.S.; Alhashemi, A.H. Determination of trace elements concentration in wet and dry atmospheric deposition and surface soil in the largest industrial city, southwest of Iran. *Int. J. Environ. Res.* **2014**, *8*, 335–346. [CrossRef]

soil systems

Article

Phytoextraction of Heavy Metals by Various Vegetable Crops Cultivated on Different Textured Soils Irrigated with City Wastewater

Iftikhar Ahmad [1], Saeed Ahmad Malik [1], Shafqat Saeed [2], Atta-ur Rehman [3] and Tariq Muhammad Munir [4,*,†]

[1] Department of Botany, Institute of Pure and Applied Biology, Bahauddin Zakariya University, Multan 60800, Pakistan; rocky_91294@yahoo.com (I.A.); saeedbotany@yahoo.com (S.A.M.)

[2] Department of Entomology, Muhammad Nawaz Sharif University of Agriculture, Multan 60500, Pakistan; shafqat.saeed@mnsuam.edu.pk

[3] Soil and Water Testing Laboratory, Vehari 60660, Pakistan; aacvehari@gmail.com

[4] Department of Geography, University of Calgary, 2500 University Dr. NW, Calgary, AB T2N 1N4, Canada

* Correspondence: tmmunir@ucalgary.ca; Tel.: +1-403-971-5693

† Current address: Department of Geography and Planning, University of Saskatchewan, 117 Science Place, Saskatoon, SK S7N 5C8, Canada.

Abstract: A challenging task in urban or suburban agriculture is the sustainability of soil health when utilizing city wastewater, or its dilutes, for growing crops. A two-year field experiment was conducted to evaluate the comparative vegetable transfer factors (VTF) for four effluent-irrigated vegetable crops (brinjal, spinach, cauliflower, and lettuce) grown on six study sites (1 acre each), equally divided into two soil textures (sandy loam and clay loam). Comparisons of the VTF factors showed spinach was a significant and the best phytoextractant, having the highest heavy metal values (Zn = 20.2, Cu = 12.3, Fe = 17.1, Mn = 30.3, Cd = 6.1, Cr = 7.6, Ni = 9.2, and Pb = 6.9), followed by cauliflower and brinjal, while lettuce extracted the lowest heavy metal contents (VTF: lettuce: Zn = 8.9, Cu = 4.2, Fe = 9.6, Mn = 6.6, Cd = 4.7, Cr = 2.9, Ni = 5.5, and Pb = 2.5) in response to the main (site and vegetable) or interactive (site * vegetable) effects. We suggest that, while vegetables irrigated with sewage water may extract toxic heavy metals and remediate soil, seriously hazardous/toxic contents in the vegetables may be a significant source of soil and environmental pollution.

Keywords: phytoremediation; phytoextraction; heavy metal; wastewater; sewage water; pollution; sustainability; spinach; cauliflower; lettuce

Citation: Ahmad, I.; Malik, S.A.; Saeed, S.; Rehman, A.-u.; Munir, T.M. Phytoextraction of Heavy Metals by Various Vegetable Crops Cultivated on Different Textured Soils Irrigated with City Wastewater. *Soil Syst.* **2021**, *5*, 35. https://doi.org/10.3390/soilsystems5020035

Academic Editors: Matteo Spagnuolo, Paola Adamo and Giovanni Garau

Received: 24 May 2021
Accepted: 17 June 2021
Published: 18 June 2021

Publisher's Note: MDPI stays neutral with regard to jurisdictional claims in published maps and institutional affiliations.

Copyright: © 2021 by the authors. Licensee MDPI, Basel, Switzerland. This article is an open access article distributed under the terms and conditions of the Creative Commons Attribution (CC BY) license (https://creativecommons.org/licenses/by/4.0/).

1. Introduction

A 1–3 °C increase in global warming is predicted by the 2050s [1]; the warming in the South Asian region is expected to be higher (2.2–3.3 °C) [2,3], with surface warming as high as 4.2 °C predicted to be in the northern regions of Pakistan [4] under the RCP8.5 emission scenario. Subsequently, enhanced glacier melts [5,6], soil surface drying, and water table lowering [7] are broadly expected, which may result in an acute shortage of surface- and/or groundwater supply for irrigating crops. Therefore, Pakistan, whose economy is ~21% agriculture driven, would be one of the regions most severely affected by climate change [6]. Some of these impacts are already being observed. That is why some of the suburban crops are irrigated using dilutes of city wastewater [8,9]. Likewise, approximately 20 million hectares of vegetable or cereal crops grown in a total of 50 countries are also being supplied with substandard waters, including ~80% untreated or partially treated wastewater of household or industrial nature [10,11] to cope with the issue of food security [12,13]. However, given the use of untreated wastewater for growing vegetables or cereal crops for human or animal consumption, human and soil health are at risk.

Wastewaters potentially contain a large variety of pollutants [9], including, but not limited to, unknown chemicals (organic, inorganic or biological nature) and/or salts, metals and metalloids, pathogens and hosts, residual drugs and pesticides, endocrine-disrupting chemicals, or active ingredients of human care products. On one hand, these pollutants can impair soil and environmental health [14]; on the other hand, they can be taken up by the growing field crops resulting in buildups of toxic levels of heavy metals in the vegetable biomass. The toxic crops (especially vegetables) when consumed by humans may put their health at risk. Therefore, urban and farming communities' environments and human and animal health are at risk, which indirectly poses an even greater risk of global food security.

A heavy metal is defined as a chemical element of 500% higher specific gravity than that of normal water [15]. Continuous use of heavy metals containing wastewater for vegetable crop irrigation results in heavy metal accumulation in soil [16] and subsequent transfer to vegetable plants above the safe limits [17,18]. The plant accumulation concentration divided by soil accumulation concentration is called the vegetable transfer factor (VTF) [19], which shows the vegetable accumulation rate concerning soil accumulation concentration [20]. The soil–plant transfer of heavy metals is largely dependent on the plant species and is evaluated using the soil–plant TF [21]. The transfer factor is further controlled by several factors: plant age and species, crop variety, heavy metal concentration and its physical and chemical properties, and duration of effect [22].

Cd toxicity, even at low levels, has been attributed to its bioaccumulation and long half-life of ~30 years [8,23,24]. Cd also is known for its high mobility across the soil–water–plant–environment continuum [25]. A few major toxic plant effects include, but are not limited to, leaf chlorosis, stunted growth, and limited uptake of essential nutrients and protein synthesis [24,26]. Cr toxicity also reduces crop yield via impaired leaf and root hair growth, reduced enzymatic dynamics, and mutagenesis [27]. Toxic soil–water—plant concentrations are reported to impair overall plant growth and reproduction [28]. Excessive amounts of soil Zn together with soil Cu may decrease overall plant growth but increase TF, and ingestion of these higher TF vegetables result in acute depression symptoms in humans [29]. Excess levels of Cu alone in human blood showed acute stomachache and subsequent liver damage in many patients [30].

While the accumulation of heavy metals by wastewater-irrigated crops has been studied over several soil types, these investigations were made using one crop and one soil type at a time. Thus, various crops have not been compared for their comparative VTFs using the same or different soil types. Phytoremediation of the contaminated soils irrigated with wastewaters is an environmentally friendly and green technique [31,32] to remediate the soil–water–air continuum and quantify the translocation of these heavy metals by calculating the VTF.

Thus, an evaluation of the comparative phytoextraction efficiencies of various crops grown on the same and different soil types and irrigated with several wastewater types, or comparison of the VTFs, remained largely unexplored in Pakistan. Therefore, our unique study aims at identifying and assessing the potential sources of contamination to the soil, water, or plant, and evaluating the comparative VTFs of various crops grown on the same and different textured soils irrigated with a variety of wastewaters. The conclusions of this study will be helpful to suggest the necessary mitigation measures and inform policy development.

We hypothesize that the comparative VTF evaluation of various crops grown on the same or different textured soils irrigated with a variety of wastewaters will differ. Therefore, the present study was carried out to quantify the heavy metal (Zn, Cu, Mn, Fe, Cd, Cr, Ni, and Pb) accumulation in four vegetables (spinach, brinjal, lettuce, and cauliflower) grown on the same and different soil types irrigated with different wastewaters, and to evaluate the suitability of the wastewater used for growing these vegetables.

The specific objectives of the present study are (1) to quantify the contents of heavy metals in two different soil types and their irrigation wastewater samples, both collected from six different study sites of the Multan suburban area; and (2) to monitor the comparative accumulation of heavy metals between the edible portions of the vegetables grown on the same and different textured soils irrigated with wastewaters, by quantifying the vegetable transfer factor. The VTF will also be related to the soil concentrations of heavy metals.

2. Materials and Methods

2.1. Study Sites

To determine the vegetable transfer factors (VTF) in the effluent-irrigated vegetable crops, a total of six study sites were chosen in the vicinity of the WASA disposal stations within the suburban area of Multan city. The sites had several open and covered drainage channels that fed the vegetable crops. Each site was around one acre in size. We divided the sites (6) into two major soil texture types: sandy loam (3) and clay loam (3). Each of the texture types was irrigated with three types of water: normal, waste (sewage), and normal + waste. The brinjal (*Solanum melogena* L.) and spinach (*Spinacia oleacea* L.) crops were sown during January 2016 on a quarter of each site, randomly, while the cauliflower (*Brassica oleracea* L.) and lettuce (*Lactuca sativa* L.) were also sown on the rest of the quarters of each site during September 2015. Details of the study sites, their textural classes, the irrigation water types, and the heavy metal concentrations of soil and irrigation waters are shown in Table 1.

Table 1. Chemistry (EC and pH) and concentrations of the heavy metals in the study soils and their irrigation wastewaters, sampled during 2015–2016 from six major vegetable production areas in the Multan region of Pakistan †.

Study Site	Soil Texture/Water	Chemistry		Heavy Metal Concentration (mg kg^{-1})							
Soil/Water		EC_s/EC_{iw} (dSm^{-1})	pH_s/pH_{iw}	Zn	Cu	Fe	Mn	Cd	Cr	Ni	Pb
Khan village											
Soil	Sandy loam	1.4	8.3	1.80	0.94	6.80	2.90	0.66	1.10	0.42	1.40
Water	Normal	0.3	7.2	0.03	0.04	0.03	0.06	0.02	-	-	0.02
Vehari road											
Soil	Clay loam	1.5	8.4	1.66	1.28	9.64	3.68	1.40	1.62	0.46	1.74
Water	Normal	0.3	7.1	0.04	0.03	0.09	0.02	0.02	0.02	-	-
Shujabad road											
Soil	Sandy loam	3.1	8.4	2.48	1.70	12.60	3.38	4.30	2.60	1.04	2.66
Water	Sewage	2.8	6.9	0.06	0.14	0.31	0.11	0.04	0.09	0.06	0.06
Industrial estate											
Soil	Clay loam	3.9	8.4	3.90	2.58	17.34	4.04	4.76	4.36	1.72	3.38
Water	Sewage	3.6	6.8	0.10	0.11	0.34	0.19	0.06	0.08	0.09	0.11
Suraj miani											
Soil	Sandy loam	2.7	8.3	2.12	2.04	11.66	3.36	2.28	3.98	1.34	2.60
Water	Normal + Sewage	2.0	7.1	0.05	0.04	0.18	0.16	0.04	0.02	0.05	0.08
Sameeja abad											
Soil	Clay loam	3.5	8.2	1.82	0.48	7.52	2.30	1.94	1.72	0.46	1.28
Water	Normal + Sewage	2.1	7.1	0.40	0.08	0.07	0.14	0.07	0.40	0.07	0.05
UNESCAP *		–	6.1	5.00	1.00	2.00	1.50	0.10	1.00	1.00	0.50
Pescod, MD **		–		2.00	0.20	5.00	0.20	0.01	0.01	0.20	5.00

† Each value is a mean of four sample months. * Permissible limits for liquid municipal and industrial effluents in Pakistan. ** Threshold levels of trace elements in irrigation water. ECs and pHs, and ECiw and pHiw denote electrical conductivity and pH of the soil and irrigation water, respectively. After four months, the study sites (soils) were significantly different in chemistry and heavy metal concentrations (ECs: $p = 0.001$; pHs: $p < 0.001$; Zn: 0.036; Cu: $p = 0.002$; Fe: $p < 0.001$; Mn: $p < 0.001$; Cd: $p = 0.006$; Cr: $p = 0.003$; Ni: $p = 0.005$; Pb: $p < 0.001$). The variable buildup of concentrations that was observed may be due to sewage water irrigation during the experimental period. All respective irrigation waters were also significantly different except for Cr (ECs: $p = 0.019$; pHs: $p < 0.001$; Zn: 0.031; Cu: $p = 0.005$; Fe: $p = 0.012$; Mn: $p = 0.003$; Cd: $p = 0.004$; Cr: $p = 0.058$; Ni: $p = 0.002$; Pb: $p = 0.013$).

2.2. Sampling and Analysis

A total of 4 composite surface soil (0–20 cm) samples were randomly collected monthly from each of the six sites receiving wastewater regularly for irrigation, (4 samples * 6 sites * 4 months = 96 samples). However, the soil samples from the brinjal and spinach crop sites were collected during January–April 2015, compared to the cauliflower and lettuce crop site soil samples collected during September–December 2016. The soil samples were air-dried, crushed and sieved to <2 mm, and stored at room temperature before analyses of the physicochemical properties and heavy metal concentrations. Soil samples were analyzed for textural class, saturation paste electrical conductivity (ECs), and saturation paste pH (pHs) following methods described by the US Salinity Laboratory Staff following Richards [33]. Textural class of only the first batch (month) of soil samples was analyzed. To quantify the water-soluble soil Zn, Cu, Fe, Mn, Cd, Cr, Ni, and Pb concentrations, 10 g of dry soil was extracted with 50 mL deionized water following Zia et al. [34].

Irrigation wastewater samples were also collected monthly during the soil sampling campaigns. Four replicate polyethylene bottles (acid washed) of 500 mL each were filled with wastewater one by one at an interval of 10 s from an open channel flowing to the study site, for all sites. Each of the collected wastewater samples was acidified immediately with 1 mL of concentrated HCl to avoid microbial degradation of the heavy metals. The samples were placed in a cooler and transported to a soil- and water-testing laboratory in Multan. Within a week, 50 mL of the sample was digested with 10 mL of concentrated HNO_3 at 80 °C until the solution turned clear [35]. The clear solution was then filtered through a Whatman™ 42 filter, diluted back to 50 mL using distilled water, and stored for analysis.

Edible parts of the harvested vegetables were thoroughly washed sequentially with 1% HCl and deionized water (to clean/remove any dust material), air-dried in shade for 24 h, and then oven-dried at 70 °C until a constant weight. The dried matter was ground to a powder form and then sieved to <1 mm. One gram of the powder was digested with a mixture of HNO_3 and $HClO_4$ in a 2:1 ratio, respectively. The clear digest was filtered and diluted to 50 mL using deionized water and stored for analysis.

Plant total and soil and wastewater soluble Zn, Cu, Fe, Mn, Cd, Cr, Ni, and Pb concentrations were measured from the stored extracts using an Atomic Absorption Spectrophotometer (Model AAS Vario 6, Analytik Jena AG, Jena, Germany).

Transfer factors of the vegetables were calculated by dividing the vegetable total heavy metal concentration with the soil water-soluble heavy metal concentration [20], to interpret comparative bioaccumulation of heavy metals by the experimental vegetables grown on the same or different textures soils irrigated with various wastewater sources in Multan.

2.3. Statistical Analyses

All data analyses were performed using the SPSS 26.0 package (SPSS, Chicago, IL, USA). A two-way (soil/water and vegetable) multivariate (Zn, Cu, Fe, Mn, Cd, Cr, Ni, and Pb) ANOVA for quantifying the individual and interactive effects of the soil/water and vegetable factors on the response variables of the Zn, Cu, Fe, Mn, Cd, Cr, Ni, and Pb concentrations in vegetables. Regressions and correlations were also performed where needed. Data were not normally distributed; therefore, they were normalized to log10 values. Differences were significant when $p < 0.05$.

3. Results

The ECs of the Industrial estate was the highest (3.9 dSm^{-1}; Table 1) and was significantly higher than those of the Khan Village (1.4 dSm^{-1}) and Vehari road (1.5 dSm^{-1}; $p = 0.001$, $p = 0.019$, respectively) sites, in which the two sites had significantly lower ECs than the Sameej abad site (3.5 dSm^{-1}; $p = 0.016$). All other ECs comparisons did not significantly differ. The pHs of the sites were not significantly different, except for the Industrial estate pHs (8.4; Table 1), which was significantly higher than that of the Sameeja abad site (8.2; $p = 0.005$).

The ECiw of Industrial estate was also the highest (3.6 dSm^{-1}; Table 1) and was significantly higher than those of the Khan Village (0.3 dSm^{-1}) and Vehari road (0.3 dSm^{-1}; $p = 0.001$, $p = 0.017$, respectively) sites, in which the two sites had significantly lower ECiw than the Shujabad site (2.8 dSm^{-1}; $p = 0.017$). All other comparisons did not significantly differ. The pHiw of the sites were not significantly different with the exceptions of the Khan Village pHiw (7.2; Table 1), which was significantly higher than those of the Industrial estate (6.8; $p = 0.001$) and Shujabad (6.9; $p = 0.017$) where the Shujabad site had significantly lower pHiw than that of the Soraj miani site (7.1; $p = 0.017$).

Overall, the study sites (soils) were significantly different in chemistry and heavy metal concentration (Table 1; ECs: $p = 0.001$; pHs: $p < 0.001$; Zn: 0.036; Cu: $p = 0.002$; Fe: $p < 0.001$; Mn: $p < 0.001$; Cd: $p = 0.006$; Cr: $p = 0.003$; Ni: $p = 0.005$; and Pb: $p < 0.001$). All respective irrigation waters were also significantly different, except for Cr (ECs: $p = 0.019$; pHs: $p < 0.001$; Zn: 0.031; Cu: $p = 0.005$; Fe: $p = 0.012$; Mn: $p = 0.003$; Cd: $p = 0.004$; Cr: $p = 0.058$; Ni: $p = 0.002$; and Pb: $p = 0.013$). It is to be noted that Mn is not exactly a heavy metal but toxic when absorbed or present in excessive amounts.

Soil heavy metal concentrations were significantly different ($p < 0.05$; Table 1) between sites with the highest concentrations of all metals at the Industrial estate (Zn = 3.9, Cu = 2.6, Fe = 17.3, Mn = 4.0, Cd = 4.8. Cr = 4.4, Ni = 1.7, and Pb = 3.4 mg kg^{-1}), and the lowest at the Khan Village site (Zn = 1.8, Cu = 0.9, Fe = 6.8, Mn = 2.9, Cd = 0.7, Cr = 1.1, Ni = 0.4, and Pb = 1.4 mg kg^{-1}). All other sites had a mix of higher or lower concentrations of the heavy metals compared to each other.

There were significant ($p < 0.05$) main (site/water and vegetable) and interactive (soil/water * vegetation) effects on phytoextraction of all metal concentrations. At the site level, the Cu, Mn, Cd, and Ni phytoextraction values were significantly different, except for the Cu and Cd values at Khan Village that did not differ from the corresponding values at Shujabad road and Vehari road (both; $p = 1.000$, $p = 0.073$, respectively; Figures 1 and 2), Cd concentrations at Sameeja abad did not differ from those at the Khan village and Vehari road values (both $p = 1.00$, respectively Figures 3 and 4). The phytoextraction concentrations of Mn, Cd, and Ni also significantly differed between all four vegetables; however, the lettuce Zn concentration did not differ from that in brinjal ($p = 1.00$), the cauliflower Cu value did not differ from that in spinach (0.189), and the lettuce and cauliflower values did not differ from those in brinjal and spinach (both; $p = 1.00$; Figures 1 and 2)

There were significant ($p < 0.05$) main (site and vegetable) and interactive (soil * vegetation) effects on vegetable transfer factor (VTF) for all metals, except no site/water * vegetable interaction for the Cr VTF was found ($p = 0.585$; Tables 2 and 3). VTF comparisons revealed that spinach was the best phytoextractant with the highest phytoextraction values (spinach: Zn = 20.2 ($\pm$3.6), Cu = 12.3 ($\pm$8.3), Fe = 17.1 ($\pm$8.1), Mn = 30.3 ($\pm$16.5), Cd = 6.1 ($\pm$4.5), Cr = 7.6 ($\pm$5.7), Ni = 9.2 ($\pm$3.1), and Pb = 6.9 ($\pm$1.7)—Tables 2 and 3) followed by cauliflower and brinjal, while lettuce had the lowest VTF values for phytoextraction (lettuce: Zn = 20.2 ($\pm$4.3), Cu = 12.3 ($\pm$4.0), Fe = 17.1 ($\pm$2.0), Mn = 30.3 ($\pm$2.4), Cd = 6.1 ($\pm$4.7), Cr = 7.6 ($\pm$5.1), Ni = 9.2 ($\pm$2.6), and Pb = 6.9 ($\pm$1.3)—Tables 2 and 3) under main (site and vegetable) or interactive (site * vegetable) effects. Values in brackets show the standard deviation.

The soil metal concentration–VTF correlations were significant ($p < 0.05$; Figures 3 and 4) for brinjal and lettuce for Zn; for brinjal and cauliflower for Cu; for brinjal only for Fe; all for Mn and Ni; for spinach for Cd and Cr; and for brinjal, lettuce, and spinach for Pb (Figures 3 and 4).

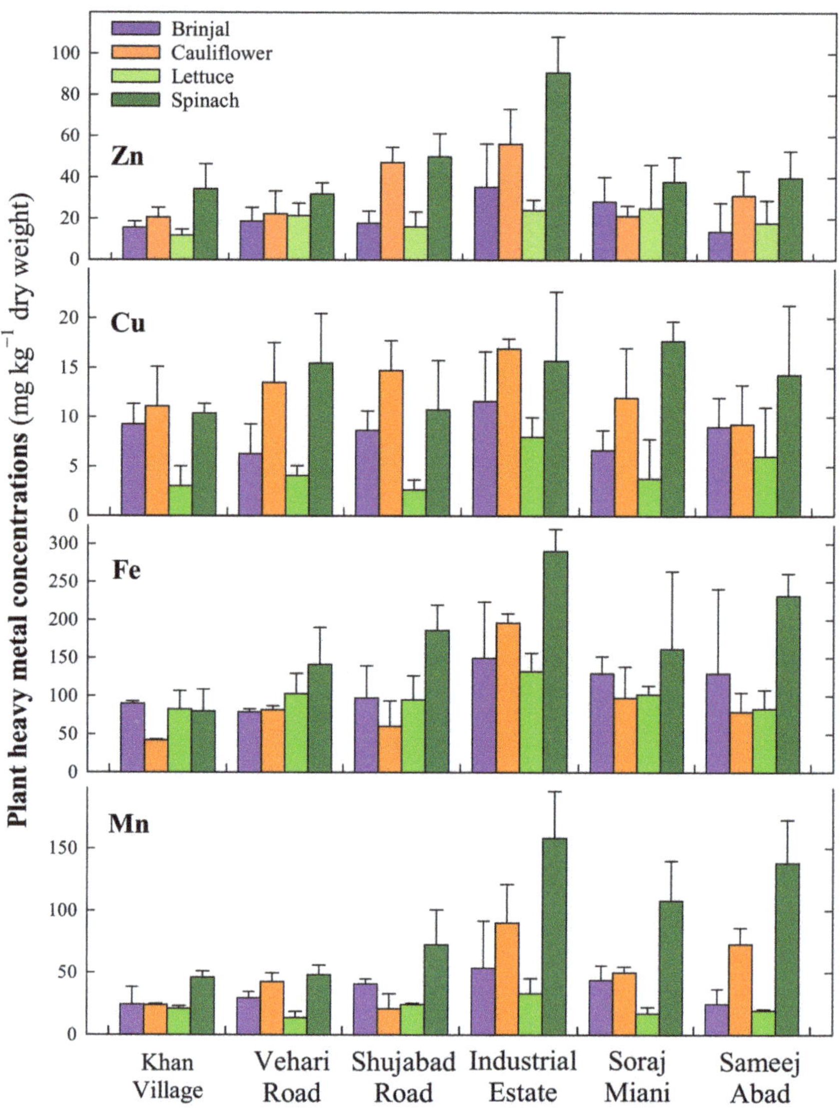

Figure 1. Zn, Cu, Fe, and Mn concentrations (mg kg^{-1} dry weight) in brinjal (*Solanum melogena*) and spinach (*Spinacia oleacea*) harvested in April 2016, and lettuce (*Lactuca sativa* L.) and cauliflower (*Brassica oleracea*) harvested in December 2015 from the Khan Village, Vehari road, Shujabad road, Industrial estate, Soraj miani, and Sameej abad study sites.

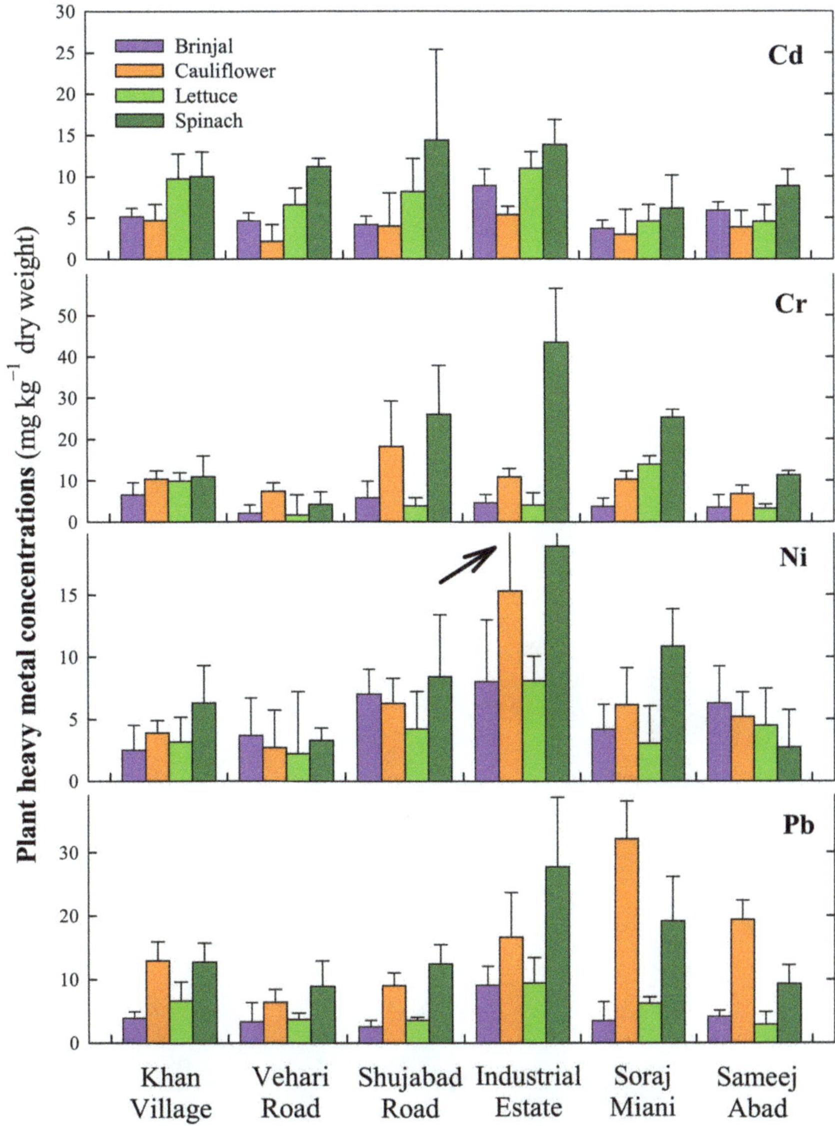

Figure 2. Heavy metal (Cd, Cr, Ni, and Pb) concentrations (mg kg^{-1} dry weight) in brinjal (*Solanum melogena*) and spinach (*Spinacia oleacea*) harvested in April 2016, and lettuce (*Lactuca sativa* L.) and cauliflower (*Brassica oleracea*) harvested in December 2015 from the Khan village, Vehari road, Shujabad road, Industrial estate, Suraj miani and Sameej abad study sites.

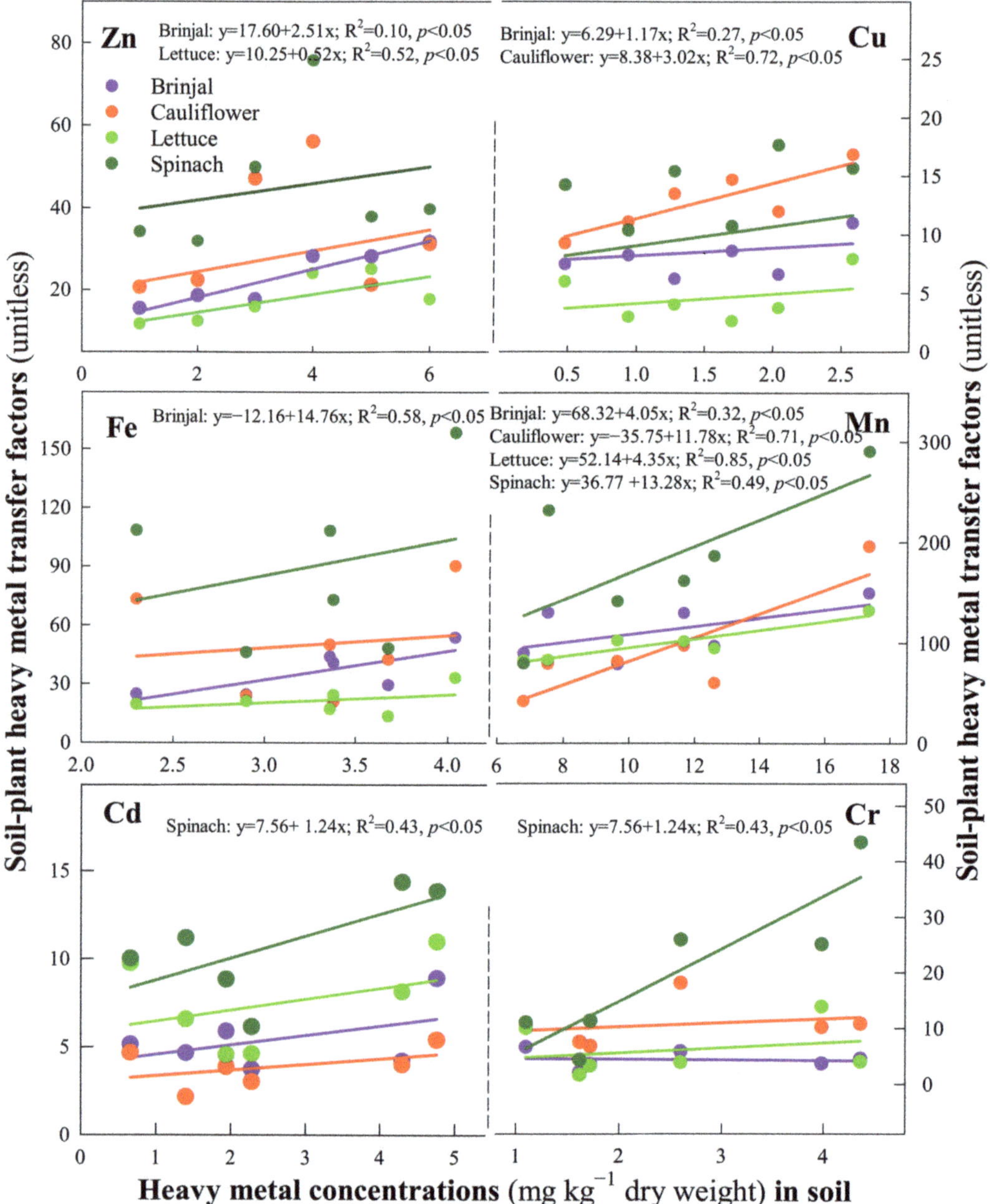

Figure 3. Relationships between the transfer factors (TF) of the vegetables (brinjal, cauliflower, lettuce, and spinach) and heavy metal (Zn, Cu, Fe, Mn, Cd, and Cr) concentrations (mg kg^{-1} dry weight) in the respective soils of the Khan Village, Vehari road, Shujabad road, Industrial estate, Soraj miani, and Sameej abad study sites. A relationship is significant at $p < 0.05$. Only significant relationships are retained to sustain the brevity of the manuscript.

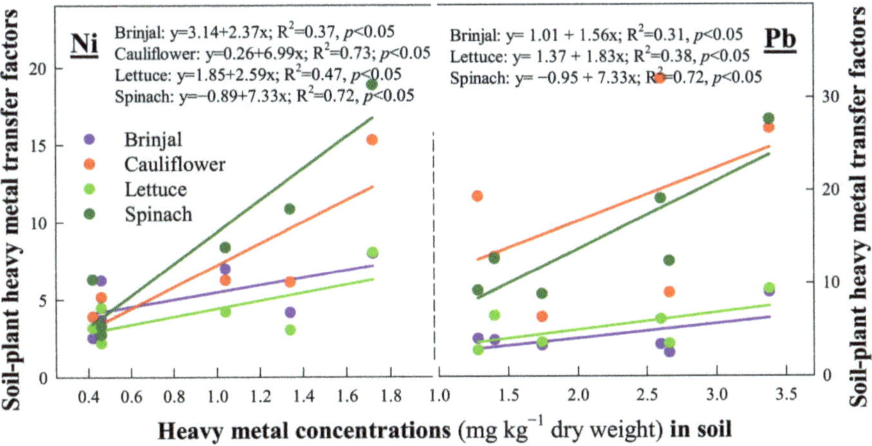

Figure 4. Relationships between the transfer factors (TF) of the vegetables (brinjal, cauliflower, lettuce, and spinach) and heavy metal (Ni and Pb) concentrations (mg kg^{-1} dry weight) in the respective soils of the Khan Village, Vehari road, Shujabad road, Industrial estate, Soraj miani, and Sameej abad study sites. A relationship is significant at $p < 0.05$. Only significant relationships are retained to sustain the brevity of the manuscript.

Table 2. Mean ($\pm$ SD) values (unitless) of the soil–plant transfer factors (vegetable leaf concentration/soil concentration) for all heavy metals across the study sites *.

Site/Vegetable	Zn	Cu	Fe	Mn	Cd	Cr	Ni	Pb
Khan Village								
Brinjal	8.6 ± 3.9	9.8 ± 0.6	13.3 ± 0.9	8.4 ± 0.5	7.8 ± 0.4	5.9 ± 5.8	6.0 ± 0.1	2.8 ± 0.2
Cauliflower	11.4 ± 1.7	11.8 ± 0.6	6.2 ± 0.9	8.2 ± 3.5	7.0 ± 5.3	9.5 ± 10.0	9.3 ± 0.5	9.2 ± 7.5
Lettuce	6.5 ± 3.8	3.2 ± 0.4	12.1 ± 0.3	7.2 ± 3.2	14.7 ± 0.4	9.0 ± 10.0	7.5 ± 1.0	4.6 ± 0.3
Spinach	19.0 ± 1.6	11.0 ± 0.1	11.7 ± 8.5	15.9 ± 1.0	15.1 ± 0.6	10.0 ± 8.8	15.0 ± 0.0	9.0 ± 0.1
Vehari road								
Brinjal	11.2 ± 2.6	4.8 ± 0.7	8.1 ± 0.9	8.0 ± 0.0	3.3 ± 1.2	1.2 ± 0.5	8.0 ± 0.1	1.9 ± 0.2
Cauliflower	13.4 ± 4.6	10.5 ± 0.7	8.5 ± 0.6	11.5 ± 0.5	1.5 ± 0.6	4.6 ± 1.3	5.9 ± 0.7	3.6 ± 0.5
Lettuce	12.9 ± 6.8	3.1 ± 0.4	10.6 ± 0.5	3.7 ± 0.5	4.7 ± 0.5	0.9 ± 0.0	4.8 ± 0.5	2.1 ± 0.2
Spinach	19.2 ± 2.2	12.0 ± 0.1	14.6 ± 0.5	13.1 ± 0.1	7.9 ± 0.0	2.6 ± 0.5	7.1 ± 0.3	5.0 ± 0.2
Shujabad road								
Brinjal	7.1 ± 1.4	5.0 ± 0.2	7.7 ± 0.5	12.0 ± 4.2	0.9 ± 0.0	2.2 ± 1.3	6.7 ± 0.5	0.9 ± 0.1
Cauliflower	19.0 ± 0.8	8.6 ± 0.8	4.8 ± 1.7	6.1 ± 0.9	0.9 ± 0.1	7.0 ± 4.8	6.0 ± 0.1	3.3 ± 3.1
Lettuce	6.4 ± 1.2	1.5 ± 0.6	7.5 ± 0.5	7.1 ± 3.2	1.9 ± 0.2	1.5 ± 1.0	4.0 ± 0.1	1.3 ± 0.5
Spinach	20.1 ± 5.5	6.3 ± 0.5	14.7 ± 6.9	21.5 ± 3.7	3.3 ± 0.5	9.9 ± 5.8	8.0 ± 0.1	4.6 ± 0.5
Industrial estate								
Brinjal	9.0 ± 0.1	4.5 ± 1.7	8.6 ± 0.5	13.3 ± 1.4	1.8 ± 0.3	1.0 ± 0.1	4.6 ± 0.5	2.6 ± 1.0
Cauliflower	14.4 ± 0.5	6.5 ± 1.0	11.3 ± 3.3	22.3 ± 2.5	1.1 ± 0.3	2.4 ± 1.0	8.8 ± 0.2	4.9 ± 2.1
Lettuce	6.1 ± 0.4	3.0 ± 2.1	7.6 ± 0.5	8.21 ± 0.5	2.3 ± 0.5	0.9 ± 0.1	4.6 ± 0.5	2.7 ± 1.5
Spinach	23.2 ± 5.2	6.0 ± 0.2	16.7 ± 1.5	39.2 ± 1.4	2.9 ± 0.2	9.9 ± 5.8	10.9 ± 0.0	8.1 ± 0.6
Soraj miani								
Brinjal	13.3 ± 0.5	3.2 ± 1.0	11.1 ± 1.6	13.0 ± 0.7	1.6 ± 0.5	0.9 ± 0.1	3.1 ± 0.2	1.3 ± 0.5
Cauliflower	10.0 ± 0.1	5.8 ± 0.6	8.3 ± 0.9	14.9 ± 0.8	1.3 ± 0.5	2.5 ± 0.5	4.5 ± 0.7	12.2 ± 1.4
Lettuce	11.8 ± 4.0	1.8 ± 0.6	8.7 ± 1.0	5.1 ± 0.6	2.0 ± 0.0	3.5 ± 5.7	2.2 ± 0.5	2.3 ± 0.8
Spinach	17.8 ± 0.6	8.6 ± 0.8	13.8 ± 5.9	32.1 ± 0.6	2.7 ± 0.5	6.3 ± 4.4	8.0 ± 0.2	7.3 ± 0.5
Sameeja abad								
Brinjal	7.5 ± 1.8	18.7 ± 1.5	17.2 ± 3.8	10.7 ± 0.9	3.0 ± 0.1	2.0 ± 0.1	13.6 ± 0.5	3.1 ± 0.5
Cauliflower	17.1 ± 1.1	19.3 ± 0.9	10.5 ± 5.3	31.8 ± 2.1	1.9 ± 0.0	3.9 ± 0.1	11.2 ± 1.0	15.1 ± 0.6
Lettuce	9.8 ± 1.6	12.5 ± 1.7	11.0 ± 2.0	8.5 ± 0.6	2.3 ± 0.5	1.8 ± 0.2	9.7 ± 0.5	2.2 ± 0.5
Spinach	21.8 ± 2.5	29.7 ± 0.5	30.7 ± 4.9	60.1 ± 0.2	4.5 ± 0.5	6.5 ± 5.0	5.9 ± 0.1	7.2 ± 0.5

* Sites comparison: soil–plant transfer of Fe was significantly higher at the industrial estate site than that at the Khan Village site ($F_{5, 24} = 2.51$, $p = 0.048$, $R^2 = 0.41$). The Ni transfer factor at the Industrial site was significantly higher than all sites except the Shujabad road site ($F_{5, 24} = 5.51$, $p = 0.003$, $R^2 = 0.61$).

Table 3. Results of a two-way MANOVA of the soil–vegetable heavy metal transfer factor for six study sites in Multan, Pakistan.

Source	df	Zn	Cu	Fe	Mn	Cd	Cr	Ni	Pb
Site									
F	5, 96	2.69	93.99	6.13	66.17	168.86	3.07	243.52	20.86
p	5, 96	**0.03**	**0.00**	**0.00**	**0.00**	**0.00**	**0.01**	**0.00**	**0.00**
Vegetable									
F	5, 96	52.50	131.85	13.59	257.04	108.95	17.64	213.05	90.90
p	5, 96	**0.00**	**0.00**	**0.00**	**0.00**	**0.00**	**0.00**	**0.00**	**0.00**
Site × Vegetable									
F	5, 96	3.31	4.20	2.59	13.13	4.30	0.88	68.92	4.55
p	5, 96	**0.00**	**0.00**	**0.00**	**0.00**	**0.00**	0.59	**0.00**	**0.00**
R^2	5, 96	0.75	0.93	0.61	0.95	0.95	0.53	0.98	0.86

Bold values are significant at the =0.05 level. R^2 shows the relationship between the soil and plant heavy metal concentrations.

4. Discussion

In this study, the phytoextractions of soil Zn, Cu, Fe, Mn, Cd, Cr, Ni, and Pb heavy metals by the four vegetables (brinjal, cauliflower, lettuce, and spinach), grown on six sites equally divided into sandy loam and clay loam soils, and where each soil was irrigated with three types of water—normal, sewage, and normal + sewage—were measured. Since comparing the phytoextraction values of the vegetables grown on different textured soils irrigated with different types of waters would not be satisfying, we additionally compared the vegetable transfer factor (VTF; Cui et al. [20]), which is equal to the heavy metal concentration in the vegetable divided by the heavy metal concentration in the soil. The VTF minimizes biases in comparative phytoextraction values owing to differences in heavy metal concentrations among the soil textures as well as the irrigation waters used in the study.

4.1. Effect of Site/Water on Phytoextraction of Heavy Metals

In this study, heavy metal phytoextraction of all heavy metals was the highest at the Industrial estate study site, which had a clay loam texture and was being irrigated with the sewage water. The overall phytoextractions of heavy metals by the four vegetables at this site were significantly higher than those at the Khan Village and Sameej abad sites, which had sandy loam and clay loam textures, respectively, and were being irrigated with normal water. Moreover, Shujabad road site (sandy loam) vegetables had also clearly higher phytoextractions than those of the crops at the sites being irrigated with normal water. These results are in line with that of Yargholi [36], who also reported significant increases in heavy metal concentrations in soil (or VTF) in response to irrigation with sewage water of high heavy metal concentrations. The toxic concentration of the heavy metals or other micro/nutrients in the irrigation water has frequently been reported, for example, [37,38] to directly concentrate in the plant roots/shoots/leaves, which food may put the health and life of animals and humans in jeopardy.

The highest phytoextraction by the industrial site vegetables was likely due to the greatest soil concentrations of the respective heavy metals owing to continuous use of sewage water at this heavy soil texture site. It also reveals that the soil particle exchange sites of this heavy texture site tend to be fully occupied and are transferring the surplus to the crops, resulting in high VTF factors. A similar phenomenon is also shown by the Shujabad road site; however, the extent of the transfer of heavy metals to the vegetables was much lower due to the light texture nature of the soil, which may, for example, accumulate or supply lesser heavy metals to the plant parts [39].

Overall phytoextraction of heavy metals by vegetable crops showed that spinach had the highest phytoextraction of all the heavy metals followed by cauliflower and lettuce, while brinjal had the lowest phytoextraction of heavy metals, as was also found by Sharma et al. [17,18]. The phyto uptake capability of the spinach crop we found is comparable to the findings of Ng et al. [40]. The highest heavy metal phytoextraction of spinach could be attributed to its leafy structure, short stature and growth cycle, and the fastest metabolism, which can uptake toxic compounds or heavy metals in toxic quantities and transform or mineralize them to simpler compounds, which can be assimilated or accumulated for the long term. There was a site * vegetable interaction for spinach at the Industrial estate site, which has a heavy soil texture irrigated with sewage water, again supporting our findings on the individual effects of site and vegetable on the phytoextraction of the heavy metals, and also support the findings of Zia et al. [34], except they used phytoextraction enhancing chemicals.

4.2. Effect of Site/Water on Vegetable Transfer Factor

In contrast to the phytoextraction values, the VTF values were found to be the highest at the Sameeja abad site of heavy texture and irrigated with a mix of normal and sewage waters (the exact ratio of the mix is not reported since it kept changing). It was found that the crops had visually much higher coverages and biomasses than those at the other experimental sites. Vegetation may attain maximum coverage and biomass when supplied with exponentially dissolved organic carbon and salts or ions in soils [41]. Therefore, a higher VTF at this site could be attributed to greater uptake of heavy metals (relative to its soil concentrations) along with essential nutrients in contrast to the Industrial estate, which had greater concentrations of heavy metals in both the soil and vegetable crop and the VTF was low due to the much higher soil heavy metal concentrations than those at the Sameeja abad site. The findings of Cui et al. [20] corroborate our explanation of the higher VTFs of vegetables due to lower soil heavy metal concentrations.

Interestingly, spinach had overall the highest VTF values of all the heavy metals, similar to its corresponding phytoextraction values. Additionally, the VTF values of cauliflower, lettuce, and brinjal were also in the order similar to those of the phytoextraction values of these vegetables. These findings complement our results on the phytoextraction capacities of the four vegetable crops used in this study.

5. Conclusions

Vegetables, one of the most important foods for humans, are also efficient phytoextractants of heavy metals in soils. When grown on heavy or light-textured soils irrigated with sewage water, vegetable crops phytoextracted the heavy metals to highly toxic levels. Spinach was the most efficient phytoextractant followed by cauliflower and lettuce, while brinjal was the least efficient of all. We suggest that soils, when irrigated with sewage water having excessive concentrations of the studied heavy metals, may transfer these metals in toxic amounts to vegetable crop plants.

Author Contributions: Conceptualization, I.A., S.S. and S.A.M.; methodology, I.A., A.-u.R.; software, T.M.M.; validation, I.A., S.S. and S.A.M.; formal analysis, T.M.M.; investigation, I.A., S.S., A.-u.R. and S.A.M.; resources, S.S.; data curation, I.A.; writing—original draft preparation, I.A.; writing—review and editing, S.S., A.-u.R., and T.M.M.; visualization, I.A., and A.-u.R.; supervision, S.S., T.M.M.; project administration, S.A.M., and S.S.; funding acquisition, S.S. All authors have read and agreed to the published version of the manuscript.

Funding: This research received funding from the Bahauddin Zakariya University (GR-2021-197), Multan, and Muhammad Nawaz Sharif University of Agriculture (GR-2021-Mul-427), Multan.

Institutional Review Board Statement: Not applicable.

Informed Consent Statement: Not applicable.

Data Availability Statement: All data analyzed in this study are available from the corresponding author upon request.

Conflicts of Interest: The authors declare no conflict of interest.

References

1. IPCC. Intergovernmental Panel on Climate Change, in Climate change 2014: Impacts, Adaptation, and Vulnerability: Working Group II Contribution to the Fifth Assessment Report of the Intergovernmental Panel on Climate Change. 2014. Available online: https://www.ipcc.ch/report/ar5/wg2/ (accessed on 18 June 2021).
2. Chaudhry, Q.U.Z. *Climate Change Profile of Pakistan*; Asian Development Bank: Mandaluyong, Philippines, 2017.
3. Krishnan, R.; Shrestha, A.B.; Ren, G.; Rajbhandari, R.; Saeed, S.; Sanjay, J.; Syed, M.A.; Vellore, R.; Xu, Y.; You, Q. Unravelling climate change in the Hindu Kush Himalaya: Rapid warming in the mountains and increasing extremes. In *The Hindu Kush Himalaya Assessment*; Springer: Berlin/Heidelberg, Germany, 2019; pp. 57–97.
4. Ali, S.; Kiani, R.S.; Reboita, M.S.; Dan, L.; Eum, H.I.; Cho, J.; Dairaku, K.; Khan, F.; Shreshta, M.L. Identifying hotspots cities vulnerable to climate change in Pakistan under CMIP5 climate projections. *Int. J. Climatol.* **2021**, *41*, 559–581. [CrossRef]
5. Barros, V.; Field, C.; Dokke, D.; Mastrandrea, M.; Mach, K.; Bilir, T.; Chatterjee, M.; Ebi, K.; Estrada, Y.; Genova, R. *Climate Change 2014: Impacts, Adaptation, and Vulnerability. Part B: Regional Aspects*; Contribution of Working Group II to the Fifth Assessment Report of the Intergovernmental Panel on Climate Change; Cambridge University Press: New York, NY, USA, 2014.
6. Asif, M. *Climatic Change, Irrigation Water Crisis and Food Security in Pakistan*; Uppsala University: Uppsala, Sweden, 2013.
7. Munir, T.M.; Perkins, M.; Kaing, E.; Strack, M. Carbon dioxide flux and net primary production of a boreal treed bog: Responses to warming and water-table-lowering simulations of climate change. *Biogeosciences* **2015**, *12*, 1091–1111. [CrossRef]
8. Lone, M. Comparison of blended and cyclic use of water for agriculture. In *Final Report University Grant Commission, Islamabad. Pakistan*; University Grants Commission: Islamabad, Pakistan, 1995.
9. Silva, H.F.; Silva, N.F.; Oliveira, C.M.; Matos, M.J. Heavy Metals Contamination of Urban Soils—A Decade Study in the City of Lisbon, Portugal. *Soil Syst.* **2021**, *5*, 27. [CrossRef]
10. Mahmood, S. Waste water irrigation: Issues and constraints for sustainable irrigated agriculture. *J. Ital. Agron.* **2006**, *3*, 12–15.
11. Drechsel, P.; Raschid-Sally, L.; Williams, S.; Weale, J. Recycling Realities: Managing health risks to make wastewater an asset. *Water Policy Brief.* **2006**, *17*, 1–7.
12. Cooper, R.C. Public health concerns in wastewater reuse. *Water Sci. Technol.* **1991**, *24*, 55–65. [CrossRef]
13. Mara, D.D.; Cairncross, S. *Guidelines for the Safe Use of Wastewater and Excreta in Agriculture and Aquaculture*; World Health Organization: London, UK, 1989.
14. Curci, M.; Lavecchia, A.; Cucci, G.; Lacolla, G.; De Corato, U.; Crecchio, C. Short-Term Effects of Sewage Sludge Compost Amendment on Semiarid Soil. *Soil Syst.* **2020**, *4*, 48. [CrossRef]
15. Thakur, I.S. *Environmental Biotechnology: Basic Concepts and Applications*; IK International: Delhi, India, 2011.
16. Matzen, S.; Fakra, S.; Nico, P.; Pallud, C. Pteris vittata Arsenic Accumulation Only Partially Explains Soil Arsenic Depletion during Field-Scale Phytoextraction. *Soil Syst.* **2020**, *4*, 71. [CrossRef]
17. Sharma, R.; Agrawal, M.; Marshall, F. Heavy metal contamination in vegetables grown in wastewater irrigated areas of Varanasi, India. *Bull. Environ. Contam. Toxicol.* **2006**, *77*, 312–318. [CrossRef]
18. Sharma, R.K.; Agrawal, M.; Marshall, F. Heavy metal contamination of soil and vegetables in suburban areas of Varanasi, India. *Ecotoxicol. Environ. Saf.* **2007**, *66*, 258–266. [CrossRef]
19. Ilic, Z.; Filipovic-Trajkovic, R.; Jablanovic, M. Transfer factor (coefficient) soil/plant as indicator concentration of heavy metals content in different vegetable species. *Contemp. Agric.* **2001**, *50*, 41–44.
20. Cui, Y.-J.; Zhu, Y.-G.; Zhai, R.-H.; Chen, D.-Y.; Huang, Y.-Z.; Qiu, Y.; Liang, J.-Z. Transfer of metals from soil to vegetables in an area near a smelter in Nanning, China. *Environ. Int.* **2004**, *30*, 785–791. [CrossRef] [PubMed]
21. Rattan, R.; Datta, S.; Chhonkar, P.; Suribabu, K.; Singh, A. Long-term impact of irrigation with sewage effluents on heavy metal content in soils, crops and groundwater—A case study. *Agric. Ecosyst. Environ.* **2005**, *109*, 310–322. [CrossRef]
22. Vassilev, A.; Yordanov, I. Reductive analysis of factors limiting growth of cadmium-treated plants: A review. *Bulg. J. Plant Physiol.* **1997**, *23*, 114–133.
23. Lone, M.I.; Saleem, S.; Mahmood, T.; Saifullah, K.; Hussain, G. Heavy metal contents of vegetables irrigated by Sewage/Tubewell water. *Int. J. Agric. Biol.* **2003**, *5*, 533–535.
24. Moral, R.; Cortés, A.; Gomez, I.; Mataix-Beneyto, J. Assessing changes in Cd phytoavailability to tomato in amended calcareous soils. *Bioresour. Technol.* **2002**, *85*, 63–68. [CrossRef]
25. Das, P.; Samantaray, S.; Rout, G. Studies on cadmium toxicity in plants: A review. *Environ. Pollut.* **1997**, *98*, 29–36. [CrossRef]
26. Mir, I.R.; Rather, B.A.; Masood, A.; Majid, A.; Sehar, Z.; Anjum, N.A.; Sofo, A.; D'Ippolito, I.; Khan, N.A. Soil Sulfur Sources Differentially Enhance Cadmium Tolerance in Indian Mustard (*Brassica juncea* L.). *Soil Syst.* **2021**, *5*, 29. [CrossRef]
27. Shanker, A.K.; Cervantes, C.; Loza-Tavera, H.; Avudainayagam, S. Chromium toxicity in plants. *Environ. Int.* **2005**, *31*, 739–753. [CrossRef]
28. Nolan, K.R. Copper toxicity syndrome. *J. Orthomol. Psychiatry* **1983**, *12*, 270–282.
29. Di Toppi, L.S.; Gabbrielli, R. Response to cadmium in higher plants. *Environ. Exp. Bot.* **1999**, *41*, 105–130. [CrossRef]
30. Rahman, M.A.; Rahman, M.M.; Reichman, S.M.; Lim, R.P.; Naidu, R. Heavy metals in Australian grown and imported rice and vegetables on sale in Australia: Health hazard. *Ecotoxicol. Environ. Saf.* **2014**, *100*, 53–60. [CrossRef]
31. Chaney, R.L.; Malik, M.; Li, Y.M.; Brown, S.L.; Brewer, E.P.; Angle, J.S.; Baker, A.J. Phytoremediation of soil metals. *Curr. Opin. Biotechnol.* **1997**, *8*, 279–284. [CrossRef]

32. Salt, D.E.; Blaylock, M.; Kumar, N.P.; Dushenkov, V.; Ensley, B.D.; Chet, I.; Raskin, I. Phytoremediation: A novel strategy for the removal of toxic metals from the environment using plants. *Nat. Biotechnol.* **1995**, *13*, 468–474. [CrossRef] [PubMed]
33. Richards, L. Diagnosis and improving of saline and alkaline soils. US, Salinity Laboratory Staff. *Agric. Handb.* **1954**, *46*, 290.
34. Zia, M.H.; Meers, E.; Ghafoor, A.; Murtaza, G.; Sabir, M.; Zia-ur-Rehman, M.; Tack, F. Chemically enhanced phytoextraction of Pb by wheat in texturally different soils. *Chemosphere* **2010**, *79*, 652–658.
35. APHA. Standard methods for the examination of water and wastewater. *Water Environ. Fed.* **2005**, *21*, 258–259.
36. Yargholi, B. *Investigation of the Firozabad Wastewater Quality-Quantity Variation for Agricultural Use*; Final Report; Iranian Agricultural Engineering Research Institute: Tehran, Iran, 2007.
37. Ali, H.; Khan, E.; Ilahi, I. Environmental chemistry and ecotoxicology of hazardous heavy metals: Environmental persistence, toxicity, and bioaccumulation. *J. Chem.* **2019**, *2019*, 6730305. [CrossRef]
38. Iqbal, Z.; Abbas, F.; Ibrahim, M.; Qureshi, T.I.; Gul, M.; Mahmood, A. Human health risk assessment of heavy metals in raw milk of buffalo feeding at wastewater-irrigated agricultural farms in Pakistan. *Environ. Sci. Pollut. Res.* **2020**, *27*, 29567–29579. [CrossRef]
39. Liang, J.; Chen, C.; Song, X.; Han, Y.; Liang, Z. Assessment of heavy metal pollution in soil and plants from Dunhua sewage irrigation area. *Int. J. Electrochem. Sci.* **2011**, *6*, 5314–5324.
40. Ng, C.C.; Rahman, M.M.; Boyce, A.N.; Abas, M.R. Heavy metals phyto-assessment in commonly grown vegetables: Water spinach (*I. aquatica*) and okra (*A. esculentus*). *SpringerPlus* **2016**, *5*, 1–9. [CrossRef] [PubMed]
41. Strack, M.; Munir, T.M.; Khadka, B. Shrub abundance contributes to shifts in dissolved organic carbon concentration and chemistry in a continental bog exposed to drainage and warming. *Ecohydrology* **2019**, *12*, e2100. [CrossRef]

Article

Spatial Analysis of Soil Trace Element Contaminants in Urban Public Open Space, Perth, Western Australia

Andrew W. Rate

School of Agriculture and Environment, The University of Western Australia, Perth, WA 6009, Australia; andrew.rate@uwa.edu.au

Abstract: Public recreation areas in cities may be constructed on land which has been contaminated by various processes over the history of urbanisation. Charles Veryard and Smith's Lake Reserves are adjacent parklands in Perth, Western Australia with a history of horticulture, waste disposal and other potential sources of contamination. Surface soil and soil profiles in the Reserves were sampled systematically and analysed for multiple major and trace elements. Spatial analysis was performed using interpolation and Local Moran's I to define geochemical zones which were confirmed by means comparison and principal components analyses. The degree of contamination of surface soil in the Reserves with As, Cr, Cu, Ni, Pb, and Zn was low. Greater concentrations of As, Cu, Pb, and Zn were present at depth in some soil profiles, probably related to historical waste disposal in the Reserves. The results show distinct advantages to using spatial statistics at the site investigation scale, and for measuring multiple elements not just potential contaminants.

Keywords: urban soil; parkland; contaminants; trace elements; metals; spatial autocorrelation; arsenic; chromium; copper; nickel; lead; zinc

Citation: Rate, A.W. Spatial Analysis of Soil Trace Element Contaminants in Urban Public Open Space, Perth, Western Australia. *Soil Syst.* **2021**, *5*, 46. https://doi.org/10.3390/soilsystems5030046

Academic Editors: Matteo Spagnuolo, Paola Adamo and Giovanni Garau

Received: 29 June 2021
Accepted: 12 August 2021
Published: 14 August 2021

Publisher's Note: MDPI stays neutral with regard to jurisdictional claims in published maps and institutional affiliations.

Copyright: © 2021 by the author. Licensee MDPI, Basel, Switzerland. This article is an open access article distributed under the terms and conditions of the Creative Commons Attribution (CC BY) license (https://creativecommons.org/licenses/by/4.0/).

1. Introduction

A number of studies have documented the potential for contaminant additions to soils from a range of urban activities. Horticultural activities are known to leave a legacy of soil contamination related to use of fertilisers, manures, and other materials [1–3]. The disposal of metalliferous and other wastes is known to cause soil contamination with trace elements [4]. Excavation of peaty and/or sulfidic subsoils is known to result in contamination of soils with acidity and metals [5,6]. Public facilities such as vehicle and storage depots and electrical substations are also potential contaminant sources known to have caused soil pollution [7,8]. Finally, building construction is a likely source of soil, sediment, and water contamination [9].

Smith's Lake and Charles Veryard Reserves are public recreation spaces in metropolitan Perth, Western Australia (WGS84 115.8505° E, 31.9319° S), with a complex history of land use change that is typical of many urban areas worldwide [10].

Spatial statistical techniques represent useful tools for identifying and describing soil contamination. For example, the use of variogram and/or spatial autocorrelation analysis can be used to quantify the degree of spatial dependence between contaminant concentrations in soil samples [11]. In addition, use of local spatial autocorrelation statistics such as Local Moran's I can be used to categorize locations, or clusters of locations, using the statistical significance and the magnitude of the response variable, such as in Local Indicators of Spatial Association (LISA) analysis [12]. Use of these techniques to study urban soil contamination has been limited so far to citywide spatial scales (tens of kilometers), covering multiple current land uses [11,13,14]. On whole-city scales, clusters of positively autocorrelated samples with higher concentrations are interpreted to represent "regional hotspots" of contamination. Conversely, isolated samples of higher concentration showing negative autocorrelation with surrounding low-concentration samples ("high-low" points in the LISA framework) are interpreted as being "isolated hotspots", potentially caused by point sources [14]. Very few published studies have used autocorrelation statistics to analyse spatial patterns of soil contamination at scales of a few hundred metres, with a single or restricted range of land use, which are typical of environmental site investigations where contamination is suspected. At smaller spatial scales, a reasonable hypothesis is that clusters of positively autocorrelated, high concentration points ("high-high" points in LISA) are more likely to represent point source contamination, whereas isolated "high-low" points will have less significance.

This study therefore had multiple objectives. The potential contaminants of primary interest were the trace elements As, Cr, Cu, Ni, Pb, and Zn, due to their known effects on human health and ecosystem functioning. This set of elements is relevant to urban soil contamination in many cities globally, and also represents a range of geochemical behaviour with As and Cr often existing as oxyanions in soils in contrast to the cationic metals Cu, Ni, Pb, and Zn. In addition, a range of mobilities would be expected, with Cr and Pb commonly showing low mobility in soils in contrast with Zn and As which are usually more mobile. The major elements Al, Ca, and Fe, and soil pH and EC, were of interest to support and explain the trace element data. The scientific objectives, therefore, were first: to characterize the concentrations and spatial distributions of potential contaminants in soil in the Smith's Lake and Charles Veryard Reserve area. The second objective was to identify any spatial patterns in the data over scales of a few hundred metres, and match these to the known history of the sites. The final aim was to evaluate the findings from spatial analysis of surface sampling as indicators of subsoil contamination. The research approach evaluates the utility of spatial analysis to provide more quantitative evidence of zones of contamination in urban soils and should therefore be applicable to other urban soil environments at similar spatial scales.

2. Materials and Methods

2.1. Study Site

Smith's Lake and Charles Veryard Reserves are situated in the historical location of Three Island Lake and associated water bodies, on land that was drained from the 1870s to allow the establishment of horticulture. The area has experienced multiple land uses, including market gardening (1920s–1950s), dumping of rubbish, and recreation/parkland [15]. The Claise Brook Main Drain was changed from an open drain to an underground stormwater pipe in the mid-1970s, and the present Smith's Lake was also constructed as a stormwater-compensating basin at this time. The 1970s were also a time of substantial residential development in this area. The 2000s saw urban gentrification and infill occurring in the area; a local government depot to the east of Smith's Lake Reserve was redeveloped to residential land in 2000–2002. A large sports centre building in the south of Smith's Lake Reserve was demolished and rehabilitated to parkland in 2008, with conversion of some larger residential lots to allow construction of higher density housing also occurring around 2008.

While the land around Smith's Lake Reserve is currently residential, several previous facilities and activities adjacent to the current reserve had the potential to modify fluxes of a range of contaminants. These include the land uses described above, and: the burial of the Claise Brook Main Drain, the Council Works Depot and its subsequent redevelopment, the demolition and rehabilitation of the sports centre, and the electrical supply substation constructed in the 1960s (and upgraded in 2001) adjacent to the north-west corner of the reserve. Smith's Lake itself was rehabilitated by a community group in 1999 [16], with construction of a path approximating the line of the underground main drain in 2010–2011. Floodlight pylons were installed in Charles Veryard Reserve in 2016, involving excavation of potential acid sulfate soil material (based on maps from [17]).

2.2. Sampling

Sampling of soil at Smith's Lake and Charles Veryard Reserves was conducted on two occasions: in March 2017 (a grid of surface samples) and March 2018 (profile sampling at selected locations). In 2017, surface soil sampling locations were pre-selected prior to sampling using a randomised-within-grid sampling strategy (Figure 1) using a 52 × 52 m grid to maximise site coverage, and two samples per grid square, with the objective of sampling the grassed areas within the Reserves without statistical bias. In the field, preselected sample locations were located using handheld GPS. If randomised sample coordinates were too close to paved surfaces, water, structures, etc., the location was moved by approximately 5 m and the revised coordinates were recorded. For subsequent analysis, the sampling area was divided into zones based on exploratory data analyses (Figure 2).

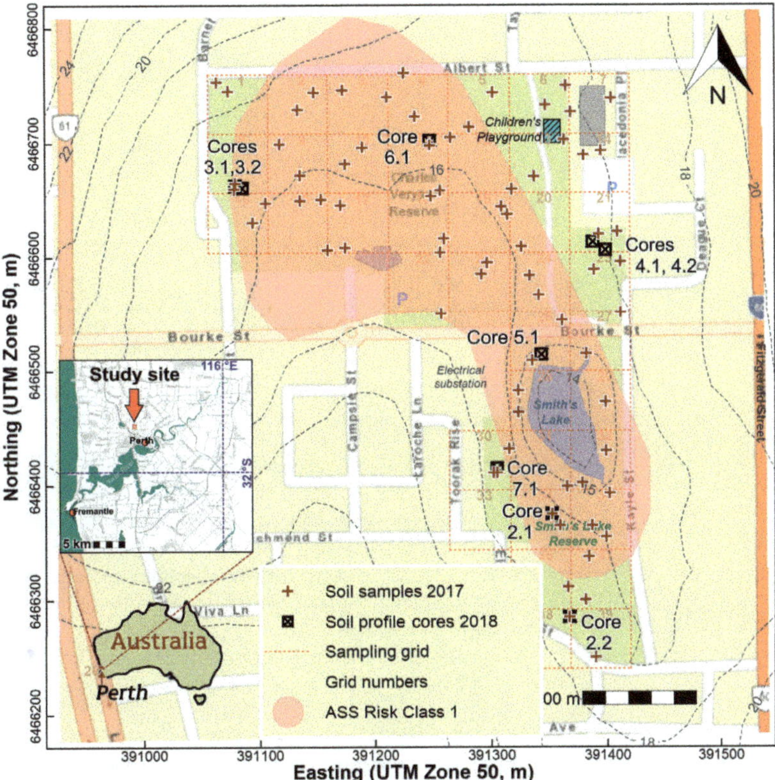

Figure 1. Street map showing soil sampling locations, and the initial grid used for sampling design, at Smith's Lake and Charles Veryard Reserves. Contours are land elevations in metres; the pale red polygon shows moderate to high acid sulfate soil (ASS) risk.

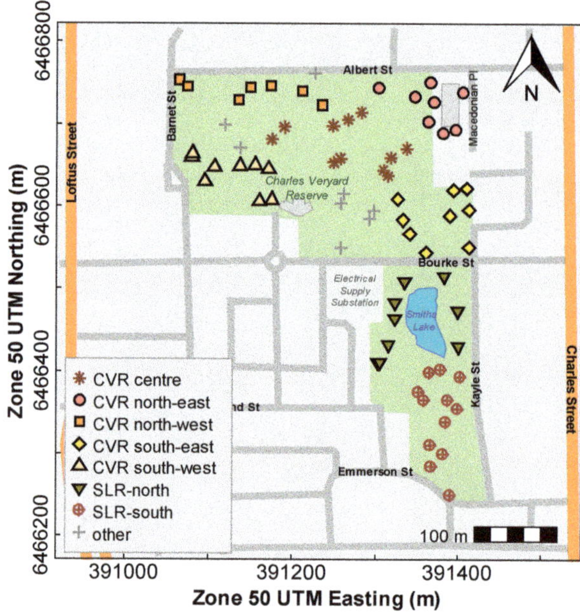

Figure 2. Simplified map showing sampling zones at Smith's Lake and Charles Veryard Reserves. CVR is Charles Veryard Reserve; SLR is Smiths Lake Reserve.

Surface soil was sampled in cylindrical cores from 0–10 cm depth using a stainless steel corer. Triplicate cores at each location were bulked to achieve a sample mass of ca. 500 g, and stored in zip-lock plastic bags prior to transport back to the laboratory. Soil sample cavities were re-filled with clean soil supplied by the sampling team.

In 2018, a second sampling of soil profiles to up to 100 cm depth was conducted using Garret-style augers. Separate samples were collected for each of eight profiles (Figure 1) in 10 cm depth increments.

Soil samples were air-dried in a laminar air-flow drying cabinet at 40 °C and sieved through a 2 mm aperture prior to analysis.

2.3. Chemical Analysis

The electrical conductivity (EC; proportional to soluble salt content) of soil samples was determined on 1:5 solid: Deionised water suspensions using a calibrated conductivity cell electrode. The pH was measured on the same suspensions using a glass-reference pH electrode after a 2-point buffer calibration [18].

The near-total concentrations of 26 elements (Al, As, Ba, Ca, Cd, Ce, Cr, Cu, Fe, Gd, K, La, Mg, Mn, Mo, Na, Nd, Ni, P, Pb, S, Sr, Th, V, Y, and Zn) were measured on samples by inductively-coupled plasma optical emission spectrometry (ICP-OES) following digestion of soil in concentrated nitric and hydrochloric acids (i.e., *aqua regia*) at ca. 130 °C [19]. *Aqua regia* digestion is commonly used to determine environmentally significant concentrations, since it largely excludes elements within the matrices of silicates and other recalcitrant minerals. Before acid digestion, samples were ground to ≲50 μm using ceramic mortars and pestles. Reagent blanks, and grinding blanks composed of acid-washed silica sand, were included in analytical runs to check for contamination. The standard reference stream sediment material STSD-2 [20] was analysed identically to samples to assess analytical accuracy. Measurement precision was assessed using analytical duplicates on ca. 10% of samples.

The lower limits of analytical detection were calculated, where possible, from $3\times$ the standard deviation of multiple reagent blank concentrations [21]. Concentrations lower than mean blank values, or below calculated lower detection limits, or both, were deleted from the dataset.

2.4. Statistical and Numerical Analysis

Data management and transformation of variables was conducted using Excel® (Version 2016, Microsoft, Redmond, WA, USA) Statistical and graphical analyses of data were performed in the statistical computing environment 'R' [22] and associated packages. Skewed variables (identified with the Shapiro-Wilk test for normality) were $\log_{10}$-transformed, or power-transformed based on the Box-Cox algorithm and re-checked for normality.

A general inability of variables to be transformed to yield normal distributions dictated the use of the non-parametric Spearman correlations, and Wilcoxon or Kruskal-Wallis tests for mean comparisons. If Kruskal-Wallis tests showed a significant difference, the R package 'PMCMR' [23] was used to apply the post-hoc Conover's test for pairwise comparisons of mean rank sums. Simple regression models were fitted using the $\log_{10}$-transformed variables. The potentially misleading effects of compositional closure were addressed using transformations to centred log-ratios [24], which were used for principal components analyses. Principal components analyses were conducted using only variables having minimal or no missing observations.

Distribution maps were constructed using the 'OpenStreetMap' package [25] with elevation contours interpolated from a dense grid of land elevations from Google [26] generated using the R package 'googleway' [27] and interpolated using the R package 'akima' [28]. Spatial autocorrelations were assessed using global and local Moran's I statistics, calculated using the R package 'lctools' [29]. Local Moran's I values showing significant association ($p \leq 0.05$) were categorised using high-low notation, based on the point measurement relative to the median and the sign of the Local Moran's I statistic. Spatial interpolations were achieved using an inverse distance weighting method using the R packages 'sp' [30] and 'gstat' [31]. Preliminary analysis showed that inverse-distance interpolation gave similar results to simple kriging, but kriging interpolation was not used, based on the requirement of ≥ 100 observations to generate a reliable experimental variogram [32].

A composite estimate of soil contamination was calculated from the concentrations of As, Cu, Pb, and Zn as the Integrated Pollution Index, IPI [33], shown in Equation (1):

$$\mathrm{IPI} = \left(\sum_{i=1}^{n} \left(\frac{C_i}{S_i} \right) \right) / n, \tag{1}$$

In Equation (1), $\sum$ means the sum of terms 1 to n, C_i = the measured concentration of th i-th element, S_i = the background concentration of the i-th element, n = the number of elements. The S_i values used (in mg/kg: As = 1.5, Cr = 10, Cu = 2, Pb = 5, Zn = 6) were published ambient background concentrations for the Perth region [34], but this report suggests a zero-background concentration for Ni. In this study 1 mg Ni/kg was used for background, which is the lowest (most conservative) 25th percentile concentration among similar datasets (e.g., [35]).

3. Results

3.1. Bulk Chemical Analyses of Surface Soil

Substantial variability in the concentrations of major elements and basic soil properties (Table 1), including potential trace element contaminants (Table 2), was observed in surface soils at Smith's Lake and Charles Veryard Reserves. In most cases, the maximum concentrations were at least five times the minima. Much greater variability was observed for calcium (Ca, maximum/minimum $\approx$ 108); cadmium (Cd), copper (Cu), magnesium

(Mg), manganese (Mn), neodymium (Nd), nickel (Ni), phosphorus (P), lead (Pb), sulfur (S), strontium (Sr), and zinc (Zn) all had maximum/minimum ratios between 20 and 90. Soluble salt content measured by EC showed a maximum/minimum ratio of $\approx$29, and there was a relatively large, $\approx$3.4 units range in pH across the Reserves. Except for zinc, which exceeded the interim Ecological Investigation Level (EIL; National Environment Protection Council, 1999) in three samples, no other soil thresholds were exceeded by any element (Table 2).

Table 1. Summary of pH, EC, and major element concentrations in surface (0–10 cm) soil and in vertical soil profile samples at Smith's Lake and Charles Veryard Reserves. EC and pH were measured in 1:5 solid: deionised water suspensions; element concentrations were measured using *aqua regia* digestion followed by ICP-OES.

Statistic	pH	EC	Major Element Concentration (mg/kg)							
		(µS/cm)	Al	Ca	Fe	K	Mg	Na	P	S
Surface soil (random-in-grid samples, 2017)										
Mean	6.92	183	2667	5083	2695	161	425	130	244	309
Std. Dev.	0.63	146	758	7082	939	75.3	285	69	123	186
Minimum	5.28	29.1	396	283	1205	41.1	57.3	27.8	18.5	43
Median	6.84	151	2637	1813	2530	150	347	118	221	280
Maximum	8.63	835	4722	30,472	5640	424	1471	410	596	1293
No of valid analyses	73	58	68	68	68	68	68	68	68	68
Soil profiles to $\leq$ 100 cm (2018)										
Mean	7.67	116	3214	11,061	4430	173	548	148	146	518
Std. Dev.	1.96	61	2168	14,254	3358	100	505	100	137	334
Minimum	4.48	7.92	149	38	218	26	14	11	4	48
Median	7.98	83	2880	7894	3593	134	450	127	140	293
Maximum	9.40	860	10,353	57,265	22,180	655	2365	513	386	4929
No of valid analyses	84	83	85	85	85	85	85	85	85	83

Table 2. Summary of minor/trace element concentrations in surface (0–10 cm) soil and in vertical soil profile samples at Smith's Lake and Charles Veryard Reserves. Element concentrations were measured using *aqua regia* digestion followed by ICP-OES.

Statistic	Concentrations (mg/kg)											
	As	Ba	Cd	Cr	Cu	Mn	Mo	Ni	Pb	Sr	V	Zn
Surface soil (random-in-grid samples, 2017)												
Mean	2.58	15.7	0.11	7.33	8.69	39.5	0.23	2.58	25.7	27.8	5.38	55.2
Std. Dev.	1.09	7.38	0.1	2.29	11	23.7	0.1	2.83	31.7	40.3	1.66	56.5
Minimum	0.9	6.2	0.02	1.17	2.13	2.68	0.07	0.25	3.23	2.4	1.28	5.59
Median	2.3	14.5	0.08	7.32	5.25	34.5	0.22	2.13	14.7	10.3	5.35	34.7
Maximum	5.86	41.7	0.66	13.8	67.5	127	0.6	21.3	174	186	10.6	304
No of analyses	68	68	62	68	68	68	68	68	68	68	68	68
No > HIL(C) [1]	0	-	0	0	0	0	-	0	0	-	-	0
No > EIL [3]	0	0	0	0	0	0	-	0	0	-	0	3
Soil profiles to $\leq$ 100 cm (2018)												
Mean	3.28	25.2	0.18	9.09	41.4	28.0	0.48	5.39	72.7	62.3	8.38	117
Std. Dev.	2.79	26.2	0.31	6.96	66.6	21.7	0.81	20.7	98.8	85.1	5.56	191
Minimum	0.6	1.9	0.01	0.2	2.8	1.5	0.06	0.6	3.5	1.8	0.3	3.3
Median	2.5	15.3	0.06	8.1	15.1	24.3	0.21	1.97	32.2	34.3	7.2	43.2
Maximum	14.6	139	1.6	34.4	356	97.9	5.63	181	568	391	30.4	1155
No of analyses	83	85	79	85	67	85	78	75	84	80	85	84
No > HIL(C) [1]	0	-	0	0	0	0	-	0	0	-	-	0
No > EIL [3]	0	0	0	0	8	0	-	1	1	0	0	14

[1] Health Investigation Level C (Recreational) [36]; [3] Ecological Investigation Level (interim urban) [37].

3.2. Spatial Distributions in Surface Soil

The measurements of primary interest (pH, EC, Al, As, Ca, Cu, Fe, Pb, Zn) generally showed significant overall spatial patterns across the study area, shown by *p* values $\leq$ 0.05 for Global Moran's I. The exceptions were Al, Ca, Cr, and Ni for which the Global Moran's

I values were close to zero (Table 3). The spatial patterns and clusters of points with significant local autocorrelation are shown in Figures 3–6, and summarised in Table 3.

Table 3. Global Moran's I, *p*-values simulated by Monte-Carlo randomization, and information on local spatial autocorrelation for the variables of principal interest in surface soil at Smith's Lake and Charles Veryard Reserves. Variables except pH were $\log_{10}$-transformed before calculation. IPI is integrated pollution index (Equation (1)).

Variable	Moran's I	*p*-Value	Number of Points with Significant Local Moran's I	Location (and Number of Points) of High-High LOCAL Moran's I Clusters [a]
pH	0.440	>0.001	13	CVR-SW (2), CVR-SE (3), SLR-S (4)
EC	0.157	0.035	7	SLR-S (3)
Al	0.014	0.691	4	CVR-SE (2) [b]
As	0.228	0.002	6	CVR-SE (4)
Ca	0.074	0.254	6	SLR-S (2)
Cr	0.016	0.672	4	CVR-SE (1)
Cu	0.331	>0.001	9	CVR-NE (3), CVR-SE (5)
Fe	0.142	0.044	11	CVR-NE (1), CVR-SE (5)
Ni	0.047	0.426	5	CVR-SE (1) [b]
Pb	0.324	>0.001	11	CVR-SE (5), SLR-N (2)
Zn	0.209	0.004	10	CVR-NE (2), CVR-SE (4)
IPI	0.257	>0.001	10	CVR-NE (1), CVR-SE (5)

[a] CVR is Charles Veryard Reserve; SLR is Smiths Lake Reserve; NE is north-east; NW is north-west; SE is south-east; SW is south-west; N is north; S is south; E is east; [b] not associated with main CVR-SE cluster.

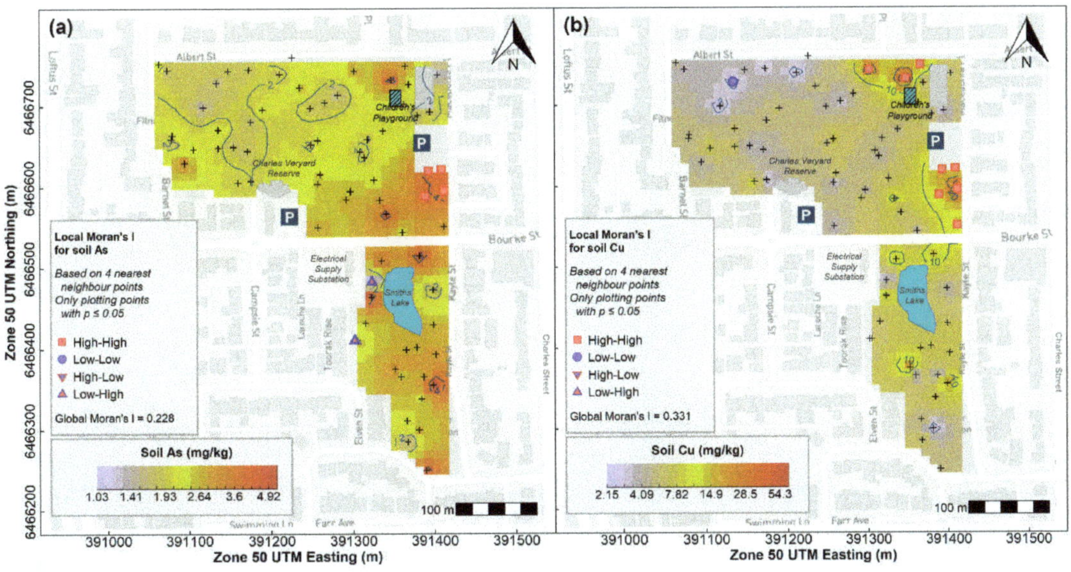

Figure 3. Spatial distributions of (**a**) As and (**b**) Cu in soil across Smith's Lake and Charles Veryard Reserves, interpolated by inverse distance weighting with significant local spatial autocorrelations (Local Moran's I ≤ 0.05) indicated with filled symbols. All sample points shown by + symbols.

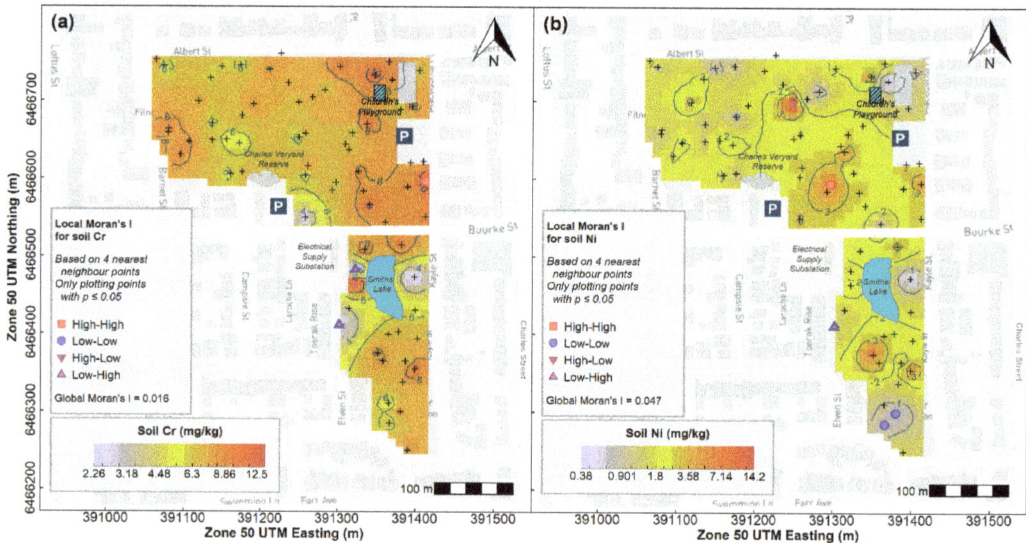

Figure 4. Spatial distributions of (**a**) Cr and (**b**) Ni in soil across Smith's Lake and Charles Veryard Reserves, interpolated by inverse distance weighting with significant local spatial autocorrelations (Local Moran's I ≤ 0.05) indicated with filled symbols. All sample points shown by + symbols.

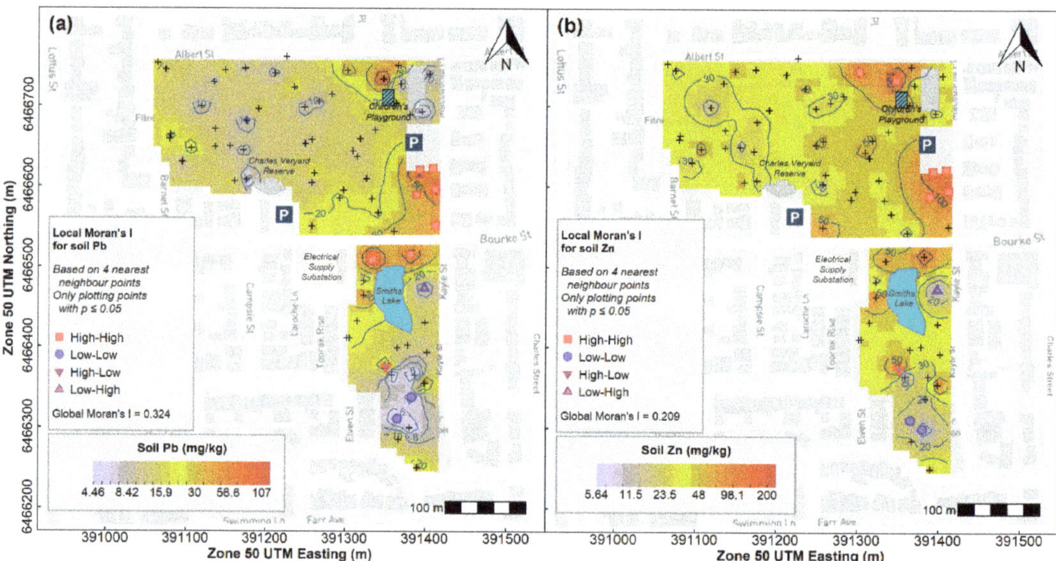

Figure 5. Spatial distributions of (**a**) Pb and (**b**) Zn in soil across Smith's Lake and Charles Veryard Reserves, interpolated by inverse distance weighting with significant local spatial autocorrelations (Local Moran's I ≤ 0.05) indicated with filled symbols. All sample points shown by + symbols.

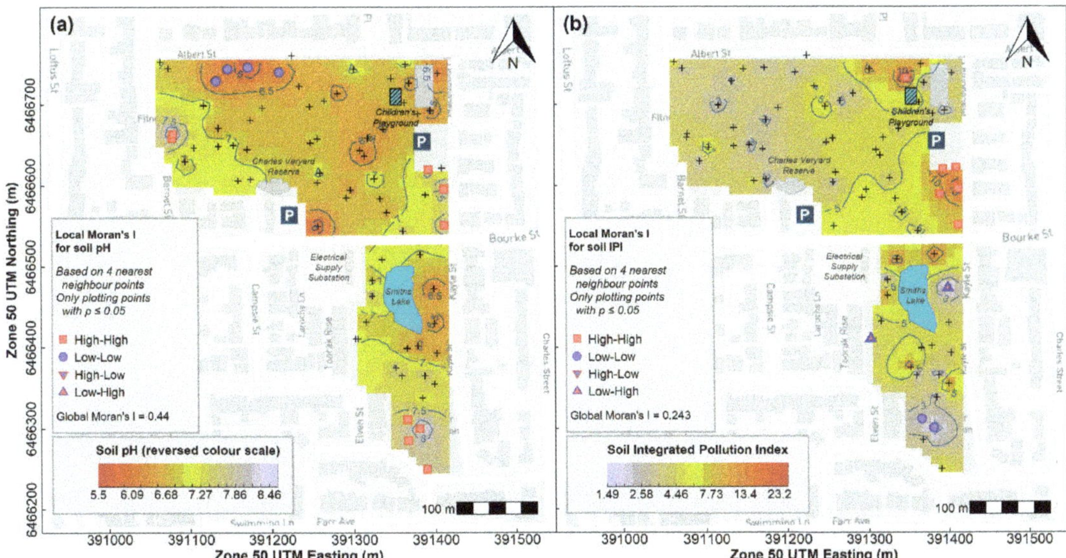

Figure 6. Spatial distributions of (**a**) pH and (**b**) Integrated Pollution Index (IPI) in soil across Smith's Lake and Charles Veryard Reserves, interpolated by inverse distance weighting with significant local spatial autocorrelations (Local Moran's I ≤ 0.05) indicated with filled symbols. All sample points shown by + symbols.

Arsenic (As) showed a broad concentration peak in soil in the south-east of Charles Veryard Reserve, with scattered local maxima in a few other locations (Figure 3). The As peak in the south-east of Charles Veryard Reserve was co-located with samples having significant ($p \leq 0.05$) high-high local Moran's I. Two points in Smith's Lake Reserve had significant low-high local Moran's I (i.e., isolated low As concentrations).

Copper (Cu) showed peaks in the north-east and south-east of Charles Veryard Reserve, with no other obvious maxima (Figure 3). Samples in both peaks in Cu concentration were significantly spatially autocorrelated (high-high, local Moran's I $p \leq 0.05$). One point in north-west Charles Veryard Reserve had significant low-high local Moran's I (i.e., an isolated low Cu concentrations).

Lead (Pb) showed peaks in concentration in soil in the south-east of Charles Veryard Reserve and the north of Smith's Lake Reserve, with scattered local maxima in Pb concentration in a few other locations (Figure 5). The Pb peak in the south-east of Charles Veryard Reserve was co-located with samples having significant ($p \leq 0.05$) high-high local Moran's I. A broad area of low Pb concentrations in Smith's Lake Reserve coincided with significant ($p \leq 0.05$) low-low local Moran's I, and instances of significant high-low and low-high local Moran's I values represented isolated high and low Pb concentrations.

Zinc (Zn) showed peaks in the north-east and south-east of Charles Veryard Reserve, with a few other subtle maxima (Figure 5). Samples in both clear peaks in Zn concentration were significantly spatially autocorrelated (high-high, local Moran's I $p \leq 0.05$). An area of low Zn concentrations in Smith's Lake Reserve coincided with significant ($p \leq 0.05$) low-low local Moran's I. Similar to Pb, instances of significant high-low and low-high local Moran's I values in Smith's Lake Reserve represented isolated high and low Zn.

Soil pH showed a cluster of lower values in the north-west of Charles Veryard Reserve with significant low-low spatial autocorrelation (Moran's I $p \leq 0.05$). In contrast, significant clusters of higher pH values were present in the west of Charles Veryard Reserve, the south-east of Charles Veryard Reserve, and the south of Smith's Lake Reserve (Figure 6).

Finally, the derived Integrated Pollution Index (IPI) had a maximum in the south-east of Charles Veryard Reserve, with a minor maximum in the north-east (Figure 6). A cluster of samples in the south-east IPI peak were significantly spatially autocorrelated (high-high, local Moran's I $p \leq 0.05$). An isolated low IPI value (low-high local Moran's I, $p \leq 0.05$) was present to the east of Smith's Lake.

A comparison of mean values of the variables of interest between samples from the different zones in Figure 2 reinforced the qualitative results from spatial interpolation (Table 4). The highest mean values for several potential contaminants (As, Cr, Cu, and Pb) were observed in the south-east of Charles Veryard Reserve (no significant effect of sampling zone was found for Ni or Zn).

Table 4. Comparison of pH, EC (1:5 soil: water extract), element concentrations, and IPI in distinct Zones of Smith's Lake (SLR) and Charles Veryard (CVR) Reserves. Mean values in a *row* are different if no superscript letters are shared ($p \leq 0.05$, Conover's pairwise test with Holm's correction).

Variable	P (K-W) [1]	Mean in Each Zone [2]							
		CVR-Centre	**CVR-NE**	**CVR-NW**	**CVR-SE**	**CVR-SW**	**SLR-N**	**SLR-S**	**Other**
pH	0.0001	6.58 [a]	6.65 [ab]	6.24 [a]	7.34 [bc]	7.35 [bc]	6.88 [abc]	7.38 [c]	6.66 [abc]
EC	0.013	155 [ab]	90.7 [ab]	273 [a]	159 [ab]	232 [a]	68.1 [b]	265 [ab]	143 [ab]
Al	0.028	2530 [ab]	2857 [ab]	2285 [a]	3373 [b]	2621 [ab]	1960 [ab]	3011 [ab]	2489 [ab]
As	0.004	1.91 [a]	2.46 [ab]	1.96 [a]	3.79 [b]	2.17 [a]	2.99 [ab]	3.04 [ab]	2.38 [ab]
Ca	0.004	1766 [abc]	3511 [abc]	4534 [abc]	6710 [abc]	7472 [ab]	3570 [ac]	11,450 [b]	1343 [c]
Cr	0.020	6.98 [ab]	8.47 [ab]	6.62 [a]	9.42 [b]	7.18 [ab]	5.85 [ab]	7.51 [ab]	6.20 [a]
Cu	0.010	5.10 [ab]	16.7 [ab]	3.57 [a]	17.6 [b]	5.36 [ab]	8.43 [ab]	8.34 [ab]	4.52 [ab]
Fe	0.041	2377 [ab]	2936 [ab]	2071 [a]	3623 [b]	2418 [ab]	2680 [ab]	2861 [ab]	2535 [ab]
Ni	0.300	*ns*	*ns*	*ns*	*ns*	*ns*	*ns*	*ns*	*ns*
Pb	0.043	11.7 [a]	31.2 [ab]	15.4 [ab]	56.9 [b]	15.3 [ab]	47.8 [ab]	15.7 [a]	16.5 [ab]
Zn	0.255	*ns*	*ns*	*ns*	*ns*	*ns*	*ns*	*ns*	*ns*

[1] Overall *p*-values from Kruskal-Wallis test; [2] CVR is Charles Veryard Reserve; SLR is Smiths Lake Reserve; NE north-east; NW north-west; SE south-east; SW south-west; N north; S south.

3.3. Relationships between Soil Elements

Several significant positive correlations existed between elements across the soil data from Charles Veryard and Smith's Lake Reserves (Table S1, Supplementary Material). Calcium, Mg and Sr were very highly correlated (r = 0.80–0.96), and Ca and Sr were the only elements significantly correlated (r = 0.66) with pH. The major elements Na, K, and Mg were all highly correlated (r $\geq$ 0.7), as were P, K, Mn, and S. High correlations also existed between iron (Fe) and As, Ba, Cu, Pb, and V. Many potential contaminants were also highly correlated with one another, e.g.,: Cu with Ba, Pb, and Zn; Pb with Cd, Cu, and Zn; Cr with V.

Principal components analysis (Figure 7) showed grouping of Cu, Pb, and Zn in PC1-PC2 space, associated with some of the samples from north-east and south-east where peak concentrations of these elements were observed (Figures 3 and 5). Arsenic plots at similar values of PC2, but has an association with Fe and Cr at small positive PC1 values. No obvious element associations were observed using PCA for Ni. The principal components analysis also identifies an association of Ca with Sr and Ba, an association of nutrient elements (S, P, K) with the central Charles Veryard Reserve samples, and clustering of rare-earth and related elements (Ce, Gd, La, Nd, Y). Additional results derived from principal components analysis are available in Tables S2–S4 in the Supplementary Materials.

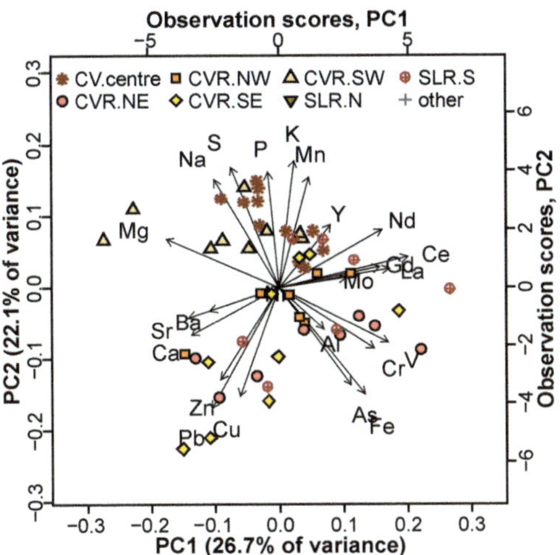

Figure 7. Principal components biplot for the first two principal components based on elemental composition in surface soil at Charles Veryard and Smith's Lake Reserves. Observation scores are identified by sampling Zone (see Figure 2). Concentrations were transformed using centered log-ratios before PCA, to avoid spurious effects of compositional closure.

In surface soil, there was a weak negative relationship between lead and vanadium concentrations and minimum (Euclidean) distance from any road surrounding or bisecting the reserves (Figure 8). No other contaminant of primary interest showed a significant trend in relation to distance from roads.

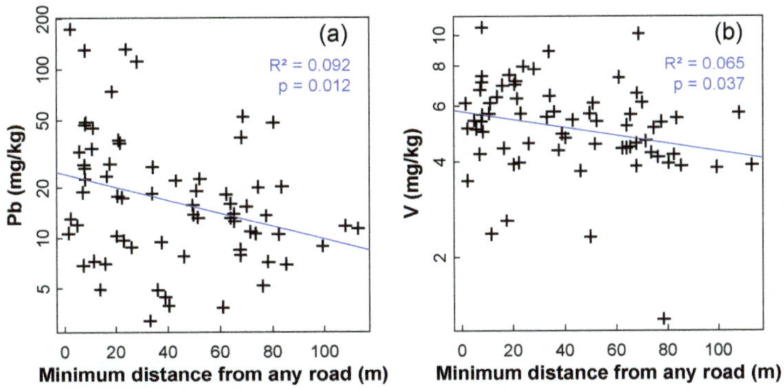

Figure 8. Weak trends in (**a**) Pb and (**b**) V concentrations in surface soil with distance from roads in Charles Veryard and Smith's Lake Reserves. Solid blue lines are log-linear models.

3.4. Depth Distributions of As, Cr, Cu, Ni, Pb, and Zn

Depth profile plots for As, Cr, Cu, Ni, Pb, and Zn are presented in Figures 9–11. Depth profiles of As, Cr, Cu, Ni, Pb, and Zn frequently showed maximum concentrations in subsurface soil samples. High maximum concentrations of Pb (376 mg/kg) and Zn (1155 mg/kg) were measured at 30–40 cm in core 4.2 (Figure 11), on the eastern side of Charles Veryard Reserve south of the Macedonia Place car park. Core 3.2 also contained 568 mg/kg Pb at 50–60 cm. Core 4.2 also contained the maximum concentration of As (14 mg/kg at 20–30 cm), Cd, Mn, and Ni. The greatest concentration of Cu (356 mg/kg) was observed in the adjacent core 4.1 (Figure 9).

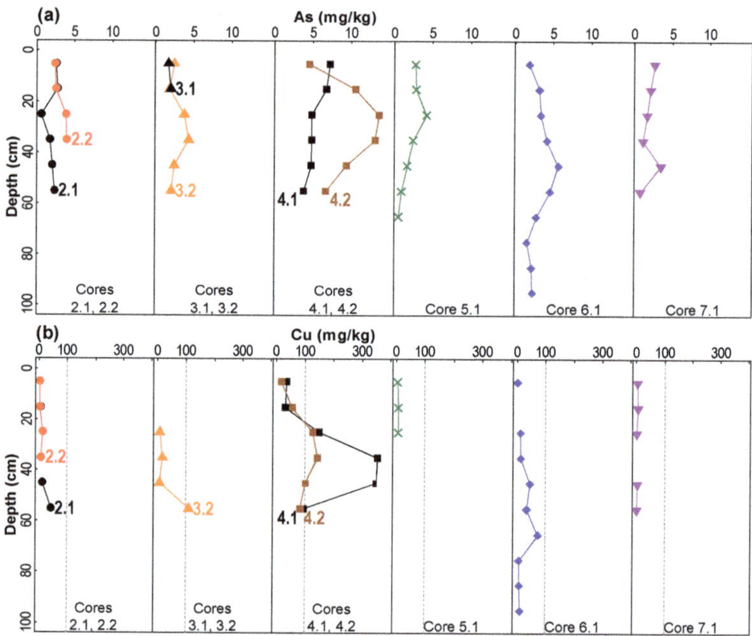

Figure 9. Depth profiles of (**a**) arsenic (As) and (**b**) copper (Cu) in soil cores collected from Smith's Lake and Charles Veryard Reserves, City of Vincent, Western Australia. Core locations from Figure 1. Ecological investigation limits (EIL) are shown as vertical dashed lines where relevant.

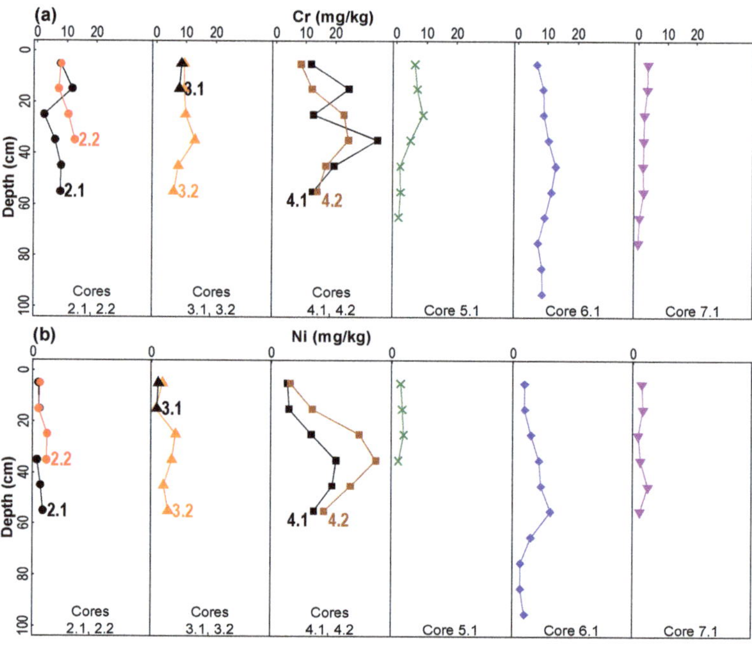

Figure 10. Depth profiles of (**a**) chromium (Cr) and (**b**) nickel (Ni) in soil cores collected from Smith's Lake and Charles Veryard Reserves, City of Vincent, Western Australia. Core locations from Figure 1.

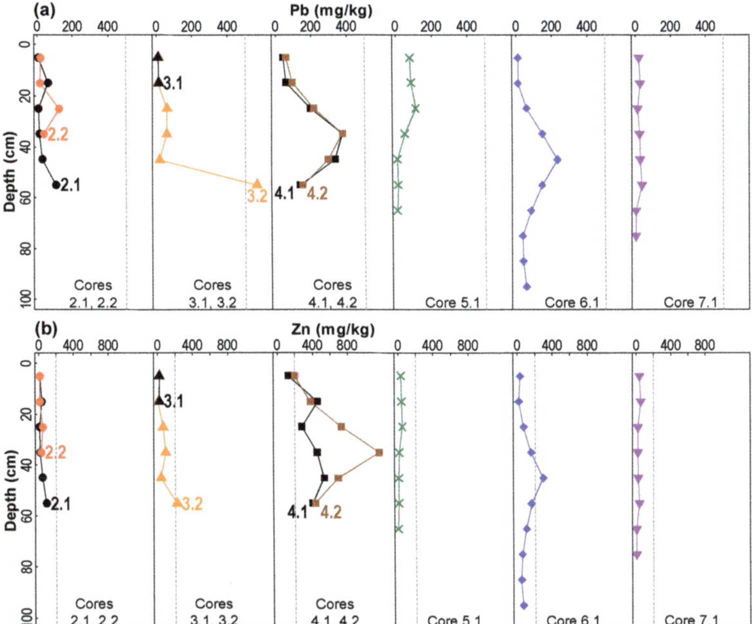

Figure 11. Depth profiles of (**a**) lead (Pb) and (**b**) zinc (Zn) in soil cores collected from Smith's Lake and Charles Veryard Reserves, City of Vincent, Western Australia. Core locations from Figure 1. Ecological investigation limits (EIL) are shown as vertical dashed lines where relevant.

There was a tendency for pH to increase with increasing depth, and EC to decrease with increasing depth, and the trends in Fe with depth were very similar to those for As (Figure S1, Supplementary Material).

4. Discussion

4.1. Concentrations of Potential Contaminants

Surface soils in the Charles Veryard and Smith's Lake Reserves were largely uncontaminated with any of the elements measured. No potential contaminant in surface soil or subsoil exceeded any relevant human health guideline for recreational/public open space land use (Table 2). This may reflect rehabilitation of the site to parkland using techniques as simple as covering with clean fill, which is known to suppress the surface expression of soil contamination [38]. In surface soil, the only element to exceed a guideline value was Zn, which had concentrations greater than the 200 mg/kg EIL threshold [37] in the southeast (two samples) and northeast (one sample) of Charles Veryard Reserve. More subsoil than surface soil samples exceeded EIL thresholds (Table 2).

The exceedance of EIL guideline values by Zn in a few surface soil samples and several subsoil samples in the eastern zones of Charles Veryard Reserve (Figures 9–11) reflects the common occurrence of zinc in urban environments, especially building materials and road traffic, and export of Zn into soil environments [39,40]. Very few toxicological studies exist on the effects of zinc on typical sports turf plants or the microbial ecology in these environments; the likelihood that this Zn represents anthropogenic additions means that the bioavailability of Zn would therefore also be expected to be greater than for native soil Zn [36,41].

4.2. Spatial Patterns of Potential Contaminants in Surface Soil

Although the incidence of actual surface soil contamination was low at Charles Veryard and Smith's Lake Reserves, the zone in which most enrichment of potential contaminants (As, Cr, Cu, Pb, and Zn) occurred coincided with the greatest subsoil concentrations of these elements (Figures 9–11). This was the CVR-SE zone, in which the greatest number of point-variable combinations had significant high-high local Moran's I statistics (Table 3), confirming the visualizations generated by inverse-distance interpolation (Figures 3–6). Based on these analyses, the south-east corner of Charles Veryard reserve, an area approximately 40 m N-S and 20 m E-W (ca. 800 m^2), is contaminated with As, Cu, Pb, and Zn, a finding supported by the Integrated Pollution Index (Figure 6b). Based on a single sample with significant high-high Local Moran's I, and weak evidence of subsoil enrichment, Cr contamination may also be present in this location. However, the Global Moran's I statistics for Cr and Ni could not reject the null hypothesis of no spatial pattern, and no local Moran's I values were significant for Ni, so it is unlikely that either Cr or Ni have been added by contamination processes at the study site. Both Cr and Ni also had significant isolated high concentrations (significant high-low local Moran's I; Figure 4)) which were not co-located; no isolated high concentrations were observed for As, Cu, Pb, or Zn.

The CVR-SE zone was the location of soil cores showing the greatest concentrations of As, Cr, Cu, Ni, and Zn, and the most exceedances of each element's ecological investigation limit (EIL) concentrations (Figures 9–11). Pb concentrations in subsoil at this location were also high, although the greatest concentration occurred in Core 3.2 in the west of Charles Veryard Reserve. The co-location of surface soil contamination identified by spatial analysis with subsoil contaminant maxima confirms the south-east Charles Veryard Reserve area as the only location of significant contamination.

4.3. Associations of Potential Contaminants

Despite the minimal surface soil contamination, the identification of distinct soil zones based on their geochemical properties (Figures 3–6) suggested that these different zones may represent the signatures of past activities or construction at the Charles Veryard and Smith's Lake Reserves. The element associations identified in the soil zones were supported by the Principal Components Analysis (Figure 7).

The associations identified by Principal Components Analysis are consistent, and also make geochemical sense. The nutrient elements K, P, and S probably represent a common source from historical horticulture [42]. The grouping of Ca, Mg, Sr, and Ba includes elements which are all commonly associated with carbonates and/or cement-based materials [35]. The metal contaminants Cu, Zn, and Pb often have a common source such as building materials or roads and traffic [43]. Finally, Fe, As, Cr, and V reflect the commonly-observed associations of As, Cr, and V with iron oxides in soils [44], and Cr and V are used with Fe in manufacture of some steel products [45].

The association of Cu, Pb, and Zn in PC1-PC2 space, and to some extent As and Cr in the second principal component dimension, validated the calculation of the Integrated Pollution Index (IPI) from these elements. The IPI values are unusually high (range 6–28), reflecting the somewhat low values used for background concentrations. Reliable background concentrations for trace elements in soils of the Swan Coastal Plain around metropolitan Perth are still subject to uncertainty and obtaining these should be a priority for local research.

The low pH and low concentrations of many elements in the north-west of Charles Veryard Reserve most likely reflect a very sandy (i.e., poorly buffered) soil material which has been subject to minor acidification. This acidification may have originated from historical or recent disturbance of the underlying peaty acid sulfate soil material (e.g., by light pylon installation), given the classification of much of the Charles Veryard and Smith's Lake Reserves area as being high to moderate risk of acid sulfate soil within 3 m of the land surface (Figure 1).

Elevated concentrations of Cu, Pb, and Zn in the north-east of Charles Veryard Reserve most likely represent contributions from construction (the Macedonian Centre, buildings on Albert Street) and possibly road traffic [39,46]. A road traffic origin for Pb is supported by the significant negative relationship of Pb with distance from roads (Figure 8). Construction and historical waste disposal sources are likely to have contributed Cu, and Pb to the south-east of Charles Veryard Reserve. The Charles Veryard Reserve south-east soil zone also has elevated pH, Al and Fe, however, so background concentrations may be naturally higher due to greater clay and/or iron oxide content of soils [44]. The greater concentrations of arsenic are most likely due to retention on Fe oxides, since there are no obvious sources of contamination and As concentrations are generally low. The greater concentrations of Al and Fe may themselves represent contamination from disposal of metalliferous wastes.

Soil in the south-west of Smith's Lake Reserve is characterized by higher pH and concentrations of Ca, Sr, Na, and P (and possibly K, S, and Mn). The high pH and elevated Ca and Sr are likely to represent additions of limestone or cement-based building materials [47]. Such additions are plausible given the relatively recent (2008) demolition of the Len Fletcher Sports Pavilion in the south of Smith's Lake Reserve.Enrichment with the nutrient elements P, K, and S, and also Na, may reflect historical market gardening at the site and associated use of fertilisers, or organic amendments such as composts or manures [1].

The weak but significant trend in lead and vanadium concentrations as a function of distance from roads (Figure 8) suggests that road traffic was a significant source of these elements, in agreement with previous studies [48]. Since leaded fuels are no longer used in Australia and numerous other countries, the inputs of Pb are likely to represent a historical legacy of Pb accumulation in roadside soils. The abrasion of road surfaces by traffic is a potential source of vanadium from bituminous materials used as asphalt binders [49].

Concentrations of Cu, Pb, and Zn in soil profiles exceeded Ecological Investigation Limits (EILs) in several samples, especially for Zn (Figures 9 and 11). Most of these higher concentrations, however, were in deeper subsoil samples, so the risk to biota (mainly plant and microbial uptake) would therefore be expected to be minimal.

The existence of subsoil maximum concentrations at some locations may represent burial of waste material or drain sediment, or an evaporation/redox front resulting in accumulation of some elements. Given that that waste disposal at the Smith's Lake and Charles Veryard Reserves site is known to have been widespread [50,51], waste material would seem the most likely source. The relatively high subsoil concentrations of trace elements may represent a health risk, for example if dust is generated during excavation [52]. The potential risk should be considered in the context of a children's playground adjacent to the most contaminated surface soils and soil profiles.

5. Conclusions

An important conclusion from the initial concentration data is that the surface soil and subsoil sampled in this study at Smith's Lake and Charles Veryard Reserves is not contaminated with As, Cr, Cu, Ni, Pb, or Zn levels of concern from a human health perspective. There was, however, multiple exceedance of ecological investigation trigger limits (EIL) for Zn in surface soil and Cu, Pb, and Zn in subsoil. The spatial analysis showed that, on the basis of global and local Moran's I, distributions of most elements were not random but showed clustering. In line with the initial hypothesis, this significant clustering of adjacent higher concentrations in surface soil allowed identification of a specific area which, at the scale of sampling design, represented inputs from a point source of As, Cu, Pb, and Zn. At this site, the specific area of surface soil contamination was co-located with the most significant subsoil contamination, but this may not be a general result.

The combination of multivariate geochemical analysis with spatial information allowed both identification of realistic associations of elements, including potential contaminants. In particular, there was a consistent association of the dominant contaminants (Cu, Pb, and Zn) in the south-east of Charles Veryard Reserve which could be deduced from univariate spatial autocorrelation analysis, a composite contamination index (IPI), and multivariate principal components analysis.

In this study, the location and significance of potential contamination in the soil of urban public open space has been assessed thoroughly by measurement of multiple parameters, and rigorous spatial and statistical analysis. It is recommended that any such study uses a similar approach if soil contamination is suspected, especially given the global tendency for urban populations to increase and for redevelopment of, and increased population density in, inner-city precincts.

Supplementary Materials: The following are available online at https://www.mdpi.com/2571-878 9/5/3/46/s1. Figure S1: 'Depth profiles of pH, EC, and Fe in soil cores collected from Smith's Lake and Charles Veryard Reserves, City of Vincent, Western Australia'; Table S1: 'Matrix of Spearman correlation coefficients for pH, EC, and elemental composition of soil samples from Smith's Lake and Charles Veryard Reserves. Values in bold type indicate a significant correlation ($p \leq 0.05$, using Holm's adjusted p-values for multiple comparisons)'. Table S2. 'Component Loadings for PC1-PC8'. Table S3. 'Summary of Principal Components'. Table S4. 'Eigenvalues (variances) for the first 8 components'.

Funding: This research received no external funding.

Institutional Review Board Statement: Not applicable.

Informed Consent Statement: Not applicable.

Data Availability Statement: The raw data and metadata have been submitted to the PANGAEA repository at https://www.pangaea.de/.

Acknowledgments: The City of Vincent, Western Australia, granted permission to sample soil in the reserves. The principal investigator is extremely grateful to the ENVT3361 classes of 2017 and 2018 at The University of Western Australia, who did most of the work to generate the data: collecting samples, conducting the laboratory analyses, and uploading analytical results. Kirsty Brooks and Emielda Yusiharni from the School of Agriculture and Environment at The University of Western Australia made substantial efforts by managing the teaching laboratories, and organising field gear. Michael Smirk conducted the ICP-OES measurements.

Conflicts of Interest: The author declares no conflict of interest.

References

1. Demiguel, E.; Degrado, M.J.; Llamas, J.F.; Martindorado, A.; Mazadiego, L.F. The overlooked contribution of compost application to the trace element load in the urban soil of Madrid (Spain). *Sci. Total Environ.* **1998**, *215*, 113–122. [CrossRef]
2. Chen, T.B.; Wong, J.W.C.; Zhou, H.Y.; Wong, M.H. Assessment of trace metal distribution and contamination in surface soils of Hong Kong. *Environ. Pollut.* **1997**, *96*, 61–68. [CrossRef]
3. Gaw, S.K.; Wilkins, A.L.; Kim, N.D.; Palmer, G.T.; Robinson, P. Trace element and ΣDDT concentrations in horticultural soils from the Tasman, Waikato and Auckland regions of New Zealand. *Sci. Total Environ.* **2006**, *355*, 31–47. [CrossRef]
4. Tarzia, M.; De Vivo, B.; Somma, R.; Ayuso, R.A.; McGill, R.A.R.; Parrish, R.R. Anthropogenic vs. natural pollution: An environmental study of an industrial site under remediation (Naples, Italy). *Geochem. Explor. Environ. Anal.* **2002**, *2*, 45–56. [CrossRef]
5. Appleyard, S.; Wong, S.; Willis-Jones, B.; Angeloni, J.; Watkins, R. Groundwater acidification caused by urban development in Perth, Western Australia: Source, distribution, and implications for management. *Aust. J. Soil Res.* **2004**, *42*, 579–585. [CrossRef]
6. Orndorff, Z.W.; Daniels, W.L.; Fanning, D.S. Reclamation of acid sulfate soils using lime-stabilized biosolids. *J. Environ. Qual.* **2008**, *37*, 1447–1455. [CrossRef]
7. Grundy, S.L.; Bright, D.A.; Dushenko, W.T.; Dodd, M.; Englander, S.; Johnston, K.; Pier, D.; Reimer, K.J. Dioxin and furan signatures in northern Canadian soils: Correlation to source signatures using multivariate unmixing techniques. *Chemosphere* **1997**, *34*, 1203–1219. [CrossRef]
8. Riemann, U. Impacts of urban growth on surface water and groundwater quality in the City of Dessau, Germany. In *Impacts of Urban Growth on Surface Water and Groundwater Quality*; Ellis, B., Ed.; IAHS-AISH Publication No. 259; International Association of Hydrological Sciences Press: Wallingford, UK, 1999; pp. 307–314.
9. Liu, E.; Yan, T.; Birch, G.; Zhu, Y. Pollution and health risk of potentially toxic metals in urban road dust in Nanjing, a mega-city of China. *Sci. Total Environ.* **2014**, *476–477*, 522–531. [CrossRef]
10. Henderson, F.M.; Xia, Z.G. SAR applications in human settlement detection, population estimation and urban land use pattern analysis: A status report. *IEEE Trans. Geosci. Remote Sens.* **1997**, *35*, 79–85. [CrossRef]
11. Huo, X.N.; Zhang, W.W.; Sun, D.F.; Li, H.; Zhou, L.D.; Li, B.G. Spatial pattern analysis of heavy metals in Beijing agricultural soils based on spatial autocorrelation statistics. *Int. J. Environ. Res. Public Health* **2011**, *8*, 2074–2089. [CrossRef]
12. Anselin, L. Local Indicators of Spatial Association—LISA. *Geogr. Anal.* **1995**, *27*, 93–115. [CrossRef]

13. Hojati, S. Use of spatial statistics to identify hotspots of lead and copper in selected soils from north of Khuzestan Province, southwestern Iran. *Arch. Agron. Soil Sci.* **2019**, *65*, 654–669. [CrossRef]

14. Zhang, C.; Luo, L.; Xu, W.; Ledwith, V. Use of local Moran's I and GIS to identify pollution hotspots of Pb in urban soils of Galway, Ireland. *Sci. Total Environ.* **2008**, *398*, 212–221. [CrossRef] [PubMed]

15. City of Vincent. *Thematic History*; City of Vincent: Leederville, WA, Australia, 2008.

16. Claise Brook Catchment Group. Restoration of Smith's Lake. Available online: http://www.cbcg.org.au/projects_smiths.html (accessed on 31 October 2017).

17. Western Australian Land Information Authority. Landgate Map Viewer Plus. Available online: https://maps.landgate.wa.gov.au/maps-landgate/registered/. https://www0.landgate.wa.gov.au/maps-and-imagery/imagery/aerial-photography/aerial (accessed on 29 June 2021).

18. Rayment, G.E.; Lyons, D.J. *Soil Chemical Methods—Australasia*; CSIRO Publishing: Clayton, VIC, Australia, 2010.

19. U.S. EPA. Method 3050B: Acid digestion of sediments, sludges, and soils test. In *Methods for Evaluating Solid Waste, Physical/Chemical Methods*; EPA Publication SW-846; U.S. EPA: Washington, DC, USA, 2007.

20. Lynch, J. Additional provisional elemental values for LKSD-1, LKSD-2, LKSD-3, LKSD-4, STSD-1, STSD-2, STSD-3 and STSD-4. *Geostand. Newsl.* **1999**, *23*, 251–260. [CrossRef]

21. Long, X.X.; Yang, X.E.; Ni, W.Z.; Ye, Z.Q.; He, Z.L.; Calvert, D.V.; Stoffella, J.P. Assessing zinc thresholds for phytotoxicity and potential dietary toxicity in selected vegetable crops. *Commun. Soil Sci. Plant Anal.* **2003**, *34*, 1421–1434. [CrossRef]

22. R Core Team. *R: A Language and Environment for Statistical Computing (Version 4.0.3)*; R Foundation for Statistical Computing: Vienna, Austria, 2020. Available online: https://www.R-project.org (accessed on 11 August 2021).

23. Pohlert, T. PMCMRplus: Calculate Pairwise Multiple Comparisons of Mean Rank Sums Extended (R Package). 2018. Available online: https://cran.r-project.org/web/packages/PMCMRplus/index.html (accessed on 11 August 2021).

24. Reimann, C.; Filzmoser, P.; Garrett, R.G.; Dutter, R. *Statistical Data Analysis Explained: Applied Environmental Statistics with R*, 1st ed.; John Wiley & Sons: Chichester, UK, 2008; p. 343.

25. Fellows, I. OpenStreetMap: Access to Open street Map Raster Images, Using the JMapViewer Library by Jan Peter Stotz. 0.3.3; (R Package Version 0.3.4). 2019. Available online: http://CRAN.R-project.org/package=OpenStreetMap (accessed on 11 August 2021).

26. Google. Getting Started | Google Maps Elevation API | Google Developers. Available online: https://developers.google.com/maps/documentation/elevation (accessed on 9 October 2017).

27. Cooley, D. Googleway: Accesses Google Maps APIs to Retrieve Data and Plot Maps; R Package Version 2.0.0; 2017. Available online: https://cran.r-project.org/web/packages/googleway/index.html (accessed on 11 August 2021).

28. Akima, H.; Gebhardt, A.; Petzoldt, T.; Maechler, M. akima: Interpolation of Irregularly Spaced Data. R Package Version 0.5–11. 2013. Available online: http://CRAN.R-project.org/package=akima (accessed on 11 August 2021).

29. Kalogirou, S. lctools: Local Correlation, Spatial Inequalities, Geographically Weighted Regression and Other Tools. R Package Version 0.2–8. Available online: https://CRAN.R-project.org/package=lctools (accessed on 28 May 2021).

30. Pebesma, E.; Bivand, R. *sp: Classes and Methods for Spatial Data*; R Package Version 1.4-4; R Foundation for Statistical Computing: Vienna, Austria, 2020.

31. Pebesma, E.J.; Graeler, B. *gstat: Spatial and Spatio-Temporal Geostatistical Modelling, Prediction and Simulation*; R Package Version 2.0-7; R Foundation for Statistical Computing: Vienna, Austria, 2021.

32. Webster, R.; Oliver, M.A. How large a sample is needed to estimate the regional variogram adequately? *Geostat. Troia'92* **1993**, *1*, 155–166.

33. Sun, Y.; Zhou, Q.; Xie, X.; Liu, R. Spatial, sources and risk assessment of heavy metal contamination of urban soils in typical regions of Shenyang, China. *J. Hazard. Mater.* **2010**, *174*, 455–462. [CrossRef]

34. DWER. *Final Report: Review of the Uncontaminated Fill Thresholds in Table 6 of the Landfill Waste Classification and Waste Definitions 1996 (as Amended 2018)*; Department of Water and Environmental Regulation, Government of Western Australia: Joondalup, WA, Australia, 2019.

35. Rate, A.W. Multielement geochemistry identifies the spatial pattern of soil and sediment contamination in an urban parkland, Western Australia. *Sci. Total Environ.* **2018**, *627*, 1106–1120. [CrossRef]

36. National Environment Protection Council. Schedule B (1): Guideline on the investigation levels for soil and groundwater. In *National Environment Protection (Assessment of Site Contamination) Measure (Amended)*; Commonwealth of Australia: Canberra, Australia, 2013.

37. National Environment Protection Council. Schedule B (1): Guideline on the investigation levels for soil and groundwater. In *National Environment Protection (Assessment of Site Contamination) Measure*; Commonwealth of Australia: Canberra, Australia, 1999.

38. Rimmer, D.L.; Younger, A. Land reclamation after coal-mining operations. In *Contaminated Land and Its Reclamation*; Hester, R.E., Harrison, R.M., Eds.; Royal Society of Chemistry: Cambridge, UK, 1997; pp. 73–90.

39. Callender, E.; Rice, K.C. The urban environmental gradient: Anthropogenic influences on the spatial and temporal distributions of lead and zinc in sediments. *Environ. Sci. Technol.* **2000**, *34*, 232–238. [CrossRef]

40. Charlesworth, S.; de Miguel, E.; Ordóñez, A. A review of the distribution of particulate trace elements in urban terrestrial environments and its application to considerations of risk. *Environ. Geochem. Health* **2011**, *33*, 103–123. [CrossRef]

41. Smolders, E.; Oorts, K.; van Sprang, P.; Schoeters, I.; Janssen, C.R.; McGrath, S.P.; McLaughlin, M.J. Toxicity of trace metals in soil as affected by soil type and aging after contamination: Using calibrated bioavailability models to set ecological soil standards. *Environ. Toxicol. Chem.* **2009**, *28*, 1633–1642. [CrossRef]
42. Pietrzak, U.; McPhail, D.C. Copper accumulation, distribution and fractionation in vineyard soils of Victoria, Australia. *Geoderma* **2004**, *122*, 151–166. [CrossRef]
43. Harrison, R.M.; Laxen, D.P.H.; Wilson, S.J. Chemical associations of lead, cadmium, copper, and zinc in street dusts and roadside soils. *Environ. Sci. Technol.* **1981**, *15*, 1378–1383. [CrossRef]
44. Hamon, R.E.; McLaughlin, M.J.; Gilkes, R.J.; Rate, A.W.; Zarcinas, B.; Robertson, A.; Cozens, G.; Radford, N.; Bettenay, L. Geochemical indices allow estimation of heavy metal background concentrations in soils. *Glob. Biogeochem. Cycles* **2004**, *18*. [CrossRef]
45. Tanner, P.A.; Ma, H.L.; Yu, P.K.N. Fingerprinting metals in urban street dust of Beijing, Shanghai, and Hong Kong. *Environ. Sci. Technol.* **2008**, *42*, 7111–7117. [CrossRef]
46. Davis, A.P.; Shokouhian, M.; Ni, S. Loading estimates of lead, copper, cadmium, and zinc in urban runoff from specific sources. *Chemosphere* **2001**, *44*, 997–1009. [CrossRef]
47. Jim, C.Y. Urban soil characteristics and limitations for landscape planting in Hong Kong. *Landsc. Urban Plan.* **1998**, *40*, 235–249. [CrossRef]
48. Mielke, H.W.; Laidlaw, M.A.S.; Gonzales, C. Lead (Pb) legacy from vehicle traffic in eight California urbanized areas: Continuing influence of lead dust on children's health. *Sci. Total Environ.* **2010**, *408*, 3965–3975. [CrossRef]
49. Béze, L.E.; Rose, J.; Mouillet, V.; Farcas, F.; Masion, A.; Chaurand, P.; Bottero, J.-Y. Location and evolution of the speciation of vanadium in bitumen and model of reclaimed bituminous mixes during ageing: Can vanadium serve as a tracer of the aged and fresh parts of the reclaimed asphalt pavement mixture? *Fuel* **2012**, *102*, 423–430. [CrossRef]
50. Conacher, J. *Historic Land Use Survey of the Claisebrook Catchment*; The University of Western Australia for the Claisebrook Catchment Group: Crawley, Australia, 2000; p. 61.
51. City of Vincent. Charles Veryard Reserve, Place Number 17957. In *inHerit—Places Database*; Heritage Council of WA: Perth, WA, Australia, 2007.
52. Ljung, K.; Maley, F.; Cook, A. Canal estate development in an acid sulfate soil-Implications for human metal exposure. *Landsc. Urban Plan.* **2010**, *97*, 123–131. [CrossRef]

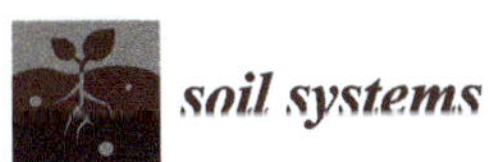

soil systems

MDPI

Article

Initial Study on Phytoextraction for Recovery of Metals from Sorted and Aged Waste-to-Energy Bottom Ash

Karin Karlfeldt Fedje [1,2,*], **Viktoria Edvardsson** [2] and **David Dalek** [2]

1   Department of Architecture and Civil Engineering, Division of Water Environment Technology, Chalmers University of Technology, SE-412 96 Gothenburg, Sweden
2   Recovery and Management, Renova AB, Box 156, SE-401 22 Gothenburg, Sweden; viktoria.edvardsson@renova.se (V.E.); david.dalek@renova.se (D.D.)
*   Correspondence: karin.karlfeldt@chalmers.se

Abstract: Sorted and aged bottom ash from Waste-to-Energy plants, i.e., MIBA (the Mineral fraction of Incinerator Bottom Ash) are potential source of metals that could be utilized to meet the increased demand from society. In this work, sunflowers (*Helianthus annuus*) and rapeseed (*Brassica napus*) were cultivated in conventional MIBA to evaluate the possibility for phytoextraction, mainly of Zn, during the period of one cultivation season in the Nordic climate. The results show that metal extraction from MIBA using rapeseed and sunflowers is workable but that neither of the used plants is optimal, mainly due to the inhibited root development and low water- and nutrient-holding capacities of MIBA. The addition of fertilizer is also important for growth. There was a simultaneous accumulation of numerous metals in both plant types, and the highest metal content was generally found in the roots. Calculations indicated that the ash from rapeseed root incineration contained about 2% Zn, and the contents of Co, Cu, and Pb were comparable to those in workable ores. This initial study shows that cultivation in and phytoextraction on MIBA is possible, and that the potential for increased metal extraction is high.

Keywords: WtE bottom ash; phytoextraction; zinc recovery; rapeseed; sunflower; MIBA

Citation: Karlfeldt Fedje, K.; Edvardsson, V.; Dalek, D. Initial Study on Phytoextraction for Recovery of Metals from Sorted and Aged Waste-to-Energy Bottom Ash. *Soil Syst.* **2021**, *5*, 53. https://doi.org/10.3390/soilsystems5030053

Academic Editors: Matteo Spagnuolo, Paola Adamo and Giovanni Garau

Received: 17 August 2021
Accepted: 26 August 2021
Published: 31 August 2021

Publisher's Note: MDPI stays neutral with regard to jurisdictional claims in published maps and institutional affiliations.

Copyright: © 2021 by the authors. Licensee MDPI, Basel, Switzerland. This article is an open access article distributed under the terms and conditions of the Creative Commons Attribution (CC BY) license (https://creativecommons.org/licenses/by/4.0/).

1. Introduction

The demand for metals is increasing in our society. The importance of a secure metal supply has been acknowledged by the European Commission, which has recommended the utilization of secondary raw materials as a way to achieve this [1]. When waste is incinerated using the most common technique, Waste-to-Energy (WtE) mass burn combustion, about 20% of the mass becomes what is known as bottom ash. In Sweden alone, almost 1 million tons of this type of bottom ash is produced annually; in the EU, Norway, and Switzerland, the corresponding amount is approximately 18 Mt/year [2,3]. These ashes contain significant amounts of metals, and the most common treatment method for bottom ash is mechanical separation, where pieces of metal are recovered and combined with natural weathering, i.e., carbonation. Ash having had this treatment is referred to as MIBA, the Mineral fraction of Incinerator Bottom Ash, to distinguish it from untreated bottom ash [3]. This treatment does not only recycle solid metals but also stabilizes the material [4,5] and opens it up for utilization. In some countries, e.g., Denmark and the Netherlands, MIBA is used for conventional road construction, whereas in others, including Sweden, Norway, and Switzerland, it is landfilled or used within landfills [3]. The variance in management approaches is a result of the different guidelines and legislation used in different countries [3]. Current guidelines and legislation are strongly correlated to the presence and potential leaching of metals, as even after mechanical metal recovery, a significant amount of metal remains in the MIBA, either bound in chemical compounds or as small pieces (typically < 1 mm); see, e.g., [6,7]. If more metals were to be removed, the potential fields of application for MIBA would increase significantly, and most importantly, these resources could be utilized instead of virgin metals needing to be mined.

Recovery of the metals remaining in MIBA cannot be carried out using physical separation, but thermal or hydrometallurgical processes could potentially be used. The content of, e.g., Cu in MIBA is comparable to workable ores, but as the MIBA matrix is different from a rock, the amounts are still too low to be of interest to the energy-intensive metal refining industry [8]. Additionally, the presence of unwanted elements such as Cl and As may harm the recovery process [8]. Instead, MIBA must be treated to generate concentrates with higher contents of interesting metals and lower amounts of unwanted elements. Hydrometallurgical processes, i.e., leaching combined with, e.g., chemical precipitation, could be a way to get such material. However, the main challenge is the huge amounts of leaching media needed for MIBA treatment. Pure water is not efficient enough to leach metals, and assuming a liquid-to-solid ratio of 4, almost 4 million m^3 leaching agents would be required for Sweden alone. Even if part of the leachate was re-circulated, the large amounts of liquid would still be problematic. In this perspective, phytoextraction is an interesting alternative for metal recovery from MIBA, as it offers a way to extract and concentrate the metals of interest. Phytoextraction uses specific plants, so-called hyperaccumulators, that cannot only survive in contaminated areas but can also extract the contaminants into their tissues. Once harvested, these plants can be incinerated, and metals can be recovered from the resulting ash. Successful laboratory-scaled Ni recovery experiments which combine phytoextraction with separate incineration of the Ni-enriched plants have been reported [9]. However, separate incineration is usually not an alternative if used on a full scale. Instead, incineration of the enriched plants in conventional full-scale WtE plants is favorable, and the metals enriched in plants would likely be found in the fly ash [10]. This is positive, as the recovery of metals from WtE fly ash has been in focus for many years. Zn especially is of high interest, and several successful initiatives for Zn recovery from ash are under development or present as full-scale processes [11–13]. This makes the production of a Zn-rich fuel generated from the growth of hyperaccumulators in MIBA especially interesting, as this would transfer the Zn from the challenging MIBA matrix into a material from which Zn can be recovered efficiently. Besides, in the future, methods for the recovery of other valuable metals in fly ash are likely to be developed as well.

Phytoextraction is used for treating contaminated soils all over the world; however, a drawback for the Nordic countries is that many of the known Zn hyperaccumulators, e.g., water hyacinth (*Eichhornia crassipes*), do not grow naturally in their climate [14]. Further, several of the potential plants have limited biomass, including alpine pennycress (*Thlaspi caerulescens*) [15]. However, sunflowers (*Helianthus annuus*), rapeseed (*Brassica napus*), corn (*Zea mays*), and haricot verts (*Phaseolus vulgaris*), as well as larger plants, such as salix trees, have also been shown to accumulate Zn [15–22]. Generally, experiments carried out so far have used contaminated soils with few contaminants, and there is a lack of studies of scientifically controlled cultivation on ash containing several potential pollutants. Most earlier studies have used coal ash [23,24]; however, there is one published study where manually sorted and washed bottom ash from WtE incineration was used [25]. The results of this study showed that Ni and Zn were extracted by Alyssum (*serpyllifolium*) and a succulent (*S. plumbizincicola*), respectively, although the biomass production was low. To the best of the authors' knowledge, no study on phytoextraction from conventional MIBA has been scientifically considered.

This project aimed to study phytoextraction, mainly of Zn, from conventional MIBA during the period of one cultivation season in the Nordic climate. The specific goals were to evaluate the survival of the plants in MIBA and to evaluate the potential to generate Zn-enriched fuel, which is to be used for metal recovery. Additionally, the simultaneous decrease in the metal content in the remaining MIBA was measured.

2. Material and method

2.1. Material

The MIBA used in this work originated from a full-scale WtE-incineration plant in southwestern Sweden. The waste, mainly municipal and industrial waste, was incinerated in four grate-fired furnaces with steam boilers that produced electricity and district heating. Bottom ash was collected from all lines, mixed, and stored in large piles that were naturally weathered, i.e., not sheltered against wind or water, for at least 6 months. After 1–3 months in storage, metal pieces (>~2 mm) were mechanically removed using mobile sorting equipment with magnets and eddy current magnets. Sub-samples from the MIBA were collected and mixed to form one sample of about 600 kg, which was used for the cultivation tests.

For the control cultivation, where sunflowers and rapeseed were cultivated under normal circumstances, a commercial "lean planting soil" generated from the biological treatment of garden waste was used (Table 1) (Reference soil) [26]. To promote plant growth, and to study the effect of fertilizer on this growth, the reference cultivation boxes and half of the cultivation boxes containing MIBA were fertilized with YaraMila Promagna 11-5-18. This fertilizer is specially developed for plants with a high need for N, which is a shortage in MIBA [27].

2.2. Cultivation Experiments

Suitable plant types were chosen based on a review of the literature, with the pre-requisites that the selected plants should be able to: grow on open land in a Nordic climate, obtain significant relative biomass within one cultivation season, and have a proven affinity for Zn. Further, the plants should preferably have a dense structure, to ensure good incineration properties in recovery processes. Sunflowers (*Helianthus annuus*) are known to be suitable for phytoremediation of Zn, but other cruciferous plants, such as rapeseed (*Brassica napus*), have also been shown to be appropriate [17,18,25,28,29]. Additionally, as rapeseed and sunflowers both have variations of taproots, these plants were chosen for the cultivation experiments.

The experiments were performed using six cultivation boxes of about 1 m^2 each. The boxes were prepared in several layers, starting with a plastic covering on the inside, to prevent uncontrolled leaching. Thereafter, an approximately 15-cm thick layer of leca (Light Expanded Clay Aggregate) spheres was placed in the bottom, followed by a ground cloth. Finally, about 30 cm of MIBA or reference soil (control cultivation) was placed on top of the ground cloth in each box. In total, 6 boxes were prepared for the cultivation experiments, 4 cultivation experiments in MIBA and 2 control cultivation experiments in reference soil:

- Sunflowers in MIBA without fertilizer (SwoF)
- Sunflowers in MIBA with fertilizer (SF)
- Sunflowers in reference soil with fertilizer (Sref)
- Rapeseed in MIBA without fertilizer (RwoF)
- Rapeseed in MIBA with fertilizer (RF)
- Rapeseed in reference soil with fertilizer (Rref)

Seeds of both sunflowers and rapeseed were supplied by the agriculture cooperative Lantmännen. The sunflowers were pre-cultivated in reference soil in a greenhouse and were about 20 cm high when planted in the boxes, while the rapeseed plants were grown directly in the boxes. To improve the growing conditions, small holes, about 2–4 centimeters deep, were made in the MIBA boxes. The holes were filled with reference soil and the plants and seeds were grown inside the holes. Eight holes, i.e., eight plants, were made in each sunflower box, while twelve holes were used for the rapeseed. Several rapeseed plants were planted in each hole. Rapeseed was also planted directly in MIBA. Fertilizer was added one week after planting, in amounts according to the recommendation from the producer. All boxes were placed in similar sun and wind conditions. Watering using ordinary tap water, management, and growth checks (ocular and height measurement) were carried out regularly during the growing season (end of April to the middle of September). Some excess water was collected in the bottom of the boxes, where it could not reach the roots. This water was not returned to the plants. Approximately 6 weeks after planting, the rapeseed plants were thinned out in some of the planting groups. This was done to study whether or not having a large number of rapeseed plants within a small area would harm their growth.

Table 1. Contents in original reference soil, MIBA and the fertilizer used. Additionally, leaching (L/S 10) from MIBA according to SS-EN-12357-3 is shown. All amounts are provided in mg/kg DS except pH, TOC, ANC, EC, and bulk density.

Element	Reference Soil Total Amounts	SD	MIBA Total Amounts	SD	Leaching (L/S 10)	SD	Fertilizer Total Amounts
Al	11,000	-[1]	59,000	7700	na[2]	-	na
As	1.8	0.5	26	7	0.05	0.02	na
B	8.5	-	200	30	na	-	500
Ba	92	4.1	1700	260	1.1	0.05	na
Be	0.6	-	1.4	0.20	na	-	na
Ca	6317	878	130,000	1000	na	-	na
Cd	0.13	0.04	2.5	0.6	0.003	0.001	<55[3]
Co	7.3	0.5	39	24	na	-	na
Cr	14	1.1	440	220	0.26	0.15	na
Cu	27	4.3	3400	1500	3.2	0.15	300
Fe	16,900	-	53,000	11,000	na	-	800
Hg	0.032	0.004	0.025	0.002	<0.001	-	na
K	5087	147	11,000	1600	na	-	176,000
Mg	4490	70	13,000	840	na	-	16,000
Mn	330	-	1100	160	na	-	2500
Mo	2.7	2.0	27	22	1.2	0.24	20
N	2250	150	na	na	na	-	110,000
Na	558	-	24,000	2100	na	-	na
Ni	10	2.5	200	100	0.09	0.03	na
P	919	269	4100	360	na	-	46,000
Pb	15	1.9	1600	1300	0.23	0.20	na
S	671	225	8900	210	na	-	100,000
Sb	0.9	0.4	80	23	0.34	0.03	na
Se	<5.0	-	na	-	0.02	0.01	na
Si	na	-	160,000	25,000	na	-	na
Sn	1.5	-	200	110	na	-	na
Sr	27	-	390	-	na	-	na
Ti	na	-	8700	720	na	-	na
V	26	1.5	48	4.4	na	-	na
Zn	103	20	5000	1100	1.6	1.4	na
Cl^-	na	-	na	-	3200	360	na
F^-	na	-	na	-	2.7	0.55	na
SO_4^{2-}	na	-	na	-	10,700	3900	na
DOC	na	-	na	-	220	64	na
pH	6.8	0.05	na	-	9.4	1.0	na
TOC [%]	6.8	1.4	1.2	0.1	na	-	na
Humus content [%]	17.1	0.6	na	-	na	-	na
ANC [mol H^+/kg DS]	na	-	2.3	0.2	na	-	na
EC_{sw}[4] [mS/m]	na	-	na	-	820	38	na
Bulk density [kg/m^3]	700	<1	1080[5]	-	na	-	na

[1] Not possible to calculate due to a single sample or parameter not analyzed; [2] Not analyzed; [3] Re-calculated from a maximum of 12 mg Cd/kg P; [4] L/S 2; [5] Re-calculated from a natural moisture content of 10%.

After 20 weeks, the plants were gently harvested manually using small spades. The plants, and any reference soil or MIBA on or close to their roots, were collected and dried at room temperature for 1 week. Due to the low growth rate, small sub-samples were collected from several cultivation spots in each box, then mixed into one MIBA sample per box. For consistency, only one sample from each reference box was collected according to the same methodology. The dried plants were separated into below-ground biomass, i.e., roots and above-ground biomass, before further analysis. The only exception was RwoF, as the biomass from this box was too small to allow separation into different plant fractions.

2.3. Chemical Analyses

Total element concentrations in the original MIBA and the solid residues after cultivation were analyzed at accredited laboratories (Eurofins, Luxembourg, and Synlab, Munich, Germany) by melting with $LiBO_2$ followed by dissolution in HNO_3 according to ASTM D3682: 2013 and ASTM D4503: 2008, or with $HNO_3/HCl/HF$ according to SS-EN 13656: 2003. Final analysis of the solutions was performed using ICP-SFMS (Inductively Coupled Plasma-Sector Field Mass Spectrometry) according to SS-EN ISO 17294-2: 2016 and EPA method 200.8: 1994. A total content analysis of the elements in the soil before and after cultivation was carried out at an accredited laboratory (Eurofins, Luxembourg) by dissolution in aqua regia according to SS-ISO 11466, followed by analysis with ICP-SFMS (SS-EN ISO 17294-1, 2 (mod) or EPA method 200.8 (mod)) or ICP- AES (Inductively coupled plasma atomic emission spectroscopy) (SS-EN ISO 11885 (mod) and EPA method 200.7 (mod)). The contents in the original MIBA and reference soil were analyzed in triplicates, while the residues after harvest, due to limited sample volume, were analyzed as singlicates.

The total organic carbon (TOC) and acid-neutralizing capacity (ANC) of the MIBA were analyzed at an accredited laboratory (Eurofins, Luxembourg) in duplicate, using SS-EN 13137: 2001 and EN 14429: 2015, respectively. The particle size distribution in MIBA was carried out as singlicate (30 g) at the Chalmers University of Technology using EN 933-1. The particle size distribution and humus content in soil were analyzed in triplicates using SS ISO-11277 (2009), while TOC was analyzed using SS-EN 15936: 2012 metodappl. A/SS-EN 13137: 2001 m and pH using DIN ISO 10390: 2005–12, all in triplicates and at accredited laboratories (Eurofins, Luxembourg). The bulk density of the soil was analyzed in triplicate (internal method at the accredited laboratory Eurofins, Luxembourg), while the corresponding value for MIBA was calculated from data in [30,31].

A 5-step sequential extraction scheme based on [32] was performed by an accredited laboratory (Synlab, Munich, Germany) on the original MIBA (single sample) to study the potential mobility and operational binding forms of selected elements [33]. The method uses 1 M NaOAc, pH 5, to extract easily adsorbed and exchangeable metals and carbonates, followed by $Na_4P_2O_7$ to extract metals weakly bound to organic matter, 0.25 M $NH_2OH\cdot HCl$ in 0.1 M HCl in 60 °C to extract metals bound to amorphic iron and manganese oxides, 1.0 M $NH_2OH\cdot HCl$ in 25% CH_3COOH in 90 °C to extract metals bound to crystalline iron oxides, and finally $KClO_3$ in 12 M HCl 4 M HNO_3 in 90 °C to extract metals bound to the stable organic forms and sulfides. The leachates were analyzed according to SS-EN ISO 17294-1,2 (mod)/EPA-method 200.8 (mod) and SS 028150-2.

To evaluate potential leaching, SS-EN-12457-3 was performed on representative original MIBA samples. The test was carried out in duplicate. The pH of the leachates was measured according to SS-EN ISO 10523: 2012. The electrical conductivity (EC) of the leachates was measured according to SS-EN 27888-1 at a liquid-to-solid (L/S) ratio of 2. All analyses were conducted at accredited laboratories (Eurofins, Luxembourg, and Synlab, Munich, Germany).

The elemental contents of the harvested plants (above- and below-ground biomass) were analyzed after microwave digestion with HNO_3/H_2O_2 using ICP-SFMS according to SS-EN ISO 17294-2: 2016 and EPA method 200.8: 1994. The analyses were done at an accredited laboratory (Synlab, Munich, Germany) in duplicate.

3. Result and Discussion

3.1. Characterisation of Original MIBA

Biomass ashes are known to act as a fertilizer and promote the growth of trees when spread in woods or used in agricultural processes [34,35]. The Swedish Forest Agency has issued recommendations on the total content of different elements in biomass ashes to be reintroduced to a forest [36]. A comparison between those values and the contents of the MIBA samples used in this study showed that of the macronutrients (Ca, K, P, Mg), Ca and Mg were present in sufficient concentrations, whereas the K and P contents were too low (Table 1). This shows that there is potential for plants to grow in MIBA and that growth could be promoted by adding NPK fertilizer. However, the amounts of the potentially toxic metals Cr, Cu, Ni, and Pb all exceeded the recommended levels, and it is well known that this can inhibit growth, especially of the roots; see, e.g., [37] and references therein. On the other hand, their mobility in water is low (Table 1) and sequential extraction has shown that of these elements, Pb was the only one with a high proportion present in exchangeable and adsorbed forms (Figure 1). In addition to Pb, Zn, Cd, and Ca are also highly mobile and were present in more than 50% of the exchangeable fraction. Chromium and Ni were mainly present in stable forms (steps 4 and 5), and are unlikely to be as available for phytoextraction. For most elements, the presence in step 2 (labile organic forms) was <10%, which was expected, as the TOC in MIBA is low (Table 1). The predicted mobility of Cu is low, and >70% was present in the sulfide fraction (step 5). The consistency of the results from the total elemental analyses for the other elements was acceptable between the methods. The mobility of anions, such as chloride and sulfate, as well as EC_{sw}, were noteworthy (Table 1), as high salinity is known to affect plant growth negatively [38].

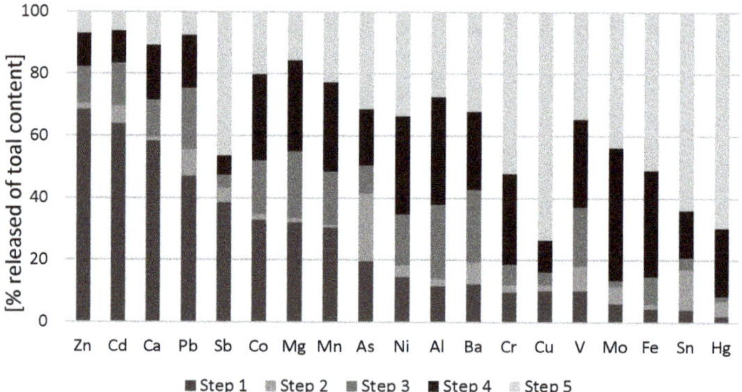

Figure 1. Sequential extraction distribution for original MIBA. Step 1: easily adsorbed and exchangeable metals and carbonates, Step 2: metals weakly bound to organic matter, Step 3: metals bound to amorphic iron and manganese oxides, Step 4: metals bound to crystalline iron oxides, and Step 5: metals bound to stable organic forms and sulfides.

3.2. Plant Growth and Biomass

The plant height growth was initially similar for all sunflower plants (Figure 2a). However, after a few weeks, the rate decreased in the MIBA boxes, and after a few more weeks, the MIBA plants started to lose their leaves and dried out. The addition of fertilizer had only a minor effect on growth, which could have been due to the presence of potentially toxic elements like Cu, Zn, and Cl that may inhibit nutrient uptake and affect plant growth negatively [37–39]. Additionally, sunflower roots are likely to be too weak to efficiently penetrate the compact MIBA and therefore don't absorb enough water and nutrients (Table 1).

Rapeseed showed a different pattern. Initially, the plants in the MIBA with fertilizer (RF) grew slower than the reference plants (Rref) (Figure 2b). However, after a few weeks, the rate increased and at harvest, the heights of the Rref and RF plants were almost the same. However, the height variation was larger for RF than for Rref, implying that cultivation in original MIBA is likely to be more diverse. Contrary to the sunflowers, the addition of fertilizer was crucial. After only a few weeks, the plants without nutrient addition (RwoF) already had a slower growth rate, and after 8 weeks, their growth stopped completely (Figure 2b). The leaves of these plants were purplish, which is a known effect of

P deficiency; see, e.g., [40]. The P content was 5 times higher in MIBA than in soil, but was not in available forms for rapeseed (Table 1). Shortage of P is more common in alkaline soils with low TOC and high Fe, and in plants with weak roots, which describes the conditions in the MIBA boxes without fertilizer (Tables 1 and 2). No difference in the growing capacity trends was observed for rapeseed plants planted at longer or shorter distances from each other, indicating that many plants can be cultivated in a limited space, giving a higher potential phytoextraction capacity.

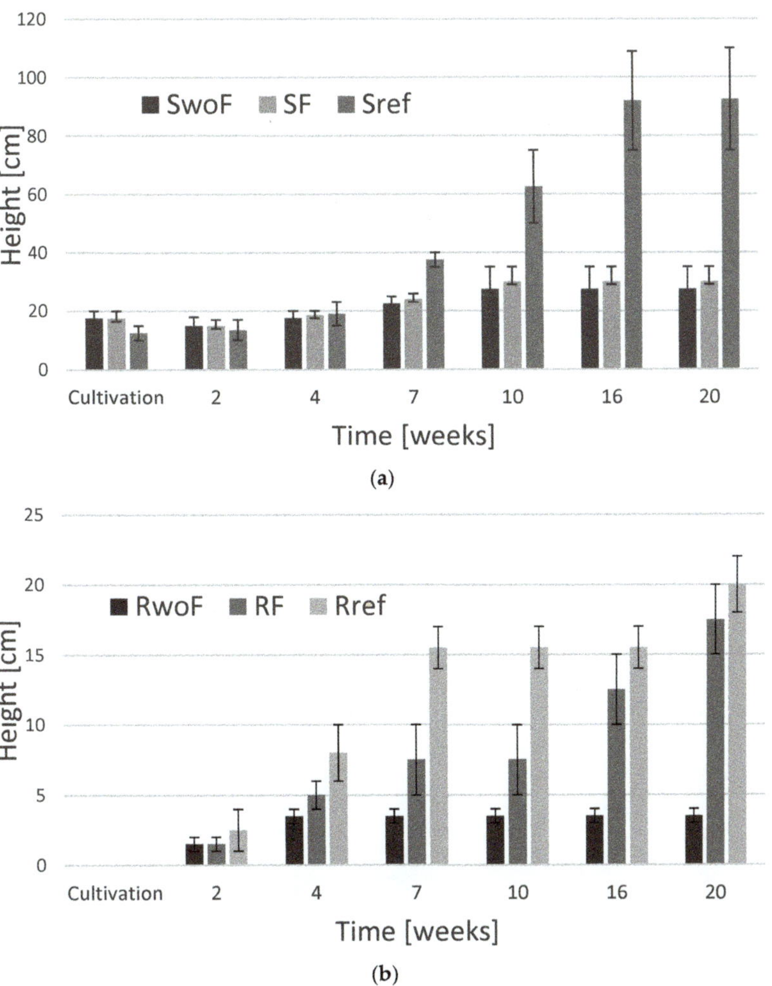

Figure 2. Average growth in cm for (**a**) sunflowers and (**b**) rapeseed cultivated in MIBA or reference soil (Sref and Rref) over 20 weeks. Error bars indicate variations in height.

The main fraction of the biomass was found in the above-ground biomass irrespective of the cultivation conditions or plant type (Table 2). However, the above-ground biomass was significantly higher for the control plants, Sref, than for the sunflowers grown in MIBA, which was due to the fact that the MIBA plants dried out and stopped growing. Both the heights and the distances between the plants in Rref and RF were similar, but the total biomass was approximately 3 times larger in the control plants, Rref (Figure 2b and Table 2). The ocular inspection confirmed that both the above- and under-ground biomass of the RF plants was weaker than that of the Rref plants. This indicates that although the taproots of rapeseed developed better than those of the sunflowers, their growth was still restricted in MIBA. There are likely several reasons for this, and except for the higher bulk density of MIBA (Table 1), which can inhibit root growth [41], another major difference was the particle size distribution (Figure 3). MIBA contains a higher fraction of larger particles, which is well known to give lower water- and nutrient-holding capacities. At harvest, it was noted that the reference soil contained more water than the MIBA samples, despite the same amount of water having percolated through all cultivation boxes. This was also noted during the growing season, when the top layer of the MIBA appeared slightly drier than the reference soil. MIBA is known to be a draining material, and much of the added water probably percolated faster through the cultivation boxes and collected in the lower parts, where the roots were unable to reach it, while the water in the reference boxes was more evenly distributed. Additionally, the organic matter (TOC) in MIBA was low (Table 1), which further affects this. Mixing the MIBA with organic matter would likely give a better material for cultivation, as the presence of organic matter not only improves the nutrient- and water-holding capacities but also provides better access to nutrients [42]. The fraction of smaller particles will increase, and organic matter can also protect plants from absorbing too much salt [42]. The latter might be of high importance in MIBA (Table 1).

Table 2. Above- and under-ground biomass yield [g] for sunflowers and rapeseed cultivated in MIBA or reference soil (Sref and Rref) with SD in brackets.

Plant	Biomass Above-Ground/Plant	Biomass Under-Ground/Plant
	[g]	[g]
SwoF	0.8 (0.1)	0.2 (0.04)
SF	0.8 (<0.01)	0.3 (<0.01)
Sref	118 (41)	4.0 (0.3)
RwoF	0.1 [1] (0.01)	-
RF	0.2 (0.02)	0.1 (<0.01)
Rref	0.6 (0.01)	0.4 (0.06)

[1] This includes the whole plant, due to insufficient biomass for division into parts.

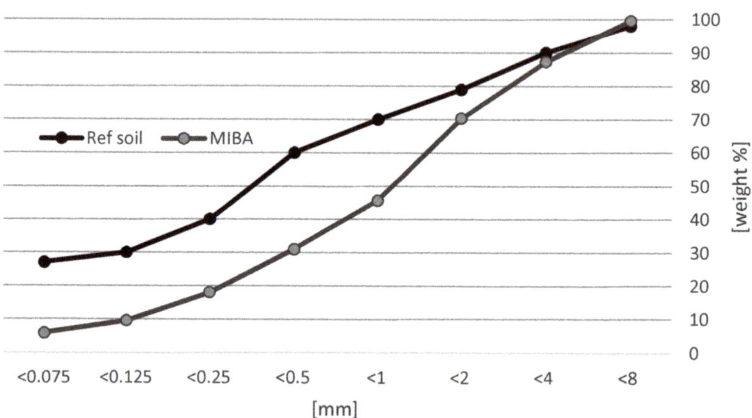

Figure 3. Cumulative particle size distribution curves for original reference soil and MIBA.

Rosenkranz et al. added compost in their study, but unfortunately, they did not include any cultivation in reference soil [25]. The mass above ground per rapeseed plant (*Brassica n.*) in their study was, however, consistent with the Rref in this study, showing the effect of the addition of organic matter to MIBA. However, the higher contents of especially Cu, Pb, and Zn in the conventional MIBA used here are also likely to have contributed to the inferior growth rate [37]. Cultivation of rapeseed in sandy soil contaminated with

Zn and Pb in amounts corresponding to those in the MIBA used here resulted in 30% less above-ground biomass, indicating that the contamination level and the density of the cultivation material both have an impact [43].

3.3. Metal Accumulation in Plants

Although most of the biomass was present in the parts above ground, the highest metal concentrations were generally found in the roots for both plant types (Table 3). This was particularly true for minor and trace elements such as Cu, Pb, and Zn in RF (Table 3). Accumulation of Pb and Zn in rapeseed roots has previously been reported and suggested to be caused by the formation of low mobility compounds between the heavy metals and organic substances, as the metals enter the plasma in the roots [29,43]. For RwoF, the distribution could not be evaluated due to insufficient biomass. The distribution trend was weaker for the sunflowers, likely due to the limited germination of the MIBA plants. Earlier research has indicated that the Zn content is highest in the leaves of the sunflower [18]. However, this was not found in the present study, neither in the sunflowers planted in MIBA nor in those planted in the reference soil (Table 3). The metal contents were generally higher in the rapeseed plants grown in MIBA than in the sunflowers (Table 3). For instance, the Zn content in the below-ground biomass was about 10 times higher for the rapeseed plants than for the sunflowers (Table 3). This is probably because the root systems of rapeseed are more suitable for cultivation in MIBA than those of sunflowers, as discussed above, although the effects of potential water and nutrient shortages are also likely to influence this.

When calculating the bioaccumulation coefficients (BAC), i.e., the concentration of metal in plant parts above ground divided by the metal concentration in original MIBA, it is obvious that none of the used plant types should be classified as hyperaccumulators for any of the elements studied, as BAC < 1. On the other hand, BAC might not be optimal for evaluating the extraction efficiency in this material, as MIBA is more complex and generally contains higher concentrations of metals than polluted soils. However, the results indicate that for most elements, the higher the pollutant content in the solid material, the higher the content in the plants. The Zn concentration was approximately 5 times higher in the above-ground sunflower parts cultivated in MIBA than in the reference soil, even though the growth was limited (Table 3). This was even more distinct for rapeseed, where the Zn concentrations detected in the RwF roots were almost 20 times higher than those in the Rref roots (Table 3). The highest enrichments compared to the reference cultivation were found for Cu (50 times), and for Sn and Pb (25 times). Marchiol et al. reported that the Zn content in rapeseed tissues cultivated in polluted soil substrate was 100 times higher than in reference soil, while the enrichment for elements like Cu and Pb was one order of magnitude higher [43]. This indicates that if the cultivation properties of MIBA are improved by, e.g., the addition of organic matter [42], the extraction efficiency in rapeseed probably will increase.

3.4. Metal Contents in MIBA after Harvest

Although the enrichment of several elements in the plants was significant, the observed decreases in the MIBA samples were limited (Figure 4). This was also confirmed by the low BACs, as calculated above. Except for elements that to a large extent form water-soluble compounds (Ca, K, Na, and S), i.e., are washed out during watering, no noteworthy changes in the MIBA composition were identified. Naturally, this was partly due to the limited growth, and better cultivation conditions would most likely result in higher extraction rates. Regardless, cultivation in MIBA should primarily be seen as a way to recover unutilized metals, rather than as a way to efficiently remediate the material, as this would take many years due to the complex contamination situation [22]. However, phytoextraction would likely decrease metal mobility, consequently enabling further utilization alternatives, even though metal compounds would still be present.

Table 3. Total amounts in parts above ground and roots from sunflower and rapeseed cultivated in MIBA, with or without fertilizer, and in reference soil. All values are shown in mg/kg DS.

	Sunflower														Rapeseed								
	Above Ground						Root								Above Ground						Root		
Element	SwoF	sd	SF	sd	Sref	sd	SwoF	sd	SF	sd	Sref	sd	RwoF [1]	sd	RF	sd	Rref	sd	RF	sd	Rref	sd
	[mg/kg DS]						[mg/kg DS]								[mg/kg DS]						[mg/kg DS]		
Al	450	46	370	41	46	13	2600	1000	1000	170	5600	640	5200	-	270	150	210	77	9100	580	3500	2300
As	0.22	<0.01	0.35	0.04	0.09	0.06	1.8	1.2	0.88	0.13	1.8	0.25	4.4	-	0.32	0.06	0.16	0.05	8.4	0.12	0.99	0.63
Ba	16	2.7	10	1	2.5	0.67	76	54	33	2.6	55	8.2	190	-	11	2.8	12	0.1	330	42	39	23
Ca	14,000	3200	21,000	3100	5100	1000	15,000	8300	9000	180	7400	10	39,000	-	30,000	300	22,000	300	61,000	2100	5100	1000
Cd	0.05	<0.01	0.07	<0.01	0.01	<0.01	0.35	0.16	0.21	0.04	0.16	<0.01	0.82	-	0.09	<0.01	0.04	<0.01	1.3	0.07	0.08	0.04
Cl	2200	140	3700	560	12,000	2000	11,000	4000	6700	390	7400	540	9100	-	12,000	1600	24,000	1000	4800	90	3000	670
Co	0.35	0.01	0.46	0.08	0.04	0.01	2.1	1.5	1.2	0.1	3.5	0.68	6.8	-	0.59	0.17	0.41	0.03	10	0.25	2.3	1.7
Cr	1.5	0.26	1.2	0.16	0.15	0.02	9.6	6.7	4.1	0.42	7.3	0.97	19	-	1.2	0.43	1.2	0.73	65	31	4.7	3.1
Cu	12	0.35	16	1.9	8	0.97	180	120	110	30	36	7.9	430	-	39	7.9	10	4.8	570	19	11	6.1
Fe	730	75	630	110	80	33	2100	1600	770	170	7900	1400	5200	-	200	77	260	72	7600	920	4700	3400
Hg	0.02	<0.01	0.02	<0.01	<0.01	<0.01	<0.01	<0.01	<0.01	<0.01	0.05	<0.01	0.01	-	<0.01	<0.01	0.02	<0.01	0.02	<0.01	<0.02	<0.02
K	4400	400	4100	790	13,000	2700	21,000	6400	17,000	4600	16,000	3800	16,000	-	30,000	2500	29,000	1500	12,000	100	11,000	990
Mg	1400	15	1600	45	940	44	2200	360	1200	55	2800	330	3000	-	2800	5	2900	130	2900	130	2400	960
Mn	27	3.6	43	4.8	8.8	1	64	46	34	3.1	190	35	200	-	52	3.9	24	3	300	2	120	79
Mo	0.78	0.06	1.3	0.03	0.15	0.06	5.1	2.6	5.2	0.9	2.2	0.6	8.1	-	3.9	0.18	11	1.6	9.5	0.16	1.7	0.23
Na	1600	45	1500	25	300	71	5700	430	5600	1400	850	340	2100	-	1500	10	1200	75	1400	160	520	250
Ni	1.2	0.08	1.3	0.11	0.19	0.05	13	11	5	0.39	5.2	0.9	31	-	3.7	0.73	2.3	0.19	52	5	4.8	1.8
P	530	7.5	650	72	2300	270	480	230	340	2.5	2400	470	2600	-	3000	190	3400	290	3700	440	3000	720
Pb	3	0.02	2.6	0.24	0.23	0.03	37	27	18	2	11	1.5	91	-	3.6	1.4	1.1	0.53	170	11	6.6	4.1
S	1200	45	2500	60	820	40	4000	65	3700	890	1400	110	9300	-	13,000	50	7500	190	6700	310	2100	55
Se	<0.01	<0.01	<0.01	<0.01	<0.05	<0.01	<0.08	<0.03	<0.05	<0.01	0.22	0.03	0.32	-	0.17	0.01	<0.09	<0.04	0.36	0.02	<0.12	<0.07
Sn	0.52	0.02	0.3	0.02	<0.02	<0.01	4.8	3.8	1.9	0.45	0.27	0.03	4	-	0.4	0.14	0.09	0.02	7.5	0.65	0.15	0.06
V	1.3	0.18	1.1	0.22	0.13	0.06	2.8	2.1	1.2	0.05	18	0.75	5.6	-	0.25	0.09	0.45	0.1	9	0.27	9.8	6.7
Zn	59	1.2	88	7.5	16	0.35	460	28	110	18	88	8.2	700	-	100	1.7	38	8.9	1000	10	56	26

[1] Whole plant due to limited biomass.

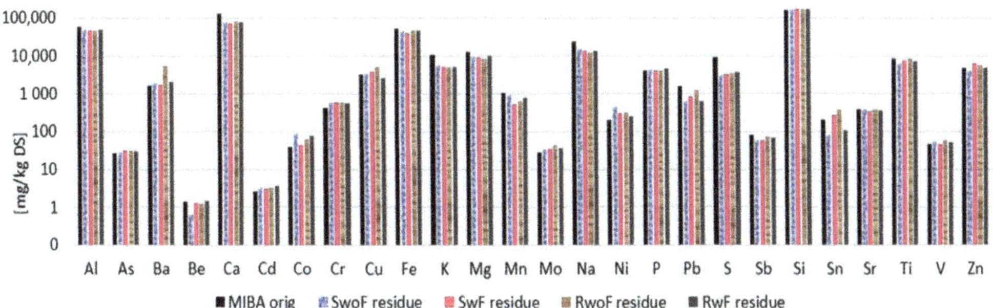

Figure 4. Contents (mg/kg DS) of selected elements in MIBA before and after cultivation of sunflowers and rapeseed.

3.5. Potential for Recovery of Zn and Other Metals from MIBA

From the results, it is obvious that the direct cultivation of sunflowers and rapeseed in conventional MIBA is not straightforward. However, although both growth and metal reduction were limited, a glimpse of the potential for using plants grown in MIBA as a metal-enriched biofuel is shown. Plants from phytoextraction could be incinerated in WtE plants, and the metals could be recovered from the fly ash, as discussed in the Introduction. Calculations assuming 6% of fly ash mass after incineration of rapeseed plants [44] provide ash with interesting quantities of Co, Cu, Pb, and Zn, for which the estimated amounts are comparable to workable ores (Table 4) [45]. Additionally, Co has been identified by the EC as a critical raw material [1]. However, for this to be interesting on a larger scale, metal enrichment in these plants must increase. In addition to improving important cultivation parameters of MIBA, such as the particle size distribution and the content of organic matter, the choice of more suitable plants for phytoextraction in this kind of material is urgent. As the metal content in MIBA is high, it would probably be more efficient to use perennial plants, as one cultivation season is too short to reach efficient extraction. For instance, plants in the salix family could be an alternative, as they are known to grow in highly polluted soils in the climate of northern Europe and have a high capacity for extracting Zn and other metals [20–22]. In a study using Salix caprea, Zn was enriched in the leaves, while Co, Cu, and Ni had mainly accumulated in the roots [20]. By collecting the leaves regularly, Zn can be recovered continuously, whereas metals mainly enriched in the roots can only be recovered when the whole plant is harvested. An additional advantage with plants in the salix family is their high biomass; this makes them suitable for incineration, which makes them interesting from an energy perspective as well, as they may be capable of displacing fossil fuels.

Table 4. Calculated indicative contents of Co, Cu, Pb, and Zn in ash after incineration of RF plants. For comparison, contents in workable ores are shown [45]. All amounts are given in %.

Element	Whole Plant Ash	Root Ash	Workable Ore
	[%]	[%]	[%]
Co	0.007	0.02	0.01–0.08
Cu	0.42	0.95	0.14–2.6
Pb	0.12	0.29	0.30–9.4
Zn	0.78	1.7	0.31–12

4. Conclusions

This paper discusses for the first time, to the knowledge of the authors, the potential of cultivating annual plants in conventional MIBA in order to generate metal-enriched biomass fuel, which is to be used for metal recovery.

The results show that cultivation of rapeseed and sunflowers in conventional MIBA is challenging. The main explanations for this are the inhibited root development and low water- and nutrient-holding capacities caused by unsuitable properties such as the high fraction of large particles, the low TOC content, and the high density of MIBA. Rapeseed plants with more developed taproots grew better than sunflowers, but the addition of fertilizer was crucial.

There was an accumulation of numerous metals in both plant types, and the highest metal content was generally found in the roots. Calculations indicated ash from rapeseed root incineration contained 2% Zn, and the contents of Co, Cu, and Pb were comparable to those in workable ores. The metal reduction in MIBA after cultivation was limited.

The results indicate that to promote phytoextraction on MIBA, the material itself should be refined, but also the choice of plants is important. Perennials with roots that efficiently penetrate the compact MIBA material, e.g., members of the salix family, are theoretically suitable. This initial study shows that cultivation in and phytoextraction on MIBA is possible, but further research is needed.

Author Contributions: Conceptualization, K.K.F., V.E. and D.D.; Investigation, K.K.F.; Methodology, K.K.F., V.E. and D.D.; Resources, K.K.F., V.E. and D.D.; Visualization, K.K.F.; Writing—original draft, K.K.F. All authors have read and agreed to the published version of the manuscript.

Funding: This research was funded by "the Ash program", managed by Energiforsk (2018-113) and Avfall Sverige (F226), grant number (2018-113) and (F226).

Institutional Review Board Statement: Not applicable.

Informed Consent Statement: Not applicable.

Acknowledgments: This work was supported by the "Ash program", managed by Energiforsk (2018-113) and Avfall Sverige (F226), which is gratefully acknowledged. Sandra Dassoum is recognized for her assistance with cultivation management.

Conflicts of Interest: The authors declare no conflict of interest.

References

1. European Commission. *Report on the Critical Raw Materials for the EU, Report of the Ad Hoc Working Group on Defining Critical Raw Materials*; European Commission: Brussels, Belgium, 2014; p. 41.
2. Avfall Sverige. *Svensk Avfallshantering 2018*; Avfall Sverige: Malmö, Sweden, 2019.
3. Blasenbauer, D.; Huber, F.; Lederer, J.; Quina, M.; Blanc-Biscarat, D.; Bogush, A.; Bontempi, E.; Blondeau, J.; Chimenos, J.M.; Dahlbo, H.; et al. Legal situation and current practice of waste incineration bottom ash utilisation in Europe. *Waste Manag.* **2020**, *102*, 868–883. [CrossRef] [PubMed]
4. Arickx, S.; Van Gerven, T.; Vandecasteele, C. Accelerated carbonation for treatment of MSWI bottom ash. *J. Hazard. Mater.* **2006**, *137*, 235–243. [CrossRef]
5. Freyssinet, P.; Piantone, P.; Azaroual, M.; Itard, Y.; Clozel-Leloup, B.; Guyonnet, D.; Baubron, J. Chemical changes and leachate mass balance of municipal solid waste bottom ash submitted to weathering. *Waste Manag.* **2002**, *22*, 159–172. [CrossRef]
6. Klymko, T.; Dijkstra, J.J.; Van Zomeren, A. *Guidance Document on Hazard Classification of MSWI Bottom Ash in ECN-E-17-024*; Confederation of European Waste-to-Energy Plants: Düsseldorf, Germany, 2017.
7. Tiberg, C.; Sjöstedt, C.; Fedje, K.K. Speciation of copper and zinc in MSWI bottom ash studied by XAS and geochemical modelling. *Waste Manag.* **2021**, *119*, 389–398. [CrossRef] [PubMed]
8. Fedje, K.K.; Strömvall, A.-M. Enhanced soil washing with copper recovery using chemical precipitation. *J. Environ. Manag.* **2019**, *236*, 68–74. [CrossRef]
9. Barbaroux, R.; Plasari, E.; Mercier, G.; Simonnot, M.-O.; Morel, J.L.; Blais, J. A new process for nickel ammonium disulfate production from ash of the hyperaccumulating plant Alyssum murale. *Sci. Total Environ.* **2012**, *423*, 111–119. [CrossRef]
10. Delplanque, M.; Collet, S.; Del Gratta, F.; Schnuriger, B.; Gaucher, R.; Robinson, B.; Bert, V. Combustion of Salix used for phytoextraction: The fate of metals and viability of the processes. *Biomass-Bioenergy* **2013**, *49*, 160–170. [CrossRef]
11. Rasmusen, E. *Sambehandling Af RGA Og Scrubber Væske Fra Forbrændingsanlæg Med HALOSEP Processen, Miljøprojekt Nr. 1648*; Danish EPA: Copenhagen, Denmark, 2015.
12. Schlumberger, S.; Bühler, J. Metal Recovery in Fly and Filter Ash in Waste to Energy Plants. In *Ash 2012*; Strömberg, B., Ed.; Värmeforsk AB: Stockholm, Sweden, 2012.
13. Fedje, K.K.; Andersson, S. Zinc recovery from Waste-to-Energy fly ash—A pilot test study. *Waste Manag.* **2020**, *118*, 90–98. [CrossRef]

14. Odjegba, V.J.; Fasidi, I.O. Phytoremediation of heavy metals by *Eichhornia crassipes*. *Environmentalist* **2007**, *27*, 349–355. [CrossRef]

15. Zhao, L.; Yuan, L.; Wang, Z.; Lei, T.; Yin, X. Phytoremediation of zinc-contaminated soil and zinc-biofortification for human nutrition. In *Phytoremediation and Biofortification*; Yin, X., Yuan, L., Eds.; Springer Briefs in Molecular Science: Dordrecht, The Netherlands, 2012; pp. 33–57.

16. Wrobel, S. Sunflower yields as an indicator of zinc polluted soil detoxification. *Fresenius Environ. Bull.* **2010**, *19*, 330–334.

17. Belouchrani, A.S.; Mameri, N.; Abdi, N.; Grib, H.; Lounici, H.; Drouiche, N. Phytoremediation of soil contaminated with Zn using Canola (*Brassica napus* L). *Ecol. Eng.* **2016**, *95*, 43–49. [CrossRef]

18. Adesodun, J.K.; Atayese, M.O.; Agbaje, T.A.; Osadiaye, B.A.; Mafe, O.F.; Soretire, A. Phytoremediation Potentials of Sunflowers (*Tithonia diversifolia* and *Helianthus annuus*) for Metals in Soils Contaminated with Zinc and Lead Nitrates. *Water Air Soil Pollut.* **2009**, *207*, 195–201. [CrossRef]

19. Fulekar, M.H. Phytoremediation of Heavy Metals by *Helianthus annuus* in Aquatic and Soil environment. *Int. J. Curr. Microbiol. Appl. Sci.* **2016**, *5*, 392–404. [CrossRef]

20. Varga, C.; Marian, M.; Mihaly-Cozmuta, L.; Mihaly-Cozmuta, A.; Mihalescu, L. Evaluation of the phytoremediation potential of the Salix caprea in tailing ponds. *An. Univ. Oradea Fasc. Biol.* **2009**, *16*, 141.

21. Hauptvogl, M.; Kotrla, M.; Prčík, M.; Žaneta, P.; Kováčik, M.; Lošák, T. Phytoremediation Potential of Fast-Growing Energy Plants: Challenges and Perspectives—A Review. *Pol. J. Environ. Stud.* **2019**, *29*, 505–516. [CrossRef]

22. Pulford, I.D.; Watson, C. Phytoremediation of heavy metal-contaminated land by trees—A review. *Environ. Int.* **2003**, *29*, 529–540. [CrossRef]

23. Bilski, J.; Jacob, D.; McLean, K.; McLean, E.; Soumaila, F.; Lander, M. Agro-toxicological aspects of coal fly ash (FA) phytoremediation by cereal crops: Effects on plant germination, growth and trace elements accumulation. *Adv. Biores.* **2012**, *3*, 121–129. [PubMed]

24. Krgovic, R.; Trifković, J.; Milojković-Opsenica, D.; Manojlovic, D.; Markovic, M.; Mutić, J. Phytoextraction of metals by Erigeron canadensis L. from fly ash landfill of power plant "Kolubara". *Environ. Sci. Pollut. Res.* **2015**, *22*, 10506–10515. [CrossRef]

25. Rosenkranz, T.; Kisser, J.; Wenzel, W.W.; Puschenreiter, M. Waste or substrate for metal hyperaccumulating plants—The potential of phytomining on waste incineration bottom ash. *Sci. Total Environ.* **2017**, *575*, 910–918. [CrossRef]

26. Renova, A.B. Jordprodukter. 2019. Available online: https://www.renova.se/foretag/produkter-och-tjanster/jordprodukter/ (accessed on 10 December 2019).

27. Granngården, A.B. Promagna 11-5-18. 2019. Available online: https://www.granngarden.se/godsel-promagna-11-5-18-25-kg/ p/1196440 (accessed on 10 December 2019).

28. Kassas, H.; Sharaf, M.; Abdou, M.M.N. Phytoremediation of zinc and copper. *J. Eniviron. Sci.* **2003**, *7*, 323.

29. Angelova, V. Potential of rapeseed (*Brassica napus* L.) for phytoremediation of soils contaminated with heavy metals. *J. Environ. Prot. Ecol.* **2017**, *18*, 468–478.

30. Arm, M. *Handbook—Bottom Ash in Road and Public Construction (In Swedish; Handbok—Slaggrus I Väg—Och Anläggning-Sarbeten)*; Institue, S.G., Ed.; Swedish Geotechnical Institute: Linköping, Sweden, 2006.

31. Serti, S.; Löfgren, M. *Guidance for Classification of Incinceration Residues Using Calculation 2018:13 (In Swedish; Vägledning for Klassificering Av Förbränningsrester Med Beräkningsmetoder)*; Management, S.A.o.W., Ed.; Swedish Association of Waste Management: Malmö, Sweden, 2018.

32. Tessier, A.; Campbell, P.; Bisson, M. Sequential extraction procedure for the speciation of particulate trace metals. *Anal. Chem.* **1979**, *51*, 844–851. [CrossRef]

33. ALS Scandinavia AB. *Sequential Extraction*, version 20-01-2018; ALS Scandinavia AB: Luleå, Sweden, 2018 (In Swedish). Available online: www.alsglobal.se (accessed on 20 January 2018).

34. Ozolinčius, R.; Varnagirytė-Kabašinskienė, I.; Stakėnas, V.; Mikšys, V. Effects of wood ash and nitrogen fertilization on Scots pine crown biomass. *Biomass-Bioenergy* **2007**, *31*, 700–709. [CrossRef]

35. Zhang, Z.; He, F.; Zhang, Y.; Yu, R.; Li, Y.; Zheng, Z.; Gao, Z. Experiments and modelling of potassium release behavior from tablet biomass ash for better recycling of ash as eco-friendly fertilizer. *J. Clean. Prod.* **2018**, *170*, 379–387. [CrossRef]

36. Drott, A.; Anderson, S.; Eriksson, H. *Rules and Recommendations for Forest Fuel Extraction and Compensation Action, Report 2019/14 (In Swedish: Regler Och Rekommendationer för Skogsbränsleuttag Och Kompensationsåtgärder, Rapport 2019/14)*; Skogsstyrelsen, S.F.A., Ed.; Swedish Forest Agency: Jönköping, Sweden, 2019.

37. Wong, M.H.; Bradshaw, A.D. A comparison of the toxicity of heavy metals, using root elongation of rye grass, lolium perenne. *New Phytol.* **1982**, *91*, 255–261. [CrossRef]

38. Oosterbaan, R.J. Cropo tolerance to soil salinity, statistical analysis of data measured in farm lands. *Int. J. Agric. Sci.* **2018**, *3*, 57–66.

39. Küpper, H.; Zhao, F.-J.; McGrath, S.P. Cellular Compartmentation of Zinc in Leaves of the Hyperaccumulator *Thlaspi caerulescens*. *Plant Physiol.* **1999**, *119*, 305–312. [CrossRef]

40. Yara, U.K. Phosphorus Deficiency-Oilseed Rape. 2020. Available online: https://www.yara.co.uk/crop-nutrition/oilseed-rape/ nutrient-deficiencies-oilseed-rape/phosphorus-deficiency-oilseed-rape/ (accessed on 7 April 2020).

41. Magdoff, F.; van Es, H. *Building Soils for Better Crops Sustainable Soil Management*, 3rd ed.; Handbook Series Book 10th ed.; Sustainable Agriculture Research and Education (SARE) and USDA's National Institute of Food and Agriculture, University of Maryland and University of Vermont: Brentwood, CA, USA, 2019.

42. Bot, A.; Benites, J. *The Importance of Soil Organic Matter Key to Drought-Resistant Soil and Sustained Food Production*; Food and Agriculture Organization of the United Nations: Rome, Italy, 2005.
43. Marchiol, L.; Assolari, S.; Sacco, P.; Zerbi, G. Phytoextraction of heavy metals by canola (*Brassica napus*) and radish (*Raphanus sativus*) grown on multicontaminated soil. *Environ. Pollut.* **2004**, *132*, 21–27. [CrossRef]
44. Strömberg, B.; Svärd, S.H. *Handbook for Fuel A08-819 (In Swedish: Bränslehandboken A08-819)*; Strömberg, B., Ed.; Värmeforsk: Stockholm, Sweden, 2012.
45. Fennoscandian Mineral Deposits Application. *Ore Deposits Database and Maps*; Geological Survey of Finland: Espoo, Finland, 2020.

soil systems

Article

Enhanced Lead Phytoextraction by Endophytes from Indigenous Plants

Ilaria Pietrini [1], Martina Grifoni [2], Elisabetta Franchi [1], Anna Cardaci [1], Francesca Pedron [2], Meri Barbafieri [2], Gianniantonio Petruzzelli [2] and Marco Vocciante [3],*

[1] Eni S.p.A., Renewable Energy & Environmental Laboratories, Via F. Maritano 26, 20097 San Donato Milanese, Italy; ilaria.pietrini@eni.com (I.P.); elisabetta.franchi@eni.com (E.F.); anna.cardaci@eni.com (A.C.)

[2] Institute of Research on Terrestrial Ecosystems, National Council of Research, Via Moruzzi 1, 56124 Pisa, Italy; martina.grifoni@iret.cnr.it (M.G.); francesca.pedron@cnr.it (F.P.); meri.barbafieri@cnr.it (M.B.); gianniantonio.petruzzelli@cnr.it (G.P.)

[3] Department of Chemistry and Industrial Chemistry, University of Genova, Via Dodecaneso 31, 16146 Genova, Italy

* Correspondence: marco.vocciante@unige.it

Abstract: Lead (Pb) is one of the most common metal pollutants in soil, and phytoextraction is a sustainable and cost-effective way to remove it. The purpose of this work was to develop a phytoextraction strategy able to efficiently remove Pb from the soil of a decommissioned fuel depot located in Italy by the combined use of EDTA and endophytic bacteria isolated from indigenous plants. A total of 12 endophytic strains from three native species (*Lotus cornicolatus*, *Sonchus tenerrimus*, *Bromus sterilis*) were isolated and selected to prepare a microbial consortium used to inoculate microcosms of *Brassica juncea* and *Helianthus annuus*. As for *B. juncea*, experimental data showed that treatment with microbial inoculum alone was the most effective in improving Pb phytoextraction in shoots (up to 25 times more than the control). In *H. annuus*, on the other hand, the most effective treatment was the combined treatment (EDTA and inoculum) with up to three times more Pb uptake values. These results, also validated by the metagenomic analysis, confirm that plant-microbe interaction is a crucial key point in phytoremediation.

Keywords: soil remediation; phytoextraction; mobilizing agents; microbial endophytes; lead pollution

Citation: Pietrini, I.; Grifoni, M.; Franchi, E.; Cardaci, A.; Pedron, F.; Barbafieri, M.; Petruzzelli, G.; Vocciante, M. Enhanced Lead Phytoextraction by Endophytes from Indigenous Plants. *Soil Syst.* **2021**, *5*, 55. https://doi.org/10.3390/soilsystems5030055

Academic Editors: Matteo Spagnuolo, Paola Adamo and Giovanni Garau

Received: 9 July 2021
Accepted: 2 September 2021
Published: 3 September 2021

Publisher's Note: MDPI stays neutral with regard to jurisdictional claims in published maps and institutional affiliations.

Copyright: © 2021 by the authors. Licensee MDPI, Basel, Switzerland. This article is an open access article distributed under the terms and conditions of the Creative Commons Attribution (CC BY) license (https://creativecommons.org/licenses/by/4.0/).

1. Introduction

The overture of remediation of polluted soils for sustainable innovation is one of the critical steps in addressing current global environmental issues.

Nowadays, several physicochemical and biological solutions have been developed for the recovery of contaminated water [1–3] and soils [4,5], making the optimal selection a problematic yet crucial step for the success of the reclamation [6,7].

However, remediation activities also have an environmental impact since they often use chemical products or processes, with consequent consumption of raw materials and energy that could compromise the sustainability of the approach or even invalidate its beneficial aspects [8].

In the light of this, environmentally unsound technologies for the remediation of heavy metal contaminated soils (chemical extraction, chemical oxidation, stabilization/solidification, solvent extraction, etc.) should be replaced by green and sustainable technologies. The aim of the latter is not only to eliminate or reduce contamination but also to minimize the environmental impact (reduction of air emissions, minimization of energy use, decrease of waste, etc.) and create synergies between different sectors and activities (ecosystems protection, circular economy, climate change and resilience). Concerning this, sustainable phytomanagement can make a significant contribution to supporting this transition towards sustainable remediation.

Remediation technologies based on the new natural-based solution (NBS) approach enable the achievement of the goals established by current environmental policies to protect natural resources [9]. Among the NBS remediation measures, a growing focus is on phytoremediation [10–12], which is the set of remediation technologies with plants as the main actors to remediate organic and inorganic contaminants in soil and other environmental matrices (sediments, water). The interest in these phytotechnologies has increased over time, given some significant advantages in terms of low cost, simplicity of operation and environmental benefits, as highlighted by some recent LCA-based studies [13,14]. In addition, the combined use with other solutions to further increase the overall sustainability is under investigation [15].

Among the different phytoremediation technologies, phytoextraction is considered a non-invasive technique to remove heavy metals from contaminated soil in an environmentally friendly and economical way [16,17]. In phytoextraction, metals are absorbed by roots from the soil solution and then transferred and accumulated in various tissues of the plant. Traditional phytoextraction requires hyperaccumulating species able to accumulate high amounts of metal without suffering physiological damage. However, these species exhibit low biomass production and slow growth rate, making phytoextraction a slow process with long implementation times [17,18].

Several studies have focused on improving phytoremediation efficiency using fast-growing, high biomass tolerant species, and the aid of chelating agents for increased metal uptake from the soil. The main chelating agents include ethylenediaminetetraacetic acid (EDTA), diethylenetriamine pentaacetate (DTPA), ethylene glycol tetraacetic acid (EGTA), hydroxyethyliminodiacetic acid (HEIDA), and ethylenediamine disuccinic acid (EDDS). These compounds can accelerate the release of heavy metals bound to soil particles into the soil solution, thus increasing the phytoavailable metal fraction [19]. Indeed, the only metals that plants can absorb are those in bioavailable form, i.e., present in soluble forms in the soil solution [20,21].

Among the various assisted-phytoextraction approaches to maximize the technique's effectiveness, an exciting alternative is the use of plant growth-promoting rhizobacteria (PGPR) [22]. This strategy involves rhizobacteria that can stimulate plant growth by both facilitating the bioavailability of soil nutrients and modulating the production of phytohormones (including auxins, cytokines, gibberellic acid, 1-aminocyclopropane-1-carboxylate deaminase—ACCD) and modulating plant hormone levels [22,23]. In addition, through microbial processes active in the rhizosphere, PGPR can also promote the mobility and bioavailability of metals in the soil, increasing their uptake by plants [23,24].

This work aimed to investigate the single and synergistic effect of PGPR and EDTA on the growth and Pb uptake in two tolerant species, *Brassica juncea* L. (Indian mustard) and *Helianthus annuus* L. (sunflower), to evaluate the phytoremediation potential of a Pb-contaminated site.

The selection of bacterial strains capable of improving the efficiency of phytoremediation is a fundamental step. In this study, the addition of a microbial consortium with indigenous endophytic bacteria allowed detectable levels of phytoextraction to be obtained even without the addition of mobilizing chemical agents.

2. Materials and Methods

2.1. Site Description and Soil Sampling

The contaminated site under investigation is located in an area of 40,000 m^2 near an urban area in Italy, previously affected for many years by a deposit for the storage of industrial wastes, active until the beginning of the 90s.

Being adjacent to a pond and at a distance of about 1 km from the sea, the site is characterized by sediments of marine and alluvial origin (silts, clayey silts, fluvio-lacustrine and marshy clays, arenaceous and conglomeratic, as well as alluvial deposits) and by a superficial layer (4–5 m depth) mainly composed of homogeneous artificial backfill soils of clayey, silty sand nature.

Previous chemical analyses performed on the soil of the area under examination revealed a relevant Pb contamination, which exceeded the contamination threshold concentrations (CSC) established by the Italian regulation (D.Lgs 152/2006) for sites intended for industrial use (1000 mg kg^{-1}) [25]. The highest Pb concentration found in the area was approximately 2200 mg kg^{-1}.

For the soil sampling to be addressed for the laboratory activities, 6 sampling points (SP1 to SP6) were chosen within the site. The selection was based on the geological and morphological characteristics of the soils and favoring the points where previous analytical campaigns had already detected the presence of Pb.

Approximately 50 kg of soil was collected from each sampling point with an excavator to a maximum depth of 2.5 m.

The soil samples were sieved at 2 cm on-site and then transported in special containers to the laboratories for the soil characterization analysis. Based on the results of characterization, which showed that the soil samples were homogeneous, a single soil sample was prepared for the phytoextraction tests, obtained by mixing the soil aliquots (~20 kg) from each soil sample. On this soil (Pb-soil), the determination of the total and bioavailable lead content was performed again.

2.2. Soil Characterization and Pb Analysis

The soil samples were air-dried, ground, and sieved (0–2 mm) before analysis. The physicochemical characteristics of the soil were determined following the procedures reported in Methods of soil analysis [26]. Soil pH and electrical conductivity (EC) were determined by glass electrodes with a soil/water ratio of 1:2.5 and 1:2, respectively. The cation exchange capacity (CEC) was measured by exchange with barium acetate (pH 8.1) and titration with EDTA (0.05 N). Particle size distribution (sand, silt, and clay) was evaluated by the pipette method [27].

For the determination of total Pb content, soil samples were digested in a mixture of HNO_3 (65%, v/v) and H_2O_2 (30%, v/v) using a microwave oven (FKV-ETHOS 900), according to EPA method 3051-A [28].

Potentially bioavailable concentrations of Pb in soil were determined according to an extraction procedure that sequentially involves the use of H_2O (to extract soluble Pb), KNO_3 1 M (to extract exchangeable Pb), and EDTA 1% (to extract Pb retained with bonds also of covalent nature) with a soil/extractant ratio of 1:5 and extraction time 5 h [29,30].

EDTA was selected because it is one of the most effective mobilizing agents for Pb and can be applied effectively in various soil types [31–33].

A single extraction with EDTA was also performed. The concentration selected was 2 mM, frequently used in phytoextraction tests [15,31,34]. Higher concentrations of EDTA would mobilize excessive quantities of metal, which could give rise to leaching phenomena towards the aquifer [18,19,35].

The extraction was carried out by shaking the soil and extractant (ratio of 1:5) for 2 h. All extracts were centrifuged at 15,000 rpm for 15 min and filtered.

All analyses were performed in triplicate, and the mean value was recorded.

2.3. Microcosm Experimental Design

To study the role of bacterial inocula on phytoremediation processes, alone or in combination with the EDTA mobilizing agent, a microcosm experimental campaign was performed.

Each microcosm was prepared by adding 300 g of Pb-soil and sowing *B. juncea* var. Scala or *H. annuus* var. Paola, at doses of 0.5 g and 5 seeds per pot, respectively.

The experimentation lasted about 30 days and was conducted in a growth chamber (CCL300BH-AS S.p.A., Perugia, Italy). The growth conditions were the following: photoperiod of 14 h light at 25 °C and 10 h dark at 19 °C, photosynthetic photon flux density (PPFD) of 200 μmol m^{-2} s^{-1}, and relative humidity 65%.

The experiment consisted of the following treatments: Pb-soil treated with 10^8 CFU (Colony-Forming Unit) per gram of soil of bacterial consortium SG_1 (PGPR) prepared as described in Section 2.6, Pb-soil treated with PGPR and EDTA 2 mM (PGPR + EDTA), and Pb-soil without any treatment (CT, control).

The study design consisted of a completely random design with three replicates per treatment. Bacterial inocula were added approximately 9 days after planting. EDTA addition was carried out after approximately 15 days, splitting the total dose over 5 days to avoid possible Pb toxic effects on plant growth resulting from rapid release and high metal mobilization [36].

Pots were thoroughly watered with tap water at field capacity, maintaining this condition throughout the experiment without fertilizer addition.

Plants were harvested 15 days after the addition of EDTA. Plant organs (roots and shoots) were washed with deionized water and then dried at 40 °C to constant weight to determine their Pb content. Roots were further treated in an ultrasonic bath (Branson Sonifier 250 ultrasonic processor, Branson Ultrasonic Corporation, Brookfield, CT, USA) to remove any residual soil.

Dried plant tissues were weighed, dry weight (DW) recorded, and powdered. Total Pb content was determined in the dried vegetal samples after overnight acid pre-digestion with HNO_3 (65%, *v/v*) and H_2O_2 (30%, *v/v*) according to US-EPA 3052 [37]. Quantification of Pb in soil, extract, and vegetal samples was performed by inductively coupled plasma optical emission spectroscopy (ICP-OES) Liberty AX, Varian.

2.4. Test of Phytotoxicity

Soil toxicity was assessed by a phytotoxicity screening test based on germination inhibition and root extension.

The assay was performed in a Petri dish (10 cm diameter) filled with 10 g of Pb-soil moistened with deionized water to saturation, over which was placed a Whatman #1 filter and 10 seeds of *Lepidium sativum* L. [38]. *L. sativum* is a biological indicator of soil toxicity screening highly sensitive to phytotoxic compounds.

Hydrated quartz sand was used as a negative control. Five replicates were performed for each soil (Pb-soil and negative control). The closed Petri dishes were placed in a germination chamber in the dark at 25 ± 1 °C. After 72 h, the number of germinated seeds was counted, and the root length was measured. The germination index (*GI%*) and the radical extension inhibition (*Inh%*) were estimated by combining the values of seed germination and root elongation:

$$GI\% = \frac{G_s * L_s}{G_c * L_c} * 100 \tag{1}$$

$$Inh\% = \frac{L_c - L_s}{L_c} * 100 \tag{2}$$

G_s and G_c are the average numbers of seeds germinated in the contaminated soil samples and in the negative control, respectively. L_s and L_c are the radical lengths (mm) for the contaminated soil sample and negative control, respectively.

2.5. Endophyte Isolation

The contaminated site is rich in spontaneous species that have developed on the reported soil, most of which were probably already present as seed banks in the fill soil. The prevalent herbaceous vegetation has the homogeneous characteristics of the surrounding area.

For this reason, endophytic bacteria were isolated and characterized from three of the most commonly represented spontaneous species: *Lotus cornicolatus*, *Sonchus tenerrimus*, and *Bromus sterilis*. The soil around the roots was removed by repeatedly rinsing with tap water. To sterilize the root surface, they were treated with 70% EtOH for 5 min, with NaClO for 2 min and again 70% EtOH for another 5 min. They were then thoroughly rinsed at least three times with sterile H_2O. The roots, finely chopped with a sterile scalpel, were placed in sterile flasks containing TYEG (Trypticase Yeast Extract Glucose) medium and incubated for 16 h at 30 °C. Serial dilutions (10-4, 10-6, 10-8) of the obtained suspension were prepared, and 100 µL of each dilution (in triplicate) was spread over R2A Agar (Merck®) plates. Then 100 µL of the third rinsing water was plated to confirm the efficiency of sterilization. Endophyte colonies appeared after 4–5 days.

About twenty phenotypically different colonies for each root type were isolated and pure cultures were used for DNA extraction and taxonomic classification as already described in Franchi et al. [23].

2.6. PGPR Characterization

Endophytes isolated were subjected to a series of in vitro assays to assess their plant growth-promoting potential. The production of auxin indole-3-acetic acid (IAA) was estimated following the method proposed by Shahab et al. [39], siderophore molecules release was determined as described by Milagres et al. [40]. Their ability to solubilize mineral phosphorus (P) was determined by growing the strains in NBRIP (National Botanic Research Institute's Phosphate) according to the protocol developed by Nautiyal [41]. Production of exopolysaccharide (EPS) was estimated using a modified Weaver mineral medium enriched with sucrose [42]. Proteolytic activity was determined as described by Nielsen and Sørensen [43]. The production of ammonia was determined in peptone water (5 g L^{-1} peptone and 5% NaCl, pH 7.2), following the method proposed by Kifle and Laing [44].

The isolated strains were also tested for their capacity to form biofilm in vitro, inoculating them in glass tubes with 7 mL of LB (*Luria Broth*) medium. The tubes were incubated at 30 °C for 7 days without agitation. The formation of a visible layer (pellicle) at the interface between medium and air indicated a potential capability to produce biofilms. The potential capacity to fix atmospheric nitrogen was evaluated by growing the isolates with a specific nitrogen-free medium (NFb) [45]. The strains that showed at least three growth-promoting properties were selected. With the 12 potentially most promising (Figure 1), a microbial consortium (SG_1) was prepared in the form of lyophilizate and was then used as an inoculum in the phytoremediation tests.

Endophytes features													
Bacterial isolates	Closest described relative [BLAST search]	Origin	Acc. N°	Family	IAA	Siderophores	N_2 fix	iP Solub.	NH_3	EPS	Pellicle	Proteases	SCORE
SMV297	*Pseudomonas lactis*	*Lotus corniculatus*	MN538912	Pseudomonadaceae	–	–	+	+	+	+	+	+	6
SMV298	*Pseudomonas baetica*		MN538913	Pseudomonadaceae	+	+	–	+	+	–	+	+	6
SMV303	*Pseudomonas putida*		MN538914	Pseudomonadaceae	+	–	–	+	+	–	+	–	4
SMV304	*Pseudomonas azotoformans*		MN538915	Pseudomonadaceae	–	–	+	+	+	+	+	+	6
SMV305	*Bacillus toyonensis*		MN538916	Bacillaceae	+	–	–	–	+	–	+	+	4
SMV311	*Lysinibacillus macroides*	*Sonchus tenerrimus*	MN538917	Bacillaceae	+	+	+	–	+	–	+	+	6
SMV313	*Pseudomonas plecoglossicida*		MN538918	Pseudomonadaceae	–	+	+	+	+	–	–	–	4
SMV316	*Micrococcus aloeverae*		MN538919	Bacillaceae	–	+	–	+	+	–	–	+	4
SMV321	*Pseudomonas koreensis*	*Bromus sterilis*	MN538920	Pseudomonadaceae	–	+	+	+	+	–	–	+	5
SMV326	*Pseudomonas stutzeri*		MN538921	Pseudomonadaceae	–	–	–	+	–	–	+	+	3
SMV328	*Pseudomonas taiwanensis*		MN538922	Pseudomonadaceae	–	–	–	+	–	+	+	+	4
SMV329	*Lysinibacillus macroides*		MN538923	Bacillaceae	+	+	+	–	+	–	+	–	5

Figure 1. The 12 selected endophytes are listed. In vitro, PGP properties and Genbank Accession number are shown. The + sign indicates the presence of the PGP property, the − sign its absence. Each strain was assigned a score calculated considering the number of PGP properties displayed.

2.7. Next-Generation Ion Torrent Sequencing (NGS)

An amount of 3 ng of the genomic DNA, obtained by extraction of 500 mg of soil samples and about 200 mg of roots samples through the Fast DNA® Spin Kit for Soil (MP Biomedicals, Irvine, CA, USA) and quantified by Qubit® 2.0 fluorometer (Invitrogen, Waltham, MA, USA), was amplified using the 16S Metagenomics Kit (Thermo Fischer Scientific, Waltham, MA, USA).

The amplification program was set up as follows: 95 °C for 10 min, followed by 25 cycles at 95 °C per 30 s, 58 °C for 30 s, and 72 °C for 20 s, a final hold time for 7 min at 72 °C and a cooling step at 4 °C.

The subsequent purification of the amplicons, the preparation, and the sequencing of the libraries followed the protocols for the Ion OneTouch™ 2 System, the Ion OneTouch™ ES, and the Ion PGM™, respectively, as previously described in Conte et al. [46].

The run was based on the workflow Metagenomics 16S w1.1 handling the Database Curated microSEQ®16 S and the reference Library 2013.1. The primers detected both ends to obtain 250 base pairs sequences. Alignment in Torrent Suite™ Software (version 5.8) was performed using the Torrent Mapping Alignment Program (TMAP). The sequences that occurred only once in the entire dataset were removed, and the representative sequences were defined with a 97% similarity cut-off.

After classifying the Operational Taxonomic Unit (OTU) representative sequences, the output was elaborated to obtain a relative abundance (%) of each OTU in the total amounts of the entire sample.

2.8. Quality Assurance and Quality Control

Quality assurance and quality control were performed by testing two standard solutions (0.5 and 2 mg L^{-1}) for every 10 samples. CRM ERM—CC141 for soil and CRM ERM—CD281 for plants were used as certified reference materials. The values obtained for Pb were 31.8 ± 1.2 mg kg^{-1} for CRM ERM—CC 141 and 1.69 ± 0.20 mg kg^{-1} for CRM ERM—CD281, in agreement with the certified values of 32.2 ± 1.4 mg kg^{-1} and 1.67 ± 0.11 mg kg^{-1}, respectively.

The detection limit for Pb was 5 µg L^{-1}, and the recovery of samples with spikes (5%) ranged from 93% to 101% with a relative standard deviation (RSD) of 1.92% of the mean.

2.9. Statistical Analysis

Data referred to the phytoextraction test are reported as the mean of three replicate microcosms, and analyses were performed in triplicate, recording the mean value ± standard deviation (±SD). Data statistical analyses were performed using Statistica version 6.0 (StatSoft, Inc., Tulsa, OK, USA). Treatment effects were analyzed using a one-way analysis of variance (ANOVA, San Francisco, CA, USA). Differences between means were compared, and a post hoc analysis of variance was performed using Tukey's honestly significant difference test ($p < 0.05$).

3. Results

3.1. Soil Analysis

Based on the data (Table 1), the soil samples under study (SP1-SP6) were homogeneous with each other, with a CEC in the range of 16–20 Cmol$_{(+)}$ kg^{-1} and an alkaline pH (~8.3 pH), which suggested a low Pb availability. Soil texture was also rather similar among the soil samples, with a predominance of the sandy fraction. The high conductivity values (on average 585 ± 54.5 µS cm^{-1}) raised a particular concern initially due to possible effects on plant growth, which may have a different sensitivity to soil salinity. However, during the phytoextraction test, no growth problems due to the high salt content in the soil were observed.

Table 1. Chemical properties in individual samples collected at the site under investigation (SP1-SP6). Values are reported as mean ($n = 3$) ± SD.

	SP1	SP2	SP3	SP4	SP5	SP6
pH	8.22 ± 0.1	8.44 ± 0.2	8.42 ± 0.1	8.51 ± 0.2	8.38 ± 0.1	8.12 ± 0.1
EC (µS cm^{-1})	596 ± 12	644 ± 10	548 ± 13	621 ± 8.5	495 ± 11	607 ± 6.1
Clay (%)	14.4 ± 0.2	8.74 ± 0.2	12.9 ± 0.4	9.78 ± 0.3	10.2 ± 1.0	13.1 ± 1.1
Silt (%)	18.5 ± 1.5	13.2 ± 0.9	17.4 ± 1.4	15.6 ± 1.1	16.2 ± 0.9	14.8 ± 1.3
Sand (%)	67.1 ± 1.1	78.1 ± 0.1	70.1 ± 0.3	67.8 ± 1.3	71.5 ± 0.4	72.3 ± 0.5
CEC (Cmol$_{(+)}$ kg^{-1})	18.7 ± 0.2	18.7 ± 1.2	20.5 ± 0.4	17.5 ± 1.2	16.2 ± 0.8	19.4 ± 0.5

Therefore, the homogeneity of the main characteristics of the sampled soils allowed the composition of a single sample of contaminated soil (Pb-soil) on which to perform the phytoextraction tests.

As reported in Table 2, the soils had significant amounts of Pb, with a mean value between samples SP1-SP6 of 108 ± 3.45 mg kg^{-1}. The total Pb content in the Pb-soil was in the range of values found in the original samples. These values are higher than the limit of the values for public, private, and residential green use, established by the Italian legislation [25], thus implying a potential environmental risk.

Table 2. Pb content (mg kg^{-1}) in individual samples collected at the site under investigation (SP1-SP6) and in the Pb-soil composite sample. Values are reported as mean ($n = 3$) $\pm$ SD.

	SP1	SP2	SP3	SP4	SP5	SP6	Pb-Soil
Total	106 ± 2.2	112 ± 8.1	106 ± 3.5	111 ± 6.9	107 ± 6.1	137 ± 3.5	112 ± 4.8
H$_2$O	bdl	bdl	bdl	bdl	bdl	bdl	bdl
KNO$_3$ 1 M	1.30 ± 0.2	0.77 ± 0.1	1.70 ± 0.6	0.92 ± 0.3	1.34 ± 0.6	1.04 ± 0.1	1.15 ± 0.2
EDTA 1%	21.5 ± 2.3	21.4 ± 3.1	20.3 ± 2.0	20.5 ± 1.9	20.6 ± 2.5	20.9 ± 1.7	20.8 ± 2.8
EDTA 2 mM	6.89 ± 1.8	7.27 ± 0.9	7.66 ± 1.3	6.55 ± 1.3	7.47 ± 1.1	7.72 ± 2.4	7.68 ± 2.1

bdl: below detection limit.

The sequential extraction procedure (SEP), which allowed the evaluation of potentially phytoavailable Pb samples, showed that most of the Pb was in a form not available for uptake by plants (Table 2). The Pb amounts solubilized by H$_2$O and KNO$_3$ were negligible, amounting to about 1% of the total Pb concentration. The results are ascribable to the basic pH of the soil, whereas the amount extractable in EDTA 1% was about 20%. Also, the extraction with 2 mM EDTA significantly increased the Pb phytoavailability, extracting on average 6.7% of total Pb.

3.2. Phytotoxicity Test

The phytotoxicity test was considered as a preliminary investigation to assess plant growth to provide supporting information for further investigations performed in this study.

GI% and Inh% data for Pb-soil, $85 \pm 10.8\%$ and $14 \pm 10.5\%$, respectively, suggest the lack of adverse conditions for plant growth. High GI% and low Inh% values indicate reduced phytotoxicity and good quality of the contaminated soil.

The results confirmed the low values of Pb bioavailability, which did not cause significant toxic effects in the species used. Moreover, thanks to the phytotoxicity test, it was possible to verify that the high values of EC did not affect the plant germination.

3.3. Effect of PGPR and EDTA on Plant Growth and Pb Phytoextraction Efficiency

Small-scale phytoremediation experiments, performed under controlled conditions (microcosm), are essential for evaluating both the species performance and the effectiveness of treatments to be used.

Therefore, the first parameter to be analyzed is biomass, which provides information about plant growth under stress conditions.

B. juncea and *H. annuus* species showed no visible toxic symptoms during growth in the Pb-contaminated soil, even after applying the different treatments.

The results presented in Figure 2 indicate some significant differences in dry biomass of roots and shoots between treatments in the two species.

PGPR utilization positively affected the growth of both species, both when applied alone and in combination with EDTA.

In particular, the aerial biomass of *H. annuus* increased by 13.4% and 11.5% compared to the control when PGPRs were applied individually and combined with EDTA, respectively. In contrast, a relevant effect (36.5%) on *B. juncea* can be observed only in the combined treatment (Figure 2B). As for the roots, PGPR did not significantly influence both species' biomass production, except for the combined treatment PGPR + EDTA on *B. juncea* (Figure 2A).

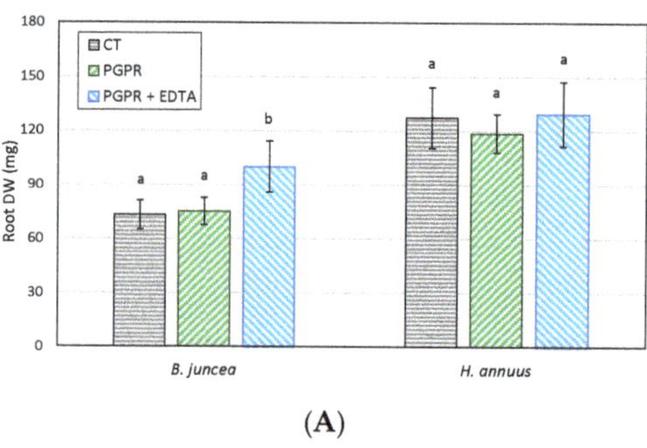 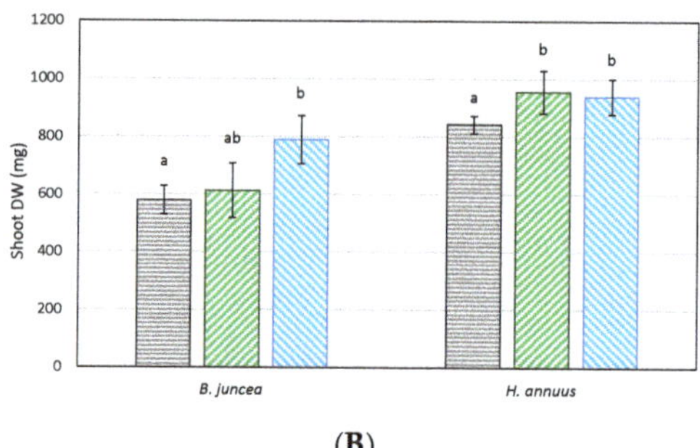

(A)   (B)

Figure 2. (**A**) Root and (**B**) shoot dry weight (mg pot^{-1}) of *B. juncea* and *H. annuus* grown on Pb-soil control (CT) and Pb-soil treated with, PGPR, and PGPR + EDTA. The values are the mean of three replicates, and the error bars show ± standard deviation. Values with different letters are significantly different at the 5% probability level (Tukey's test).

Under the experimental conditions investigated, and it is worth highlighting this, both *B. juncea* and *H. annuus* well tolerated the amounts of Pb released by EDTA.

The absence of adverse effects of EDTA on plant growth could be attributed to its ability to chelate Pb, which prevents binding between the metal and cellular components, neutralizing cytological impacts [47,48].

Pb concentrations in roots and shoots of *B. juncea* and *H. annuus* are presented in Table 3.

Table 3. Effect of PGPR and PGPR + EDTA treatment on Pb concentration in root and shoot of *B. juncea* and *H. annuus* grown on Pb-soil. The values are the mean of three replicates, and the error bars show ± standard deviation. The different letters within the same column represent different significance levels at $p < 0.05$ (Tukey's test).

Treatment	B. juncea		H. annuus	
	Root	**Shoot**	**Root**	**Shoot**
CT	1.33 ± 0.51 a	0.20 ± 0.05 a	2.14 ± 0.58 a	1.34 ± 0.25 a
PGPR	55.8 ± 4.89 c	4.82 ± 0.77 c	19.0 ± 3.46 b	1.14 ± 0.13 a
PGPR + EDTA	52.3 ± 4.74 c	2.04 ± 0.54 b	42.0 ± 5.10 c	3.46 ± 0.69 c

As suggested by SEP, which did not reveal significant amounts of metal in an immediately bioavailable form, the lowest amounts of Pb were found in the control plants of both species. These minimal concentrations could be due to the radical exudates promoting the Pb solubilization.

The application of EDTA improved Pb uptake in both species with respect to the control. Indeed, significant increases in Pb concentration in the soil solution due to the EDTA addition, as indicated by the soil bioavailability results, resulted in substantial increases in Pb concentrations in the roots and shoots of both plants.

However, concerning *B. juncea*, the highest Pb concentrations were found in the roots and shoots of plants treated with only PGPR. The shoots of *B. juncea* treated exclusively with PGPR had approximately 24 times more Pb than control shoots. The PGPR + EDTA combined treatment mainly affected *B. juncea* roots, which showed Pb values not significantly different from plants treated with PGPR alone treatment. In contrast, the PGPR + EDTA combined treatment did not increase Pb amounts in *B. juncea* shoots.

In *H. annuus*, the highest Pb concentration was found in the roots and shoots of the plants treated with PGPR + EDTA.

The phytoextraction efficiency is the relation between the metal concentration in the plants and the biomass produced by evaluating the total accumulation (total uptake), obtained by multiplying the Pb concentration in plant tissues by the corresponding biomass produced. The effects of the treatments on Pb total uptake in *B. juncea* and *H. annuus* are illustrated in Figure 3.

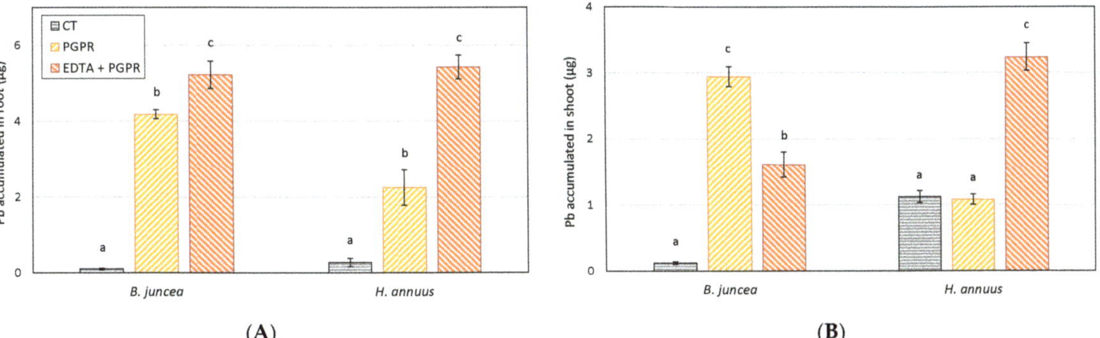

Figure 3. Effect of PGPR and PGPR + EDTA treatment on Pb uptake by (**A**) root and (**B**) shoot of *B. juncea* and *H. annuus* grown on Pb-soil. The values are the mean of three replicates, and the error bars show ± standard deviation. Values with different letters are significantly different at the 5% probability level (Tukey's test).

Our results indicated that treatment with PGPR, alone or in combination with EDTA, effectively improved Pb phytoextraction by *B. juncea* and *H. annuus*.

In *B. juncea* roots, the combined treatment PGPR + EDTA increased Pb accumulation by about 58 times compared to the control. Concerning *B. juncea* shoots, the highest Pb accumulation was observed in plants treated with only PGPR.

In contrast, in *H. annuus*, the highest Pb accumulation in shoots and roots was attributable exclusively to the PGPR + EDTA treatment. However, bacterial inocula alone increased the *H. annuus* root biomass to the levels of the combined treatment, without showing a significant difference.

3.4. NGS Analysis Results

The bacterial communities (Figure 4) were mainly constituted by the family Pseudomonadaceae (maximum value of 51% in the sample 1 R), followed by the families Caulobacteraceae (maximum value of 30% in the sample 6 R), Xanthomonadanceae (maxi-mum value of 18% in the sample 6 R) and Sphingomonadaceae (maximum value of 15% in the sample 3S). These bacteria are often correlated to hydrocarbon and metal contamination [49,50]. Considering the known PGPR strains [51] found in these communities and mainly related to Pb tolerance, the Pseudomonadaceae and Bacillaceae were isolated in the site and used for the inoculum. Pseudomonadaceae maintained a constant abundance in all the samples, while the abundance of Bacillaceae seemed to be affected by the inoculum, with values from 2% (both Controls) to 8% (sample 2S: PGPR *B. juncea*) and to 5–10% (respectively, in samples 5S: PGPR in *B. juncea* and 6S: EDTA + PGPR in *H. annuus*). Moreover, this family can be found only in the soil samples. Similarly, the family Rhodospirillaceae, mainly represented by the genus *Azospirillum sp.*, can be found, even if only with low relative abundance (maximum value 2%), in the samples belonging to *B. juncea*. Continuing to consider PGPR families, also known for their tolerance to Pb, there is a slight increase of the Alcaligenaceae in the sample's roots in the treatment added with the inoculum. In detail, these families increase their relative abundance from 0% to 2% in sample 3 R (i.e., PGPR in *B. juncea*) and from 0% to 1–2%, respectively in the samples 5 R (i.e., PGPR in *H. annuus*) and 6 R (i.e., EDTA + PGPR in *H. annuus*). The Enterobacteraceae represents another family of interest showing its increase in the root samples treated with the mobilizing agent EDTA in the presence of both *B. juncea* and *H. annuus* (i.e., increase from 0% to 4% in sample 2 R: EDTA with *B. juncea* and increase from 14% to 27% in sample 6 R: EDTA with *H. annuus*).

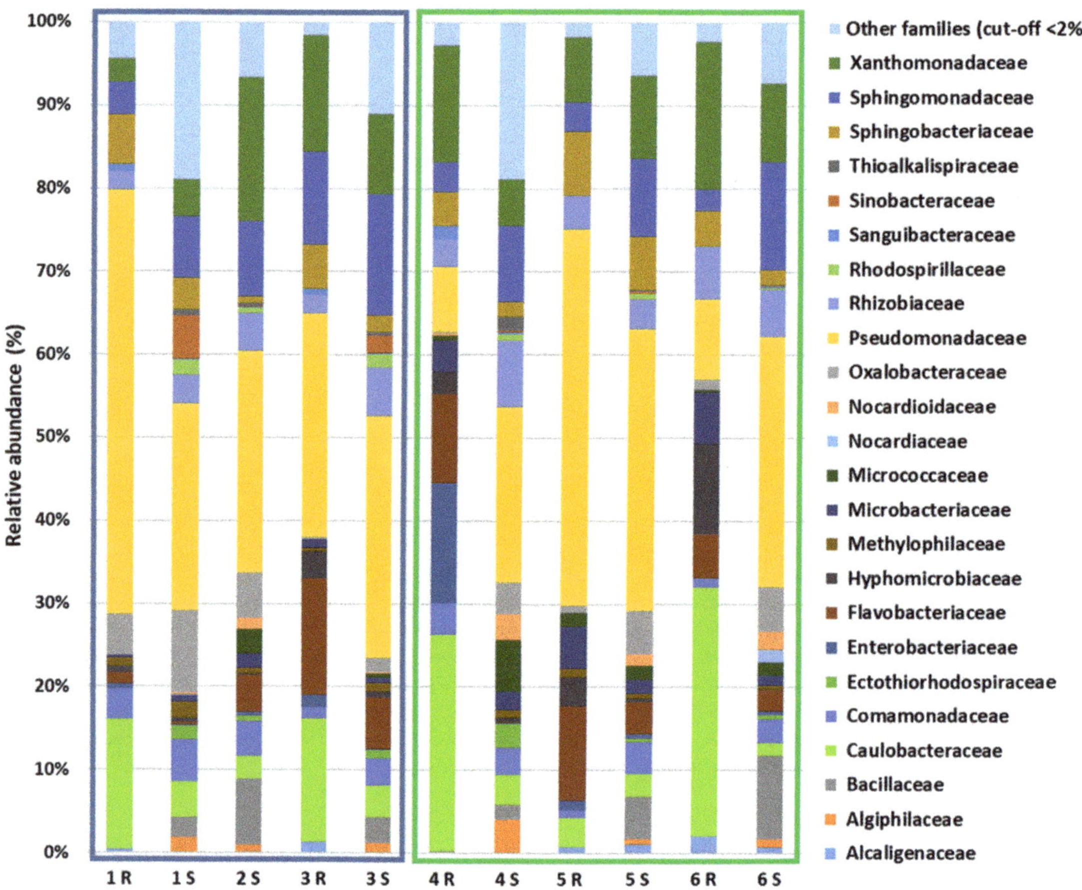

Figure 4. NGS Ion Torrent analysis of all the samples divided by the *B. juncea* (i.e., samples 1–3) and *H. annuus* (i.e., samples 5–8). The numbers represent the different treatments applied (i.e., Controls: 1 and 4; PGPR: 2 and 5; EDTA + PGPR: 3 and 6), and S stands for Soils and R for Roots. Sample 2 R was not included due to the low quality of the data required. Due to the high number of results obtained, the calculations, at the family level, were carried on by setting a cut-off < 2%.

Expanding the attention to other PGPR families, not known for their Pb tolerance, it can be observed that both the families Flavobacteriaceae and Nocardioacea were mainly represented in the microcosms with *H. annuus*; the first one with higher values in the root samples, while the Nocardioacea was present almost exclusively in soil samples.

4. Discussion

Among soil contaminants, Pb is of particular concern due to its ability to biomagnify through the food chain, threatening food safety [47]. Lead is highly toxic for most plants, causing severe physiological damage [52]. However, some plant species have proven to be particularly resistant to lead toxicity and are widely used in the phytoremediation of Pb-contaminated soils [31,47]. The plant species tested in this experiment showed a growth capacity not affected by lead contamination. The reason for the selection of these two species is twofold: in addition to their satisfactory performances, as shown in other studies on Pb phytoremediation [53–57], they have recently been recognized as energy crop species, capable of providing a high level of biomass that can be reused in the bioenergy field [58–61].

This last aspect is currently very relevant as it allows further support for phytoremediation both in economic and sustainability terms.

The phytoextraction of Pb-contaminated soils with EDTA has been extensively studied. The results obtained in our study are in agreement with previous works that have shown how the addition of this complexing agent was recognized as an effective technique to increase Pb absorption by plants [31,47,62,63].

Although several studies have reported that EDTA can promote plant growth in metal-polluted soils by stimulating auxin production and aiding nutrient translocation in plants [33,64,65], the rapid release of bioavailable metal in a short period could cause harmful transient plant phytotoxicity. In this regard, a good strategy for phytoremediation of Pb-contaminated sites could be the combined use of PGPRs and EDTA to simultaneously improve metal bioavailability in soil and plant growth under metal stress.

The stress caused in plants by the presence of heavy metals triggers complex physiological and molecular mechanisms. Among these, the production of radical exudates containing, e.g., low-molecular-weight organic acids (LMWOAs) that stimulate microbial growth of the rhizosphere and solubilize essential trace elements such as insoluble or poorly soluble (e.g., phosphorus, iron and zinc), can complex some metals such as arsenic, cadmium and lead [66].

Most metals are present in the soil in the form of insoluble and non-bioavailable salts. Chemical compounds such as EDTA, dipotassium phosphate or ammonium sulfate can separate metals from complexes bound to soil particles, favoring the absorption by the plants' roots.

PGPRs can increase the bioavailability of metals by producing microbial metabolites and siderophore molecules. Many studies show that adding chelating agents (such as EDTA) would further improve plant growth and metal uptake when combined with PGPR inoculation [67,68].

In this work, we have shown that the simple addition of the microbial inoculum (PGPR) led in *B. juncea* to a significant increase in the absorption of lead in the aerial part. A similar result was also shown in the work of He et al. [69], where the inoculation of two *Bacillus* strains, isolated from the soil, improved the rhizosphere soil environment promoting absorption of Pb by plants, enhancing the dry weight of shoots of plants growing in Pb-contaminated soil, and significantly increasing the total Pb content in aerial parts.

The possibility of avoiding the addition of chemical compounds to increase the bioavailability of the metals is undoubtedly a fascinating aspect that significantly increases the sustainability of phytoremediation.

New experiments with different plant species and endophytes isolated from metal contaminated soils are underway.

5. Conclusions

Heavy metal contamination is an important aspect of environmental pollution, and phytoextraction is one of the most effective technologies in removing many metal species, including radionuclides.

In these contamination cases, the selection of remediation technology should be directed towards the application of phytoremediation if it is to be sustainable.

The interaction between plants and microorganisms creates a fascinating ecosystem that can lead to encouraging results. As highlighted in this study, the microbial inoculation of endophytic bacteria isolated from the roots of spontaneous plants grown on the Pb-contaminated soil allowed a significant percentage of metal phytoextraction to be obtained without chemical mobilizers.

Case studies showing positive results of assisted phytoextraction only with the support of PGPRs are limited, and research in this field must undoubtedly be supported and encouraged.

Author Contributions: Conceptualization, M.G., E.F. and G.P.; methodology, M.G., F.P., M.B., E.F. and G.P.; validation, M.G., F.P., M.B., I.P., E.F. and G.P.; formal analysis, M.G., F.P., M.B. and I.P.; investigation, E.F., M.G., F.P., M.B., I.P. and G.P.; resources, I.P., M.G., F.P., M.B. and A.C.; data curation, E.F., I.P. and M.V.; writing—original draft preparation, E.F., G.P., M.G. and M.V.; writing—review and editing, M.V., G.P., I.P. and E.F.; visualization, M.V.; supervision, M.V.; project administration, G.P.; funding acquisition, E.F. All authors have read and agreed to the published version of the manuscript.

Funding: This research was supported by Eni S.p.A, Research & Technological Innovation Department, San Donato Milanese (Italy) and fully funded by Syndial S.p.A. (now Eni Rewind S.p.A.), agreement number 3500038857.

Institutional Review Board Statement: Not applicable.

Informed Consent Statement: Not applicable.

Data Availability Statement: Data is contained within the present article.

Acknowledgments: The Authors thank Irene Rosellini, IRET-CNR, for technical assistance in Phytoremediation experiments.

Conflicts of Interest: The authors declare no conflict of interest.

References

1. Farhadian, M.; Vachelard, C.; Duchez, D.; Larroche, C. In situ bioremediation of monoaromatic pollutants in groundwater: A review. *Biores. Technol.* **2008**, *99*, 5296–5308.
2. Vocciante, M.; Finocchi, A.; De Folly D'Auris, A.; Conte, A.; Tonziello, J.; Pola, A.; Reverberi, A. Enhanced oil spill remediation by adsorption with interlinked multilayered graphene. *Materials* **2019**, *12*, 2231. [CrossRef]
3. Pietrelli, L.; Francolini, I.; Piozzi, A.; Sighicelli, M.; Vocciante, M. Chromium (III) Removal from Wastewater by Chitosan Flakes. *Appl. Sci.* **2020**, *10*, 1925. [CrossRef]
4. Khan, F.I.; Husain, T.; Hejazi, R. An overview and analysis of site remediation technologies. *J. Environ. Manag.* **2004**, *71*, 95–122.
5. Pavel, L.V.; Gavrilescu, M. Overview of ex situ decontamination techniques for soil cleanup. *Environ. Eng. Manag. J.* **2008**, *7*, 815–834.
6. Vocciante, M.; Reverberi, A.P.; Pietrelli, L.; Dovì, V.G. Improved Remediation Processes through Cost-effective Estimation of Soil Properties from Surface Measurements. *J. Clean. Prod.* **2017**, *167*, 680–686.
7. da Silva, B.M.; Maranho, L.T. Petroleum-contaminated sites: Decision framework for selecting remediation technologies. *J. Hazard. Mat.* **2019**, *378*, 120722.
8. Vocciante, M.; Dovì, V.G.; Ferro, S. Sustainability in ElectroKinetic Remediation Processes: A Critical Analysis. *Sustainability* **2021**, *13*, 770. [CrossRef]
9. Song, Y.; Kirkwood, N.; Maksimović, Č.; Zhen, X.; O'Connor, D.; Jin, Y.; Hou, D. Nature based solutions for contaminated land remediation and brownfield redevelopment in cities: A review. *Sci. Total Environ.* **2019**, *663*, 568–579. [CrossRef]
10. Grifoni, M.; Pedron, F.; Barbafieri, M.; Rosellini, I.; Petruzzelli, G.; Franchi, E. Sustainable Valorization of Biomass: From Assisted Phytoremediation to Green Energy Production. In *Handbook on Assisted and Amendments Enhanced Sustainable Remediation Technology*; Prasad, M.N.V., Ed.; John Wiley & Sons: Hoboken, NJ, USA, 2020.
11. Hou, D.; Bolan, N.S.; Tsang, D.C.W.; Kirkham, M.B.; O'Connor, D. Sustainable soil use and management: An interdisciplinary and systematic approach. *Sci. Total Environ.* **2020**, *729*, 138961. [CrossRef]
12. O'Connor, D.; Zheng, X.; Hou, D.; Shen, Z.; Li, G.; Miao, G.; O'Connell, S.; Guo, M. Phytoremediation: Climate change resilience and sustainability assessment at a coastal brownfield redevelopment. *Environ. Int.* **2019**, *130*, 104945. [CrossRef]
13. Vocciante, M.; Caretta, A.; Bua, L.; Bagatin, R.; Franchi, E.; Petruzzelli, G.; Ferro, S. Enhancements in phytoremediation technology: Environmental assessment including different options of biomass disposal and comparison with a consolidated approach. *J. Environ. Manag.* **2019**, *237*, 560–568. [CrossRef]
14. Vocciante, M.; De Follis D'Auris, A.; Franchi, E.; Petruzzelli, G.; Ferro, S. CO_2 footprint analysis of consolidated and innovative technologies in remediation activities. *J. Clean. Prod.* **2021**, 126723. [CrossRef]
15. Pedron, F.; Grifoni, M.; Barbafieri, M.; Petruzzelli, G.; Franchi, E.; Samà, C.; Gila, L.; Zanardi, S.; Palmery, S.; Proto, A.; et al. New Light on Phytoremediation: The Use of Luminescent Solar Concentrators. *Appl. Sci.* **2021**, *11*, 1923. [CrossRef]
16. Ashraf, S.S.; Ali, Q.; Zahir, Z.A.; Ashraf, S.S.; Asghar, H.N. Phytoremediation: Environmentally sustainable way for reclamation of heavy metal polluted soils. *Ecotoxicol. Environ. Saf.* **2019**, *174*, 714–727. [CrossRef]
17. Shah, V.; Daverey, A. Phytoremediation: A multidisciplinary approach to clean up heavy metal contaminated soil. *Environ. Technol. Innov.* **2020**, *18*, 100774. [CrossRef]
18. Sheoran, V.; Singh Sheoran, A.; Poonia, P. Factors affecting phytoextraction: A review. *Pedosphere* **2016**, *26*, 148–166. [CrossRef]
19. Evangelou, M.W.H.; Ebel, M.; Schaeffer, A. Chelate assisted phytoextraction of heavy metals from soil. Effect, mechanism, toxicity, and fate of chelating agents. *Chemosphere* **2007**, *68*, 989–1003. [CrossRef]

20. Pedron, F.; Grifoni, M.; Barbafieri, M.; Petruzzelli, G.; Rosellini, I.; Franchi, E.; Bagatin, R.; Vocciante, M. Applicability of a Freundlich-like model for plant uptake at an industrial contaminated site with a high variable arsenic concentration. *Environments* **2017**, *4*, 67. [CrossRef]
21. Petruzzelli, G.; Grifoni, M.; Barbafieri, M.; Rosellini, I.; Pedron, F. Sorption: Release processes in soil-the basis of phytoremediation efficiency. In *Phytoremediation*; Ansari, A.A., Gill, S.S., Gill, R., Lanza, G.R., Newman, L., Eds.; Springer: Cham, Switzerland, 2018; pp. 91–112. [CrossRef]
22. Manoj, S.R.; Karthik, C.; Kadirvelu, K.; Arulselvi, P.I.; Shanmugasundaram, T.; Bruno, B.; Rajkumar, M. Understanding the molecular mechanisms for the enhanced phytoremediation of heavy metals through plant growth promoting rhizobacteria: A review. *J. Environ. Manag.* **2020**, *254*, 109779. [CrossRef]
23. Franchi, E.; Cosmina, P.; Pedron, F.; Rosellini, I.; Barbafieri, M.; Petruzzelli, G.; Vocciante, M. Improved arsenic phytoextraction by combined use of mobilizing chemicals and autochthonous soil bacteria. *Sci. Total Environ.* **2019**, *655*, 328–336. [CrossRef]
24. Abbaszadeh-Dahaji, P.; Omidvari, M.; Ghorbanpour, M. Increasing phytoremediation efficiency of heavy metal-contaminated soil using PGPR for sustainable agriculture. In *Plant-Microbe Interaction: An Approach to Sustainable Agriculture*; Choudhary, D.K., Varma, A., Narendra, T., Eds.; Springer: Singapore, 2017; pp. 187–204. [CrossRef]
25. Legislative Decree 152/2006. Rules in environmental field. In *Italian Official Journal*; National Legislation: Italy, 2006.
26. Sparks, D.L.; Page, A.L.; Helmke, P.A.; Loeppert, R.H. *Methods of Soil Analysis Part 3—Chemical Methods*; Soil Science Society of America, American Society of Agronomy: Madison, WI, USA, 1996. [CrossRef]
27. Gee, G.W.; Bauder, J.W. Particle-size analysis. In *Methods of Soil Analysis: Part 1d—Physical and Mineralogical Methods*; Klute, A., Ed.; Soil Science Society of America, American Society of Agronomy: Madison, MI, USA, 1986; pp. 383–411.
28. EPA—United State Environmental Protection Agency. Method 3051A Microwave assisted acid digestion of sediments, sludges, soils, and oils. In *Test Methods for Evaluating Solid Waste: Physical/Chemical Methods (SW-846)*; U.S. Environmental Protection Agency: Washington, DC, USA, 1995.
29. Grifoni, M.; Petruzzelli, G.; Barbafieri, M.; Rosellini, I.; Pedron, F. Soil quality protection at heavy metal-contaminated manufactured gas plant sites: Role of biological remediation. In *Enhancing Cleanup of Environmental Pollutants*; Anjum, N., Gill, S., Tuteja, N., Eds.; Springer: Cham, Switzerland, 2017; pp. 231–260. [CrossRef]
30. Pedron, F.; Petruzzelli, G.; Barbafieri, M.; Tassi, E. Strategies to use phytoextraction in very acidic soil contaminated by heavy metals. *Chemosphere* **2009**, *75*, 808–814. [CrossRef]
31. Barbafieri, M.; Pedron, F.; Petruzzelli, G.; Rosellini, I.; Franchi, E.; Bagatin, R.; Vocciante, M. Assisted phytoremediation of a multi-contaminated soil: Investigation on arsenic and lead combined mobilization and removal. *J. Environ. Manag.* **2017**, *203*, 316–329. [CrossRef]
32. Jez, E.; Lestan, D. EDTA retention and emissions from remediated soil. *Chemosphere* **2016**, *151*, 202–209. [CrossRef]
33. Kanwal, U.; Ali, S.; Shakoor, M.B.; Farid, M.; Hussain, S.; Yasmeen, T.; Adrees, M.; Bharwana, S.A.; Abbas, F. EDTA ameliorates phytoextraction of lead and plant growth by reducing morphological and biochemical injuries in *Brassica napus* L. under lead stress. *Environ. Sci. Pollut. Res.* **2014**, *21*, 9899–9910. [CrossRef]
34. Li, F.L.; Qiu, Y.; Xu, X.; Yang, F.; Wang, Z.; Feng, J.; Wang, J. EDTA-enhanced phytoremediation of heavy metals from sludge soil by Italian ryegrass (*Lolium perenne* L.). *Ecotox. Environ. Safe.* **2020**, *191*, 110185. [CrossRef]
35. Grifoni, M.; Rosellini, I.; Petruzzelli, G.; Pedron, F.; Franchi, E.; Barbafieri, M. Application of sulphate and cytokinin in assisted arsenic phytoextraction by industrial *Cannabis sativa* L. *Environ. Sci. Pollut. Res.* **2021**, *28*, 47294–47305. [CrossRef]
36. Tassi, E.; Pouget, J.; Petruzzelli, G.; Barbafieri, M. The effects of exogenous plant growth regulators in the phytoextraction of heavy metals. *Chemosphere* **2008**, *71*, 66–73. [CrossRef]
37. EPA—United State Environmental Protection Agency. Method 3052 Microwave assisted acid digestion of siliceous and organically based matrices. In *Test Methods for Evaluating Soild Waste: Physical/Chemical Methods (SW-846)*; U.S. Environmental Protection Agency: Washington, DC, USA, 1995.
38. ISO—International Organization for Standardization. *ISO 18763:2016 Soil Quality–Determination of the Toxic Effects of Pollutants on Germination and Early Growth of Higher Plants*; International Organization for Standardization: Geneva, Switzerland, 2016.
39. Shahab, S.; Ahmed, N.; Khan, N.S. Indole acetic acid production and enhanced plant growth promotion by indigenous PSBs. *J. Plant Pathol.* **2009**, *91*, 61–63.
40. Milagres, A.M.F.; Machuca, A.; Napoleão, D. Detection of siderophore production from several fungi and bacteria by a modification of chrome azurol S (CAS) agar plate assay. *J. Microbiol. Methods* **1999**, *37*, 1–6.
41. Nautiyal, C.S. An efficient microbiological growth medium for screening phosphate solubilizing microorganisms. *FEMS Microbiol. Lett.* **1999**, *170*, 265–270. [CrossRef]
42. Santaella, C.; Schue, M.; Berge, O.; Heulin, T.; Achouak, W. The exopolysaccharide of Rhizobium sp. YAS34 is not necessary for biofilm formation on Arabidopsis thaliana and Brassica napus roots but contributes to root colonization. *Environ. Microbiol.* **2008**, *10*, 2150–2163. [CrossRef]
43. Nielsen, P.; Sørensen, J. Multi-target and medium-independent fungal antagonism by hydrolytic enzymes in *Paenibacillus polymyxa* and *Bacillus pumilus* strains from barley rhizosphere. *FEMS Microbiol. Ecol.* **1997**, *22*, 183–192. [CrossRef]
44. Kifle, M.H.; Laing, M.D. Isolation and screening of bacteria for their diazotrophic potential and their influence on growth promotion of maize seedlings in greenhouses. *Front. Plant Sci.* **2015**, *6*, 1225. [CrossRef]

45. Liba, C.M.; Ferrara, F.I.S.; Manfio, G.P.; Fantinatti-Garboggini, F.; Albuquerque, R.C.; Pavan, C.; Ramos, P.L.; Moreira-Filho, C.A.; Barbosa, H.R. Nitrogen-fixing chemo-organotrophic bacteria isolated from Cyanobacteria-deprived lichens and their ability to solubilize phosphate and to release amino acids and phytohormones. *J. Appl. Microbiol.* **2006**, *101*, 1076–1086. [CrossRef]

46. Conte, A.; Chiaberge, S.; Pedron, F.; Barbafieri, M.; Petruzzelli, G.; Vocciante, M.; Franchi, E.; Pietrini, I. Dealing with complex contamination: A novel approach with a combined bio-phytoremediation strategy and effective analytical techniques. *J. Environ. Manag.* **2021**, *288*, 112381. [CrossRef]

47. Saifullah; Meers, E.; Qadir, M.; de Caritat, P.; Tack, F.M.; Du Laing, G.; Zia, M.H. EDTA-assisted Pb phytoextraction. *Chemosphere* **2009**, *74*, 1279–1291. [CrossRef]

48. Xu, Y.; Yamaji, N.; Shen, R.; Ma, J.F. Sorghum roots are inefficient in uptake of EDTA-chelated lead. *Ann. Bot.* **2007**, *99*, 869–875. [CrossRef]

49. Li, Q.; Huang, Y.; Wen, D.; Fu, R.; Feng, L. Application of alkyl polyglycosides for enhanced bioremediation of petroleum hydrocarbon-contaminated soil using *Sphingomonas changbaiensis* and *Pseudomonas stutzeri*. *Sci. Total Environ.* **2020**, *719*, 137456. [CrossRef]

50. Rathi, M.; Yogalakshmi, K.N. Brevundimonas diminuta MYS6 associated *Helianthus annuus* L. for enhanced copper phytoremediation. *Chemosphere* **2021**, *263*, 128195. [CrossRef] [PubMed]

51. Nazli, F.; Mustafa, A.; Ahmad, M.; Hussain, A.; Jamil, M.; Wang, X.; Shakeel, Q.; Imtiaz, M.; El-Esawi, M.A. A Review on Practical Application and Potentials of Phytohormone-Producing Plant Growth-Promoting Rhizobacteria for Inducing Heavy Metal Tolerance in Crops. *Sustainability* **2020**, *12*, 9056. [CrossRef]

52. Barbafieri, M.; Morelli, E.; Tassi, E.; Pedron, F.; Remorini, D.; Petruzzelli, G. Overcoming limitation of "recalcitrant areas" to phytoextraction process: The synergistic effects of exogenous cytokinins and nitrogen treatments. *Sci. Total Environ.* **2018**, *639*, 1520–1529. [CrossRef]

53. Alaboudi, K.A.; Ahmed, B.; Brodie, G. Phytoremediation of Pb and Cd contaminated soils by using sunflower (*Helianthus annuus*) plant. *Ann. Agric. Sci.* **2018**, *63*, 123–127. [CrossRef]

54. Bassegio, C.; Campagnolo, M.A.; Schwantes, D.; Gonçalves Junior, A.C.; Manfrin, J.; Schiller, A.D.P.; Bassegio, D. Growth and accumulation of Pb by roots and shoots of *Brassica juncea* L. *Int. J. Phytoremediat.* **2020**, *22*, 134–139. [CrossRef]

55. Chauhan, P.; Rajguru, A.B.; Dudhe, M.Y.; Mathur, J. Efficacy of lead (Pb) phytoextraction of five varieties of *Helianthus annuus* L. from contaminated soil. *Environ. Technol. Innov.* **2020**, *18*, 100718. [CrossRef]

56. Gurajala, H.K.; Cao, X.; Tang, L.; Ramesh, T.M.; Lu, M.; Yang, X. Comparative assessment of Indian mustard (*Brassica juncea* L.) genotypes for phytoremediation of Cd and Pb contaminated soils. *Environ. Pollut.* **2019**, *254*, 113085. [CrossRef]

57. Rathika, R.; Srinivasan, P.; Alkahtani, J.; Al-Humaid, L.A.; Alwahibi, M.S.; Mythili, R.; Selvankumar, T. Influence of biochar and EDTA on enhanced phytoremediation of lead contaminated soil by *Brassica juncea*. *Chemosphere* **2021**, *271*, 129513. [CrossRef] [PubMed]

58. Ahmed, S.; Hassan, M.H.; Kalam, M.A.; Ashrafur Rahman, S.M.; Abedin, M.J.; Shahir, A. An experimental investigation of biodiesel production, characterization, engine performance, emission and noise of *Brassica juncea* methyl ester and its blends. *J. Clean. Prod.* **2014**, *79*, 74–81. [CrossRef]

59. Hunce, S.Y.; Clemente, R.; Bernal, M.P. Energy production potential of phytoremediation plant biomass: Helianthus annuus and Silybum marianum. *Ind. Crops Prod.* **2019**, *135*, 206–216. [CrossRef]

60. Iram, S.; Tariq, I.; Ahmad, K.S.; Jaffri, S.B. *Helianthus annuus* based biodiesel production from seed oil garnered from a phytoremediated terrain. *Int. J. Ambient Energy* **2020**. [CrossRef]

61. Pant, S.; Ritika; Komesu, A.; Penteado, E.D.; Diniz, A.A.R.; Rahman, M.A.; Kuila, A. NaOH pretreatment and enzymatic hydrolysis of Brassica juncea using mixture of cellulases. *Environ. Technol. Innov.* **2021**, *21*, 101324. [CrossRef]

62. Luo, J.; Qi, S.; Gu, X.W.S.; Wang, J.; Xie, X. An evaluation of EDTA additions for improving the phytoremediation efficiency of different plants under various cultivation systems. *Ecotoxicology* **2016**, *25*, 646–654. [CrossRef] [PubMed]

63. Mahmood-ul-Hassan, M.; Suthar, V.; Ahmad, R.; Yousra, M. Heavy metal phytoextraction—natural and EDTA-assisted remediation of contaminated calcareous soils by sorghum and oat. *Environ. Monit. Assess.* **2017**, *189*, 1–10. [CrossRef]

64. Saleem, M.H.; Ali, S.; Kamran, M.; Iqbal, N.; Azeem, M.; Tariq Javed, M.; Ali, Q.; Zulqurnain Haider, M.; Irshad, S.; Rizwan, M.; et al. Ethylenediaminetetraacetic Acid (EDTA) Mitigates the Toxic Effect of Excessive Copper Concentrations on Growth, Gaseous Exchange and Chloroplast Ultrastructure of *Corchorus capsularis* L. and Improves Copper Accumulation Capabilities. *Plants* **2020**, *9*, 756. [CrossRef]

65. Sharma, P.; Pandey, A.K.; Udayan, A.; Kumar, S. Role of microbial community and metal-binding proteins in phytoremediation of heavy metals from industrial wastewater. *Bioresour. Technol.* **2021**, *326*, 124750. [CrossRef]

66. Ma, Y.; Rajkumar, M.; Zhang, C.; Freitas, H. Beneficial role of bacterial endophytes in heavy metal phytoremediation. *J. Environ. Manag.* **2016**, *174*, 14–25. [CrossRef]

67. Franchi, E.; Fusini, D. Plant Growth-Promoting Rhizobacteria (PGPR) Assisted Phytoremediation of Inorganic and Organic Contaminants Including Amelioration of Perturbed Marginal Soils. In *Handbook of Assisted and Amendment-Enhanced Sustainable Remediation Technology*; Prasad, M.N.V., Ed.; John Wiley & Sons Ltd.: Hoboken, NJ, USA, 2021; pp. 477–500.

68. Ullah, A.; Heng, S.; Munis, M.F.H.; Fahad, S.; Yang, X. Phytoremediation of heavy metals assisted by plant growth promoting (PGP) bacteria: A review. *Environ. Exp. Bot.* **2015**, *117*, 28–40. [CrossRef]

69. He, X.; Xu, M.; Wei, Q.; Tang, M.; Guan, L.; Lou, L.; Xu, X.; Hu, Z.; Chen, Y.; Shen, Z.; et al. Promotion of growth and phytoextraction of cadmium and lead in *Solanum nigrum* L. mediated by plant-growth-promoting rhizobacteria. *Ecotoxicol. Environ. Saf.* **2020**, *205*, 111333. [CrossRef] [PubMed]

soil systems

Article

Effect of Municipal Solid Waste Compost on Antimony Mobility, Phytotoxicity and Bioavailability in Polluted Soils

Stefania Diquattro, Giovanni Garau *, Matteo Garau, Gian Paolo Lauro, Maria Vittoria Pinna and Paola Castaldi

Dipartimento di Agraria, University of Sassari, Viale Italia 39, 07100 Sassari, Italy; sdiquattro@uniss.it (S.D.); matteo_gp@libero.it (M.G.); gplauro@uniss.it (G.P.L.); mavi@uniss.it (M.V.P.); castaldi@uniss.it (P.C.)
* Correspondence: ggarau@uniss.it; Tel.: +39-079229210

Abstract: The effect of a municipal solid waste compost (MSWC), added at 1 and 2% rates, on the mobility, phytotoxicity, and bioavailability of antimony (Sb) was investigated in two soils (SA: acidic soil; SB: alkaline soil), spiked with two Sb concentrations (100 and 1000 mg kg^{-1}). The impact of MSWC on microbial activity and biochemical functioning within the Sb-polluted soils was also considered. MSWC addition reduced water-soluble Sb and favored an increase in residual Sb (e.g., by 1.45- and 1.14-fold in SA-100 and SA-1000 treated with 2% MSWC, respectively). Significant increases in dehydrogenase activity were recorded in both the amended soils, as well as a clear positive effect of MSWC on the metabolic activity and catabolic diversity of respective microbial communities. MSWC alleviated Sb phytotoxicity in triticale plants and decreased Sb uptake by roots. However, increased Sb translocation from roots to shoots was recorded in the amended soils, according to the compost rate. Overall, the results obtained indicated that MSWC, particularly at a 2% rate, can be used for the recovery of Sb-polluted soils. It also emerged that using MSWC in combination with triticale plants can be an option for the remediation of Sb-polluted soils, by means of assisted phytoextraction.

Keywords: potentially toxic element; gentle remediation options; organic amendments; Sb uptake; microbial activity

Citation: Diquattro, S.; Garau, G.; Garau, M.; Lauro, G.P.; Pinna, M.V.; Castaldi, P. Effect of Municipal Solid Waste Compost on Antimony Mobility, Phytotoxicity and Bioavailability in Polluted Soils. *Soil Syst.* **2021**, *5*, 60. https://doi.org/10.3390/soilsystems5040060

Academic Editor: Jorge Paz-Ferreiro

Received: 9 August 2021
Accepted: 24 September 2021
Published: 1 October 2021

Publisher's Note: MDPI stays neutral with regard to jurisdictional claims in published maps and institutional affiliations.

Copyright: © 2021 by the authors. Licensee MDPI, Basel, Switzerland. This article is an open access article distributed under the terms and conditions of the Creative Commons Attribution (CC BY) license (https://creativecommons.org/licenses/by/4.0/).

1. Introduction

Antimony (Sb) is an environmentally relevant potentially toxic element (PTE), usually combined, in alloys, with metals such as lead and zinc [1]. Sb, generally detected as a trace element in the Earth's crust (0.2–0.3 g per metric ton) and water (less than 1 μg L^{-1}) [2], reached worrying levels of contamination in different world areas in the last decades [3]. This has been mainly due to anthropogenic activities, such as mining and the processing of Sb-containing ores [4]. High Sb levels in soils can also be due to vehicular traffic and recreational shooting [5–7]. Moreover, Sb is extensively used as a flame retardant in plastics, and as catalyst in the production of polyesters fibers [8]. As a consequence of its ubiquity and toxicity (Sb is recognized as carcinogenic and clastogenic agent), Sb is included in the list of high-priority pollutants by the U.S. Environmental Protection Agency and the European Union [9,10]. In most natural systems, Sb mainly occurs as reduced (trivalent Sb(III)) or oxidized (pentavalent Sb(V)) inorganic species [11]. Inorganic compounds of Sb are more toxic than its organic species, and Sb(III) compounds are predicted to be 10-fold more toxic than Sb(V) [1]. However, as emphasized by Filella et al. [12], this cannot be generalized, since toxicity depends on many factors (e.g., the target organism, the route of exposure, and the presence of other pollutants). Sb oxidation state, its reactivity, and its potential bioavailability in soil are largely dependent on the pH, redox conditions, and concentrations of co-occurring reducing and oxidizing agents [13]. Sb(V) is the prevalent form in aerobic conditions, and mainly occurs as octahedral antimonate ion, Sb(OH)$_6{}^-$. This latter form is considered the most stable form, compared to the thermodynamically unstable Sb(III), which mainly occurs in anoxic conditions, as Sb(OH)$_3$ [1,12,14]. On the other hand, Sb(V) mobility is greater than Sb(III) in the 5.0–8.5 pH range, due to its negative charge (pKa HSb(OH)$_6$ = 2.55; pKa Sb(OH)$_3$ = 11.8) [1,15]. Sb in soil is mostly associated with Fe, Mn and Al (hydr)oxides and organic matter [16,17], and their occurrence and abundance in soil seems to control Sb mobility and bioavailability, e.g., [18–20].

As a PTE, Sb can affect soil microbial communities and their functionality, as well as plant growth [21,22]. Diquattro et al. [21] showed a significant Sb impact on the composition and general catabolic activity of soil microbial communities, while Yu et al. [23] recorded reduced carbon mineralization and nitrification in the presence of Sb. Moreover, in situ Sb immobilization significantly stimulated the growth of *Helichrysum italicum* [22], indirectly showing a certain impact of this PTE on plant growth, which, however, was also confirmed in other studies, e.g., [24–26].

In the search for eco-friendly materials to restore Sb-contaminated soils, and/or to limit Sb ecotoxicological impact, many organic and inorganic amendments have been studied. For instance, Garau et al. [22] observed that the addition of a municipal solid waste compost (MSWC), together with an iron(Fe)-rich water treatment residue (WTR), favored the chemical and biological recovery of a subalkaline soil that was contaminated with Sb (~110 mg kg^{-1} soil). Wang et al. [27] showed that ferrous Fe and nitrate promoted the formation of Fe plaques in rice, decreasing Sb bioavailability. Teng et al. [28] showed that Fe-modified rice husk hydrochar was effective at immobilizing Sb in soil, while the same was shown by Almas et al. [29], using Fe-rich slag in combination with FeSO$_4$.

Despite the existence of substantial literature on effective amendments that are able to immobilize Sb in soil, very limited and often conflicting information can be found on the impact of compost on Sb mobility, and its toxicity and availability for plants. Nakamaru and Peinado [30] observed an increase in Sb availability in contaminated soils that were amended with compost. Verbeeck et al. [31] observed that the complexation capacity of dissolved organic matter could increase Sb mobility under different redox conditions. Lewinska et al. [32] showed that a long incubation time with compost (i.e., 140 d) did not always lead to Sb immobilization in different shooting range soils. Finally, Clemente [33] showed that adding greenwaste compost mulch to PTE-contaminated soil increased Sb mobilization. By contrast, Abou Jaoude et al. [34] showed that compost addition to Sb-contaminated sub-alkaline soils reduced water-soluble and exchangeable Sb, and increased its residual (non-extractable) fractions. Likewise, Garau et al. [35] showed that compost reduced the concentration of labile Sb in soil. Taken together, the results from these studies highlight the need for further investigations, aimed at clarifying the impact of compost on Sb-contaminated soils, and its potential use in the remediation of polluted environments. In particular, in view of a remediation intervention, it would be useful to establish the effects of compost on soil microbial communities, enzyme activities, plant growth, and Sb uptake in Sb-contaminated soils, other than establishing its impact on Sb mobility. Accordingly, the aim of the present study was to evaluate the effect of an MSWC, added at two different rates, on Sb mobility, phytotoxicity, and bioavailability in two polluted soils, as well as its impact on soil microbial activity, catabolic diversity, and enzyme activity.

2. Materials and Methods

2.1. Soil and MSWC Origin, Characteristics and Mesocosms Set Up

Different soil samples were randomly collected (0–30 cm depth) from two soils, named SA (soil A) and SB (soil B), located in North-Western Sardinia (Italy) (soil SA: 40°56′15.7″ N 8°53′30.4″ E; soil SB: 40°43′32.77″ N 8°24′48.6″ E). These soils were selected because they had never been cultivated (they were not treated with agrochemicals), were not close to Sb-contaminated areas, did not contain Sb (i.e., Sb < 1 mg kg^{-1}), and exhibited different physical–chemical properties, which were previously reported in detail (Table S1; [21]). In the laboratory, soil samples were air dried, sieved to <2 mm and pooled together according to their origin (SA and SB). SA was a loamy coarse sand while SB was a sandy clay loam soil (USDA classification).

SA and SB soils were then spiked with Sb(V) (i.e., KSb(OH)$_6$; CAS 12208-13-8; Merck) to obtain soils with medium–low (100 mg kg^{-1}; S-100) or high (1000 mg kg^{-1}; S-1000)

Sb(V) pollution level as previously reported [21]. Spiked soils were kept at 40% of their water-holding capacity (WHC) using deionized water and equilibrated for nine months at 25 °C. During this period, soils were periodically mixed once a week and maintained at constant humidity. Afterwards, triplicate mesocosms from each soil type (SA and SB) and contamination level (S-100 and S-1000) were treated as follows: T0—polluted untreated soil; T1—polluted soil amended with 1% MSWC; T2—polluted soil amended with 2% MSWC. A total of thirty-six mesocosms (each consisting of 5 kg soil; 2 soil types × 2 contamination levels × 3 amendment treatments × 3 reps) were prepared using SA (n = 18) and SB (n = 18) soils. The MSWC rates were selected based on the specific Sb-immobilizing capabilities of compost highlighted in previous studies [17,22]. The MSWC was provided by Verde Vita Srl (Sassari, Italy) and was sieved to <2 mm before addition to mesocosm soils. The main characteristics of the MSWC were previously reported [17] and resumed in Table S2. Briefly, the MSWC had sub-alkaline pH (i.e., 7.93) and 27.3% organic matter content (OM); it had high cation exchange capacity (CEC, 92.3 $cmol_{(+)}$ kg^{-1}), dissolved organic carbon (DOC, 0.82 mg kg^{-1}) and humic acids content (14.2%). After amendment with MSWC, soils were carefully mixed, brought to 40% of their WHC and equilibrated for three months at 25 °C. During this period, soils were periodically mixed once a week and maintained at constant humidity.

2.2. Soil Characterization and Sb Mobility in Treated and Untreated Soils

After the equilibration period, selected physico-chemical properties were determined in treated and untreated polluted soils (Table 1). The pH, electric conductivity (EC), total organic carbon (TOC) and nitrogen (TN) were determined for treated and untreated soils following the national standard guidelines [36]. The DOC content was estimated as previously described by Brandstetter et al. [37]. The available phosphorus was quantified following the Olsen method (P Olsen) and the CEC was determined using the $BaCl_2$ and triethanolamine method [36].

Table 1. Chemical characteristics of Sb-polluted SA and SB soils amended or not (control) with MSWC (dry matter basis). SA/SB-100 and SA/SB-1000 denote soil type and pollution level, i.e., 100 and 1000 mg Sb kg^{-1} soil. For each soil type and Sb pollution level, different letters (e.g., a, b, c) denote statistical differences (Tukey–Kramer, $p < 0.05$) between treatments.

	pH	EC (mS m^{-1})	DOC (mg g^{-1})	CEC ($cmol_{(+)} \cdot kg^{-1}$)	P Olsen (mg kg^{-1})	TOC (g kg^{-1})	TN (g kg^{-1})
SA-100 control	5.14 ± 0.59 [b]	77.7 ± 0.78 [c]	0.81 ± 0.01 [c]	10.6 ± 0.35 [b]	30.5 ± 0.33 [b]	13.8 ± 0.31 [c]	0.96 ± 0.12 [b]
SA-100 + 1% MSWC	6.16 ± 0.09 [a]	102 ± 0.78 [b]	1.09 ± 0.03 [b]	15.2 ± 0.63 [a]	36.2 ± 0.58 [a]	17.6 ± 0.03 [b]	1.36 ± 0.00 [a]
SA-100 + 2% MSWC	6.65 ± 0.26 [a]	116 ± 2.26 [a]	1.17 ± 0.02 [a]	16.0 ± 0.06 [a]	35.9 ± 0.56 [a]	18.8 ± 0.56 [a]	1.57 ± 0.14 [a]
SA-1000 control	5.53 ± 0.71 [b]	71.8 ± 5.30 [b]	0.82 ± 0.02 [c]	11.8 ± 0.01 [b]	30.3 ± 0.39 [b]	13.5 ± 0.09 [c]	1.12 ± 0.04 [b]
SA-1000 + 1% MSWC	6.46 ± 0.06 [a]	102 ± 1.91 [a]	1.00 ± 0.03 [b]	13.3 ± 0.52 [a]	34.4 ± 0.62 [a]	16.8 ± 0.34 [b]	1.24 ± 0.01 [a,b]
SA-1000 + 2% MSWC	6.98 ± 0.16 [a]	108 ± 2.83 [a]	1.09 ± 0.01 [a]	14.2 ± 0.75 [a]	33.4 ± 0.22 [a]	18.5 ± 0.52 [a]	1.30 ± 0.10 [a]
SB-100 control	7.84 ± 0.05 [a]	57.8 ± 1.48 [b]	0.29 ± 0.00 [c]	22.4 ± 0.86 [b]	10.1 ± 0.43 [c]	12.7 ± 0.20 [c]	1.06 ± 0.02 [c]
SB-100 + 1% MSWC	7.93 ± 0.01 [a]	67.7 ± 3.82 [b]	0.40 ± 0.01 [b]	24.0 ± 0.15 [a,b]	15.9 ± 0.46 [b]	15.0 ± 0.20 [b]	1.13 ± 0.01 [b]
SB-100 + 2% MSWC	7.93 ± 0.00 [a]	88.9 ± 2.97 [a]	0.69 ± 0.02 [a]	26.2 ± 1.05 [a]	19.4 ± 0.68 [a]	16.6 ± 0.05 [a]	1.19 ± 0.01 [a]
SB-1000 control	7.92 ± 0.04 [a]	63.0 ± 0.07 [c]	0.30 ± 0.00 [b]	23.2 ± 1.83 [b]	9.73 ± 0.43 [c]	13.0 ± 0.16 [c]	1.02 ± 0.00 [b]
SB-1000 + 1% MSWC	7.94 ± 0.08 [a]	70.7 ± 2.40 [b]	0.50 ± 0.01 [a]	25.2 ± 0.58 [a,b]	15.6 ± 0.59 [b]	15.3 ± 0.19 [b]	1.21 ± 0.01 [a]
SB-1000 + 2% MSWC	7.97 ± 0.13 [a]	91.3 ± 0.85 [a]	0.51 ± 0.01 [a]	27.2 ± 1.10 [a]	20.3 ± 0.11 [a]	16.9 ± 0.22 [a]	1.22 ± 0.03 [a]

Total Sb concentration was quantified in all soils after digestion with aqua regia reverse solution (HNO_3/HCl 3:1 ratio) and microwave mineralization (Milestone MLS1200), using graphite furnace atomic absorption spectroscopy (GFAAS; PerkinElmer AAnalyst 400-HGA 900, Software AA-WinLab32). A standard reference material (NIST-SRM 2711A) was included for quality assurance and quality control. The Sb mobility in polluted (amended and not) SA and SB soils, i.e., its chemical reactivity with the soil matrix, was evaluated through the sequential extraction procedure proposed by Wenzel et al. [38] with minor modifications. Briefly, triplicate soil samples (1 g each) from each mesocosm were first treated with water (25 mL) to estimate water-soluble Sb (step 0, this was the only additional step with respect to the original procedure), then they were treated with 25 mL of a 0.05 M $(NH_4)_2SO_4$ solution to quantify the readily exchangeable Sb fraction (step 1) and 25 mL of a 0.05 M $(NH_4)H_2PO_4$ solution (step 2) to estimate surface-bound Sb, while Sb associated to amorphous and crystalline Al- and Fe-(hydr)oxides was quantified after extraction with 25 mL of a 0.2 M NH_4- oxalate solution at pH 3.25 (step 3) and with 25 mL of a 0.2 M NH_4-oxalate + 0.1 M ascorbic acid solution at pH 3.25 (step 4), respectively. After each step, soil samples were centrifuged at 3500 rpm for 10 min, filtered using Whatman filter No. 42 to separate the liquid and solid phases, and Sb concentration in the supernatant was quantified using GFAAS as previously described. A standard reference material (NIST-SRM 2711A) was included for quality assurance and quality control.

2.3. Biolog Community-Level Physiological Profiles and Soil Enzyme Activities in Treated and Untreated Sb-Polluted Soils

The activity and catabolic diversity of microbial communities inhabiting the different Sb-polluted soils (amended or not) was investigated using the Biolog community-level physiological profile (CLPP) approach using Biolog EcoPlates™ (Biolog Inc., Hayward, CA, USA). In particular, triplicate soil samples (10 g) from each mesocosm were added with (90 mL) sodium pyrophosphate solution (2 g L^{-1}) and serial ten-fold dilutions were obtained using 0.89% NaCl solution. The obtained soil suspensions were centrifuged for 5 min at 500 rpm and the clear supernatant was used to inoculate the wells of the Biolog EcoPlate™ (120 µL per well). The Biolog EcoPlates™ are ready to use 96-well microtiter plates containing a different carbon source of soil/environmental relevance in each well [39]. A total of 31 different carbon sources and a control well with no carbon (all replicated three times) were present in each Biolog EcoPlate™ [40]. Inoculated plates were incubated at 28 °C for 6 days (144 h) and purple color formation in each well, due to the reduction of a tetrazolium dye and indicative of C-source catabolism, was recorded daily by measuring the absorbance at 590 nm (OD590), using an automatic Biolog MicroStation™ reader. All OD590 data were first blanked against the absorbance at time 0 and further subtracted from the respective control well (with no carbon source). Finally, they were processed to obtain a measurement of the potential catabolic activity of the microbial community, i.e., the average well color development (AWCD). The latter was calculated as follows: AWCD = $[\Sigma (R_i - C)]/31$ where C represents the absorbance value of control well, R_i is the absorbance of each response well, and 31 is the number of carbon substrates in the plate [41]. The richness value, or the number of substrates catabolized by each microbial community, was also determined as the number of wells with OD590 >0.15 [42]. Also, the Shannon–Weaver diversity index (H') was used to estimate the catabolic diversity of microbial communities and calculated as follows:

H = $-\sum(P_i \times \ln P_i)$, where P_i is the ratio between the absorbance value in the blank subtracted ith well (1 to 31) and the total absorbance values of all wells [43].

All Biolog-derived parameters (AWCD, richness and H') refer to the 120 h incubation time as this time point provided the best discrimination between treatments.

Enzymatic activities, such as dehydrogenase (DHG), β-glucosidase (GLU) and urease (URE), were determined colorimetrically in triplicate soil samples collected from each mesocosm as described by Alef and Nannipieri [44]. Briefly, the DHG activity was spectrophotometrically quantified (OD480 nm) after incubation of soil samples (10 g) with a triphenyltetrazolium chloride solution and expressed as triphenyl formazan (TPF) formed per g soil (dry weight basis). GLU activity was determined spectrophotometrically (OD400 nm) after incubation of soil samples (1 g) with p-nitrophenyl glucoside and expressed as p-nitrophenol released per g soil (dry weight basis). Finally, URE activity was determined spectrophotometrically (OD690 nm) after incubation of soil samples (5 g) with urea and expressed as ammonia released per g soil (dry weight basis).

2.4. Sb Phytotoxicity and Bioavailability in Treated and Untreated Sb-Polluted Soils

The influence of MSWC on Sb phytotoxicity was determined using triticale plants ($\times$ *Triticosecale* Wittm. cv. Universal) grown in treated and untreated Sb-polluted soils.

This plant species was selected since in previous studies its growth was significantly affected by the presence of PTE in the growing medium [45–48]. A total of thirty-six pots each containing 1.5 kg of soil deriving from the different mesocosms were set up, i.e., 3 replicated pots × 3 amendment treatments (T0, T1, T2) × 2 Sb contamination levels (100 and 1000 mg kg^{-1}) × 1 plant species × 2 soil types (SA and SB). Seven triticale plants were planted in each pot after their germination in the dark at 25 °C. Planted pots were arranged in a completely randomized design and plants were grown over 8 weeks in a naturally lit greenhouse under controlled conditions (20–25 °C temperature, 60–70% relative humidity). At harvest, shoots were separated from roots, carefully washed with deionized water and dried at 55 °C for 72 h. Plant growth, i.e., root and shoot dry weight values, was used to estimate soil Sb phytotoxicity.

Sb bioavailability, i.e., the Sb uptaken by triticale plants, was determined after mineralization of roots and shoots with 2 mL suprapure H_2O_2 and 9 mL of HNO_3 and ultrapure H_2O (ratio 1:1), using a Microwave Milestone MLS 1200. The total Sb concentration in the mineralization solutions was determined using GFAAS as previously reported. Peach leaves were used as standard reference material (NIST-SRM 1515). The Sb translocation factor (TF_{Sb}) was calculated as the ratio between Sb concentration in shoots and that present in roots.

2.5. Statistical Analysis

All chemical analyses were performed in triplicate soil samples collected from each mesocosm and mean values ± standard errors were reported in tables and figures. For each soil type (SA or SB) and contamination level (100 or 1000 mg·kg^{-1}), the MSWC influence (T0, T1, T2) on Sb mobility, soil enzyme activities, Biolog-derived parameters, plant growth, and Sb uptake data was ascertained by one-way analysis of variance followed by a post hoc Tukey–Kramer test ($p < 0.05$). Moreover, differences between the above-mentioned parameters in SA-100 and SB-100, and SA-1000 and SB-1000, were assessed by a Student t-test ($p < 0.05$). Pearson correlation between the most labile Sb fraction (i.e., that extracted in step 0 of SEP) and biochemical, root dry weight, root length and Sb uptake (by roots) data was also determined. In this regard, the entire dataset from S-100 soils was considered since SA and SB soils showed comparable concentrations of labile Sb, while data from SA-1000 and SB-1000 were separately analyzed since labile Sb concentration in these two soils was very different (Figure 1). In all tests, differences were considered statistically significant at $p < 0.05$. All statistical analyses were carried out using the NCSS software (released 1 June 2011).

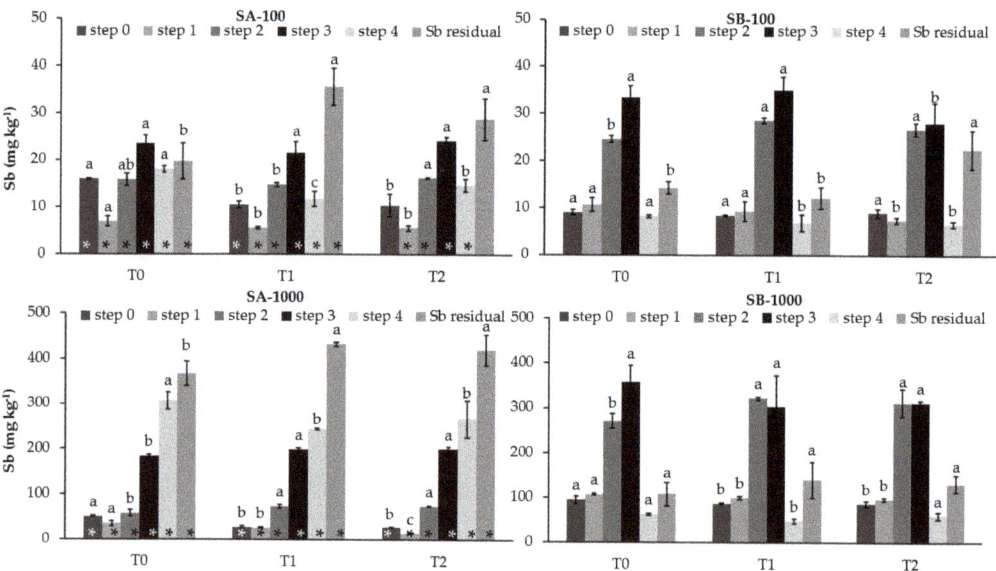

Figure 1. Sb extracted in amended and unamended Sb-polluted SA and SB soils in different SEP steps. Average values (histograms) and standard errors (bars) are reported. T0, T1 and T2 refer to MSWC addition, i.e., 0, 1 and 2%, respectively. Step 0: water-soluble Sb; step 1: exchangeable Sb; step 2: inner-sphere complexed Sb; step 3: Sb bound to amorphous Fe/Al (hydr)oxides; step 4: Sb bound to crystalline Fe/Al (hydr)oxides. For each soil and extraction step, different letters denote significant differences between treatments (Tukey–Kramer test; $p < 0.05$). For each extraction step, asterisks denote significant differences between SA-100 and SB-100, and between SA-1000 and SB-1000 (Student t-test; $p < 0.05$).

3. Results and Discussion

3.1. Influence of MSWC on Selected Chemical Properties and Sb Mobility in Polluted Soils

The treatment with MSWC caused a significant increase in pH in SA, and a significant increase in TOC, DOC, CEC, P Olsen, TN, and EC in both soils (Table 1). These results are in agreement with those reported by several researchers [49–52], and confirm the overall suitability of MSWC in improving different parameters related to soil fertility. This is particularly relevant in PTE-contaminated soils, which are often characterized by poor physico-chemical characteristics, and in which plant growth can be severely limited [53].

Sb mobility in soil after MSWC treatment was essentially evaluated through the sequential extraction procedure of Wenzel et al. [38], with minor modifications. In particular, an additional step (i.e., step 0) was added, to evaluate water-soluble Sb in soil. The water-soluble Sb (step 0) released from S-100-polluted soils was higher in SA than SB (~16 and 9% of total Sb, respectively), and was in accordance with previous findings [21], while an opposite trend was observed for S-1000 (Figure 1). The addition of MSWC did not change the water-soluble Sb fraction of SB-100, while a significant reduction was observed in SA-100 and in both S-1000 soils (i.e., −34, −9, and −47% in SA-100, SB-1000, and SA-1000, amended with 2% MSWC, respectively). These decreases suggested the occurrence of stable interactions between MSWC and Sb, which led to a reduction in water-soluble Sb. This is expected to have substantial positive implications from an ecotoxicological viewpoint, since water-soluble PTE represent the most biologically impacting fraction of contaminants (e.g., [34,54,55]). It is also important to underline that the concentration of water-soluble Sb in both soils was very high, even after MSWC addition (between 9.0 and 94.6 mg kg^{-1}), and it may represent a serious environmental hazard for soil organisms and other environmental compartments, e.g., surface and groundwater.

This is even more relevant considering that the Sb concentration threshold for Italian agricultural soils, which refers to total soil Sb, is 10 mg kg^{-1} [9].

A different trend was observed for Sb released with $(NH_4)_2SO_4$ (step 1, i.e., the relatively labile and exchangeable fraction), which was lower in SA than SB (e.g., ~3.5 and 9.6% of total Sb in SA-1000 and SB-1000, respectively). A limited, yet significant, reduction in Sb, extracted in step 1, was observed in all the amended soils, especially when 2% MSWC was applied (Figure 1). Overall, these findings highlighted a poor presence of a relatively labile and easily exchangeable Sb fraction in soils (note that sulphate can only exchange anions forming weak electrostatic bonds), and implied the prevalence of specific (inner-sphere) binding between Sb and soil/MSWC components. This was, in general, confirmed by the Sb concentrations extracted in step 2 (but also steps 3–4), which quantified the specifically adsorbed Sb; in this case, MSWC addition significantly increased such Sb fraction (e.g., by 15 and 28% in SB-1000 and SA-1000, amended with 2% MSWC, compared to the respective controls) (Figure 1). These results may have important environmental implications, since the specifically adsorbed Sb can be mobilized (becoming potentially bioavailable), as a result of a change in pH (e.g., due to plant and/or microbial activity) or phosphate increase [38].

Most of the soil Sb (~35 and 40% in SB and SA, respectively) was specifically associated with amorphous and crystalline Al- and/or Fe-(hydr)oxides (steps 3 and 4, respectively; Figure 1). After MSWC addition, the Sb fraction extracted with step 3 did not vary in SA-100 and SB-1000, significantly increased in SA-1000, and slightly decreased in SB-100, while the Sb fraction associated to crystalline Al- and Fe-(hydr)oxides significantly decreased in most soils (step 4; Figure 1). The Sb released in steps 3–4 is expected to have a limited impact on soil ecotoxicity, as it mainly represents the Sb involved in stable inner-sphere bonding with Fe- and/or Al-(hydr)oxides surfaces [38].

The residual Sb fraction was higher in treated and untreated SA soils (especially in SA-1000) than respective SB ones. This could be ascribed to the lower pH of untreated SA soil (i.e., 5.14–5.53), which could have promoted the formation of stable precipitates between Sb(V) and soluble Al or Fe, e.g., $FeSbO_4$ and $AlSbO_4$ [1,21,56–58]. The addition of MSWC favored an increase in such residual Sb in SB-100 (1.58-fold in 2% MSWC-amended soil), and in SA-100 and SA-1000 (1.46- and 1.14-fold, respectively, in 2% MSWC-amended soils). Overall, this is relevant, as residual Sb represents the very insoluble and/or occluded contaminant fraction, which can hardly be mobilized, and it is therefore expected to have a negligible impact on soil biota, at least in the medium term [17,22].

Taken together, these results show a substantial stabilizing/immobilizing effect of MSWC towards Sb, particularly in acidic soil (SA). This may have occurred by means of different processes, such as the formation of stable complexes between labile Sb and compost solid phases, such as ternary complexation, in which polyvalent metal cations (e.g., Fe, Al, and Ca) can act as bridges between the negatively charged functional groups of MSWC and the antimony oxyacid [18]. The precipitation of Sb, above all, with soluble Ca (abundantly present within MSWC; 6.3%, Table S2) may also have contributed to the increase in residual Sb in the treated soils. Additionally, the formation of mono- and di-ester bonds between $Sb(OH)_6^-$ and the hydroxyl functional groups of humic acids within MSWC could have also contributed to Sb immobilization in the amended soils [17].

In this context, the reduction in most labile Sb fractions in treated soils (steps 0–1) is expected to have relevant positive effects on soil microbial communities and their functionality, as well as plant growth. This has been proved for other PTE-contaminated soils (which, however, did not contain Sb) treated with MSWC [16,59].

3.2. Influence of MSWC on Soil Enzyme Activities in Sb-Polluted Soils

It has been shown that Sb can adversely affect microbial growth and inhibit the activity of soil enzymes, such as dehydrogenase, urease, arylsulfatase, and β-glucosidase [21,23]. Therefore, monitoring soil enzyme activities can be helpful in the assessment of the ability of MSWC to alleviate Sb microbial ecotoxicity, and to restore the biological activity and functionality of polluted soils [22]. DHG activity can provide a good estimate of the overall oxidative capacity of a soil and, at the same time, it has often been used as an indicator of soil microbial abundance and/or activity [59–62]. Overall, DHG activity was higher in SB soils, and in soils contaminated with the lower Sb rate (i.e., 100 mg kg^{-1}) (Figure 2). This likely reflected the different size (and possibly also the different community structure) of microbial communities inhabiting SA and SB soils, which were characterized by very different chemical properties (Table 1). For instance, the higher pH in SB soils (Table 1) is likely responsible for the larger bacterial populations and higher DHG values (compared to SA), as previously reported [21]. Moreover, the lower DHG values in both Sb-1000 soils (compared to the respective Sb-100 ones) clearly show a higher negative impact of Sb on soil microbial abundance and/or activity in these soils. Compost addition significantly increased DHG activity in the contaminated soils, and this increase was proportional to the rate of MSWC added (e.g., 3.50- and 4.26-fold higher in SB-1000 amended with MSWC at 1 and 2%, respectively, compared to the unamended control) (Figure 2). This could be due to the lower amount of labile Sb in the amended soils (i.e., Sb extracted in steps 0–1), and to the consequently reduced environmental pressure on soil microbial communities, as previously shown with other PTEs [63,64]. The negative and significant correlation between the most labile Sb (i.e., the water-soluble one) and soil DHG supports this view (Table 2). On the other hand, the improved physico-chemical properties of the soils after compost addition could have contributed to the higher DHG values in the amended soils, as pointed out in previous studies [53].

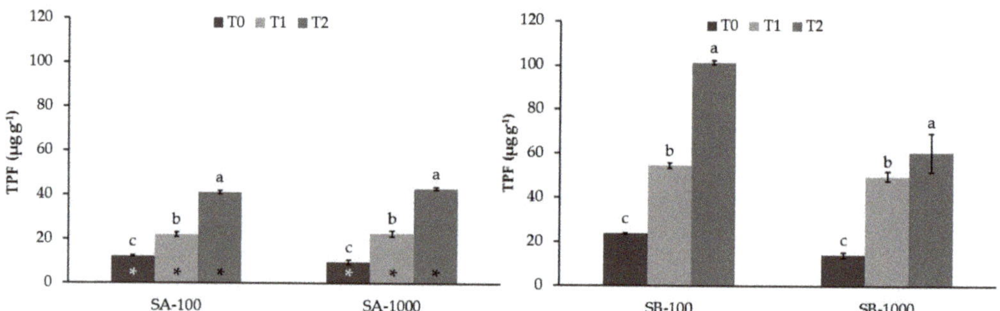

Figure 2. Dehydrogenase activity in amended and unamended Sb-polluted SA and SB soils. Average values (histograms) and standard errors (bars) are reported. T0, T1 and T2 refer to MSWC addition, i.e., 0, 1 and 2%, respectively. For each soil (SA and SB) and Sb concentration level, different letters denote significant differences between treatments (Tukey–Kramer test; $p < 0.05$). Asterisks denote significant differences between SA-100 and SB-100, and between SA-1000 and SB-1000 (Student t-test; $p < 0.05$).

Table 2. Pearson correlations between the most labile Sb fraction (i.e., Sb extracted in step 0 of SEP), soil biochemical parameters, root dry weight, root length and Sb uptake by the roots of triticale plants.

	Root Dry Weight	Root Length	Sb Uptake$_{root}$	DHG	URE	GLU	AWCD	H′	Richness
S-100	−0.62 **	−0.72 ***	0.94 ***	−0.55 *	−0.61 **	−0.76 ***	−0.85 ***	−0.92 ***	−0.80 ***
S1-1000	−0.85 **	−0.79 *	0.72 *	−0.85 **	−0.6 NS	−0.43 NS	−0.67 *	−0.88 **	−0.78 *
S2-1000	−0.74 *	−0.67 *	−0.92 ***	−0.80 **	−0.69 *	−0.98 ***	−0.73 *	−0.91 ***	−0.90 ***

* $p < 0.05$; ** $p < 0.01$; *** $p < 0.001$; NS: not significant ($p > 0.05$).

A different trend was observed for URE activity, which can provide useful information on the rate of urea hydrolysis in soil (i.e., a specific step of N cycling), and was often used as a proxy of environmental stressing conditions [65–67]. URE activity increased in both soils treated with 1% MSWC, compared to the other treatments (e.g., URE was 1.9- and 2.9-fold higher in SB-100 amended with 1% MSWC, compared with the control and 2% MSWC, respectively; Figure 3) and, as for DHG, the activity decreased at higher concentrations of added Sb.

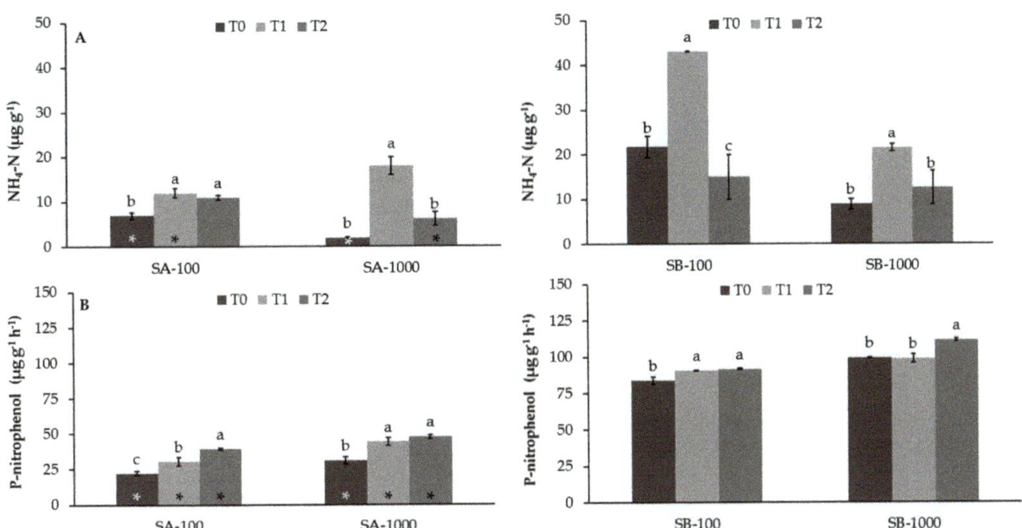

Figure 3. Urease (**A**) and β-glucosidase (**B**) activity in amended and unamended Sb-polluted SA and SB soils. Average values (histograms) and standard errors (bars) are reported. T0, T1 and T2 refer to MSWC addition, i.e., 0, 1 and 2%, respectively. For each soil (SA and SB) and Sb concentration level, different letters denote significant differences between treatments (Tukey–Kramer test; $p < 0.05$). For each enzyme activity, asterisks denote significant differences between SA-100 and SB-100, and between SA-1000 and SB-1000 (Student t-test; $p < 0.05$).

The lower URE activity recorded in soils amended with 2% MSWC (often very similar to that of respective controls) seems to be in contrast with the lower amounts of labile Sb (steps 0–1; Figure 1) recorded in such amended soils. This was reflected by the Pearson correlation values between URE and the most labile Sb fraction, which were not statistically significant for all the soils (Table 2). Previous studies showed that ureases activity can be inhibited by the binding of humic substances to the thiol group of urease [68]. This could explain the reduced URE in soils amended with the highest compost rate. Moreover, the inorganic N present in MSWC could also be involved in the inhibition of urease synthesis, which could justify the reduced URE activity in soils treated with the highest MSWC amount [69]. GLU activity, which is due to extracellular enzymes that cleave the β 1→4 bonds linking two glucose or glucose-substituted molecules, was higher in SB than SA, which is in line with what was observed for URE and DHG. The addition of MSWC (with the exception of SB-1000 amended at a 1% rate) favored a significant increase in this activity (e.g., ~1.09- and 1.77-fold higher in SB- and SA-100 amended with MSWC, respectively), and this could likely be due to an increase in the labile C pool in the amended soils [65], and/or to a decrease in labile Sb. However, as for URE, a consistent correlation between the most labile Sb and GLU was not found in all the soils (Table 2). Furthermore, it should be noted that in soils spiked with the highest Sb amount (i.e., 1000 mg kg^{-1}), GLU activities were higher, with respect to S-100 soils. This apparent stimulating effect of Sb is difficult to explain, since the relatively few studies in the literature reported that β-1,4-glucosidase activity was negatively correlated with total and bioavailable Sb [23]. Probably, as also noted by Diquattro et al. [21], the proliferation in Sb-1000 soils of Sb-resistant microbial communities, able to synthesize β-1,4-glucosidases, could explain our findings.

3.3. Influence of MSWC on Soil Microbial Activity and Catabolic Diversity in Sb-Polluted Soils

As mentioned elsewhere, ideal amendments used for environmental remediation purposes should improve soil biological and/or biochemical attributes, other than reducing the labile concentration of contaminants [53]. In this context, the role of compost in influencing the potential metabolic activity and catabolic diversity of the microbial communities of Sb-polluted soils was investigated by means of Biolog CLPP. Such an approach, which essentially detects differences in C-source utilization (if any) by microbial communities, was revealed to be particularly useful at evaluating the impact of organic and inorganic treatments on soil microbial consortia (e.g., [34,70–73]). Our results indicated a clear impact of MSWC on the potential metabolic activity of the microbial communities of amended soils, i.e., significant increases in AWCD were recorded in both S-100- and S-1000-amended soils (Figure 4). The most striking increases were recorded in SA-100- and SA-1000-amended soils, where the AWCD was 31- and 14-fold higher compared to that of respective untreated soils. Moreover, irrespective of the treatment applied, the AWCD data supported the higher metabolic activity in SB soils (compared to SA), as also indicated by DHG activity (Figure 2); essentially the same trend was detected for the richness and H' index values (Figure 4). Microbial communities of MSWC-amended soils were able to catabolize a significantly higher number of carbon sources (richness; up to ~14 and 5 in SB and SA soils, respectively), compared to those of untreated soils, while the catabolic diversity (Shannon–Weaver H' values) increased up to 3- and 30-fold in SB-1000- and SA-1000-amended soils, respectively (Figure 4).

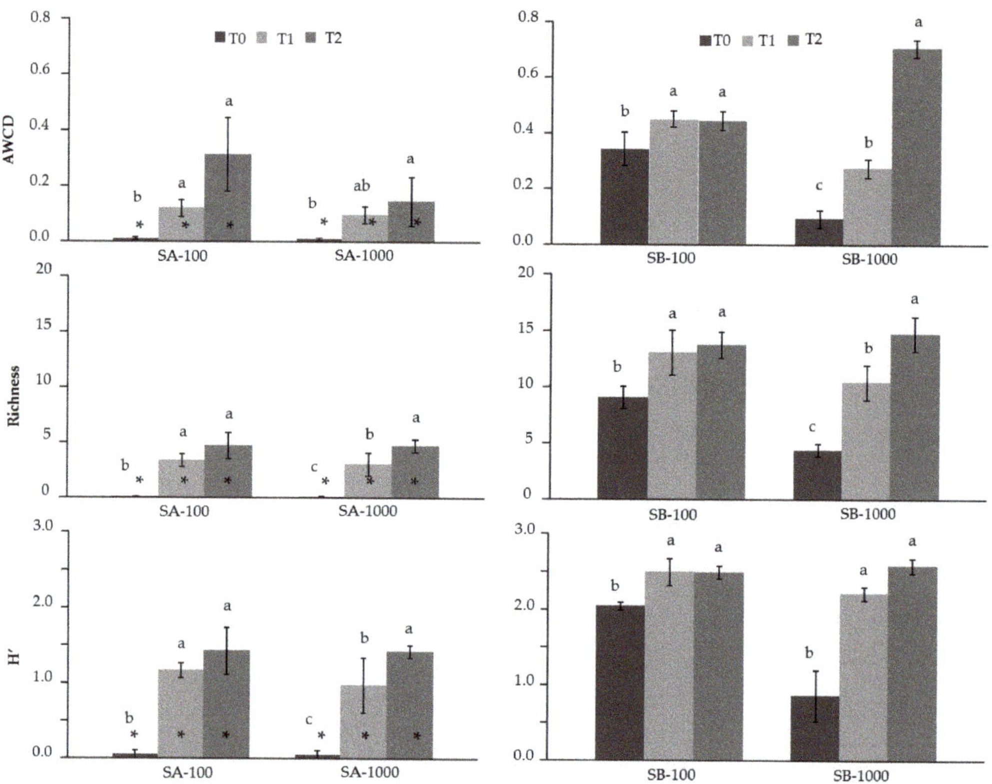

Figure 4. Biolog AWCD, richness and H' (Shannon–Weaver) index in amended and unamended Sb-polluted SA and SB soils. Average values (histograms) and standard errors (bars) relative to the 120 h incubation time are reported. T0, T1 and T2 refer to MSWC addition, i.e., 0, 1 and 2%, respectively. For each soil (SA and SB) and Sb concentration level, different letters denote significant differences between treatments (Tukey–Kramer test; $p < 0.05$). For each Biolog-derived parameter, asterisks denote significant differences between SA-100 and SB-100, and between SA-1000 and SB-1000 (Student t-test; $p < 0.05$).

Looking at the consumption of different C guilds in SB soils, the addition of MSWC greatly increased the utilization of sugar and sugar derivates by soil microbial communities, while that of amino acids substantially reduced (Figure 5). On the other hand, in SA soils, the use of the different C guilds was negligible, while MSWC addition led to significant catabolic recovery, which appeared to be proportional to the rate of MSWC addition, and was in agreement with the DHG trend (Figures 2 and 6). Taken together, Biolog EcoPlate™ data indicated a clear improvement of metabolic potentials in amended soils, which can be explained by different factors, such as the observed reduction in labile Sb in these soils (steps 0–1; Figure 1). This latter reduction likely alleviated the environmental stress posed by labile Sb towards microbial communities, leading to higher microbial abundance and/or activity, as previously reported [74]. Moreover, the reduction in labile Sb in amended soils could have favored the multiplication of Sb-sensitive bacterial strains (whose presence was negligible in the polluted untreated soils), having new catabolic capabilities, as highlighted by the increased richness and H' values. Similar findings were reported by Garau et al. [59], which showed a two-fold increase in bacterial species (and a concurrent increase in Biolog AWCD and richness) in a PTE-polluted soil amended with 4% MSWC. This view was also supported by the significant negative correlations between the most labile Sb fraction in soil and Biolog-derived parameters (Table 2).

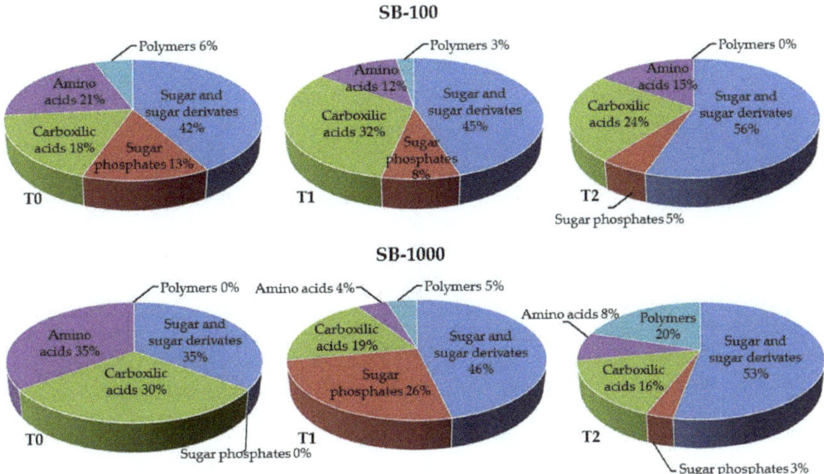

Figure 5. Use of different C-source guilds by microbial communities in amended and unamended Sb-polluted SB soils. Average percentage values relative to the 120 h incubation time are reported. T0, T1 and T2 refer to MSWC addition, i.e., 0, 1 and 2%, respectively. For overall metabolic activity of the different microbial communities see Figure 4.

However, the larger availability of organic C in amended soils, which was reflected by higher DOC values (Table 1), could have reasonably contributed to the observed increase in metabolic activity and catabolic versatility, e.g., by making new and more diverse C sources available to microbial communities and/or by means of a priming effect [75]. On the other hand, links between DOC content and bacterial abundance and/or activity were also previously reported, e.g., [59,76].

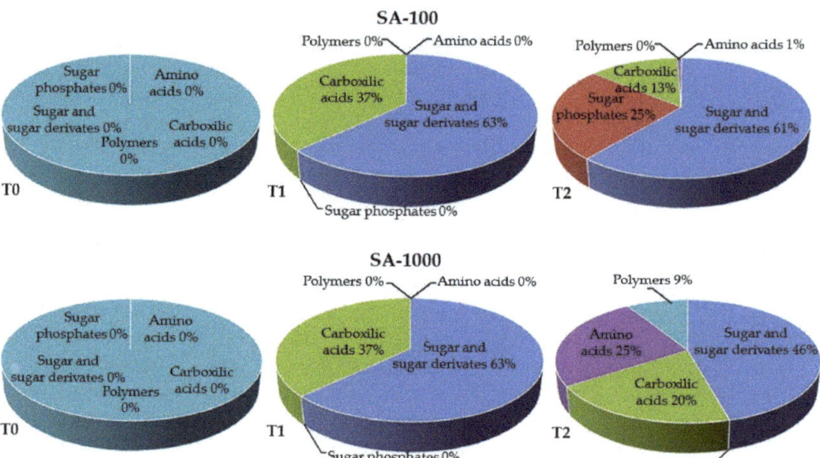

Figure 6. Use of different C-source guilds by microbial communities in amended and unamended Sb-polluted SA soils. Average percentage values relative to the 120 h incubation time are reported. T0, T1 and T2 refer to MSWC addition, i.e., 0, 1 and 2%, respectively. For overall metabolic activity of the different microbial communities see Figure 4.

3.4. Influence of MSWC on Sb Phytotoxicity and Bioavailability

The growth of triticale plants was assessed in Sb-polluted SA and SB soils, to estimate the role of MSWC in the alleviation of Sb phytotoxicity. This is an important point in the evaluation of the suitability of the amendment, since plant growth in treated soils should hopefully be increased, mainly because of PTE immobilization by the amendment and/or improved soil nutritional status. Both these factors largely contribute to reduce the phytotoxicity of contaminated soils [53]. The plant growth in the Sb-1000 soils was consistently higher than that recorded in the Sb-100 ones (Figure 7). Considering that Sb was added in the form of $KSb(OH)_6$, this was likely due to the higher availability of K in the former soils; that is why a direct comparison of plant growth in Sb-100 and Sb-1000 soils can be misleading, and was therefore avoided. Overall, plant growth was significantly stimulated in the presence of MSWC, especially when the highest rate was applied (Figure 7), suggesting reduced phytotoxicity. In particular, root dry weight (rather than shoot) was greatly increased in MSWC-amended soils, e.g., up to 24% in SB-100 and -1000, and up to 14 and 47% in SA-100 and -1000, respectively (Figure 7). More limited increases were also noted for the shoot dry weight of plants grown in SA-amended soils (Figure 7). As for soil biochemical features, these results can be explained by a reduction in labile (and potentially bioavailable) Sb concentrations in amended soils (Figure 1). This view was supported by the correlation analysis, which highlighted a negative and significant correlation in all the soils, between the most labile Sb fraction and root dry weight (Table 2).

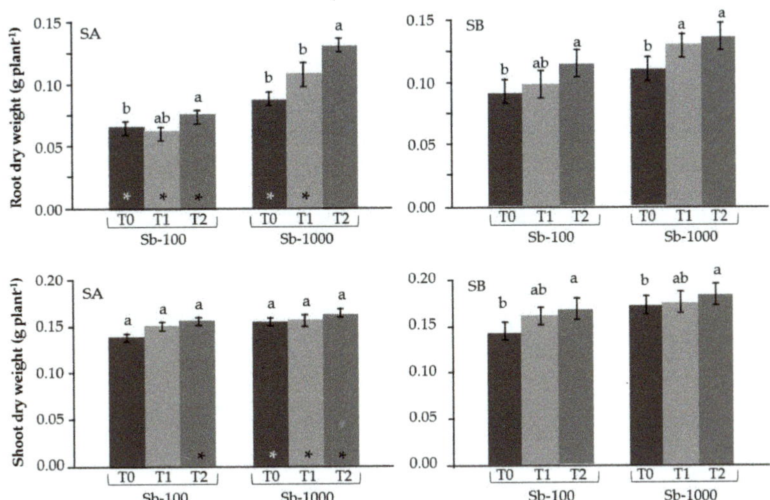

Figure 7. Root and shoot dry weight of triticale plants grown in amended and unamended Sb-polluted SA and SB soils. Average values (histograms) and standard errors (bars) are reported. T0, T1 and T2 refer to MSWC addition, i.e., 0, 1 and 2%, respectively. For each soil (SA and SB) and Sb concentration level, different letters denote significant differences between treatments (Tukey–Kramer test; $p < 0.05$). For each plant part, asterisks denote significant differences between SA-100 and SB-100, and between SA-1000 and SB-1000 (Student *t*-test; $p < 0.05$).

However, the improved nutritional status of amended soils (see, for instance, the higher DOC, TOC, P Olsen, and TN content of these latter soils; Table 1) could have contributed to the better plant growth, as also highlighted in previous studies, where *H. italicum* and maize seedlings were grown in combination with MSWC and biochar, respectively [22,77]. Regarding the impact of MSWC on Sb uptake (i.e., bioavailability) by triticale, the results showed different trends according to Sb contamination level. In soils spiked with the lower Sb concentration, MSWC addition consistently decreased Sb uptake by the roots and increased that of the shoots (Table 3). In SA-1000 soils, MSWC addition increased both Sb uptake by the shoots (especially) and roots, whereas in SB-1000, these were both reduced, even if not significantly for the shoot (Table 3). Except for SA-1000 soils, a positive and significant correlation was found between the most labile Sb and its uptake by roots (Table 2), as was also reported in other studies for different PTE (e.g., [16]). This explains the reduced Sb uptake by roots in amended soils (Table 3), where a significant reduction in labile and potentially bioavailable Sb was recorded (Figure 1; steps 0–1). However, this was not the case of SA-1000 soils, for which a significant (unexpected) negative correlation between labile Sb and its uptake by roots was found (Table 2). This could be explained by higher root activity in such soils, especially in amended ones, which likely changed the relative distribution of labile Sb in amended and unamended soils. As a result, labile Sb after plant growth could have been different from that recorded before plant growth (and that was used to calculate the Pearson correlations). In particular, a higher release of phytosiderophores and organic acids, by triticale roots, in the amended SA-1000 soils (e.g., to alleviate Fe deficiency, note that SA soil contained ~65% less Fe compared to SB soil; [21]) could be responsible for the enhanced Sb mobilization, through the (partial) dissolution of Sb binding Al and/or Fe minerals [78–81]. Reasonably, this would have also occurred in SA-100 soil, where, however, this could have had a negligible effect, due to much lower Sb fractions bound to Al and/or Fe minerals (i.e., Sb extracted in steps 3–4; Figure 1). Furthermore, a progressive increase in TF_{Sb} was recorded in amended soils, according (mainly) to the compost rate (Table 3). This suggests that the enhanced plant growth (and root activity) recorded in soils treated with MSWC (Table 3) could have possibly influenced Sb bioaccumulation by triticale, leading to a higher Sb concentration and increased translocation to shoots.

Table 3. Sb uptake, and translocation factor (TF_{Sb}), by triticale plants grown in amended and unamended Sb-polluted SA and SB soils. Average values are reported. T0, T1 and T2 refer to MSWC addition, i.e., 0, 1 and 2%, respectively. For each soil (SA and SB) and Sb concentration level, different letters (a, b and c) denote significant differences between treatments (Tukey–Kramer test; $p < 0.05$).

	SA-100			SA-1000			SB-100			SB-1000		
Sb Uptake	**T0**	**T1**	**T2**	**T0**	**T1**	**T2**	**T0**	**T1**	**T2**	**T0**	**T1**	**T2**
Shoot	1.61 [b]	1.42 [b]	2.54 [a]	3.36 [b]	5.61 [a]	6.31 [a]	0.53 [b]	0.92 [a]	0.91 [a]	12.38 [a]	10.30 [a]	9.66 [a]
Root	1.35 [a]	0.26 [b]	0.24 [b]	1.52 [b]	1.85 [a]	1.82 [a]	0.28 [a]	0.01 [b]	0.01 [b]	1.15 [a]	0.79 [b]	0.43 [c]
TF	1.2 [c]	5.5 [b]	11 [a]	2.2 [b]	3.0 [a,b]	3.5 [a]	1.9 [b]	92 [a]	91 [a]	11 [b]	13 [b]	22 [a]

Finally, it is interesting to note that triticale emerged for the first time as a phytoextracting species, with respect to Sb (i.e., $TF_{Sb} > 1$), and that MSWC could possibly be used in combination with this plant species for the remediation of Sb-polluted soils, by means of assisted phytoextraction approaches [82].

4. Conclusions

The mobility of Sb in contrasting soils, characterized by a medium–low or high pollution level, was reduced by MSWC addition and/or its residual fraction increased. Although the extent of these phenomena appeared to be dependent on the soil type and contamination level, MSWC amendment implied reduced Sb potential bioavailability, which was supported by increased soil biochemical functioning (DHG and GLU, especially) and the development of microbial communities with improved metabolic potentials and catabolic diversity. The addition of MSWC also alleviated Sb phytotoxicity towards triticale plants, and reduced, in most cases, their Sb uptake by roots. Overall, the MSWC impact on Sb-polluted soils was consistent with its use as an amendment for the recovery of such environments. The reduction in labile Sb concentration in treated soils was likely the key factor boosting soil microbial activity and diversity (and probably abundance). Reasonably, this contributed to the development of a rhizosphere environment, which favored plant growth in amended soils and stimulated Sb translocation from the roots to shoots. This is the first time that triticale has emerged as a potential species for Sb phytoextraction. In this regard, MSWC was revealed to be useful to enhance this capacity, and it appears a promising candidate, in combination with triticale, for assisted Sb (or possibly anionic PTE) phytoextraction programs. Despite a clear role of MSWC in the chemical and biochemical recovery of Sb-polluted soils, its long-lasting effect is currently unknown and should be investigated.

Supplementary Materials: The following are available online at https://www.mdpi.com/2571-8789/5/4/60/s1, Table S1: physico-chemical characteristics of SA and SB before Sb spiking; Table S2: chemical characteristics of the MSWC.

Author Contributions: Investigation, methodology, formal analysis, writing—original draft preparation, S.D.; conceptualization, methodology, resources, writing—reviewing and editing, G.G. and P.C.; data curation, software, M.G., M.V.P.; methodology, formal analysis, G.P.L. All authors have read and agreed to the published version of the manuscript.

Funding: This research was funded by the University of Sassari (Fondo di Ateneo per la Ricerca 2020).

Institutional Review Board Statement: Not applicable.

Informed Consent Statement: Not applicable.

Data Availability Statement: All data analyzed in this study are available from the corresponding author upon request.

Conflicts of Interest: The authors declare no conflict of interest.

References

1. Herath, I.; Vithanage, M.; Bundschuh, J. Antimony as a global dilemma: Geochemistry, mobility, fate and transport. *Environ. Pollut.* **2017**, *223*, 545–559. [CrossRef]
2. Pierart, A.; Shahid, M.; Sejalon-Delmas, N.; Dumat, C. Antimony bioavailability: Knowledge and research perspectives for sustainable agricultures. *J. Hazard. Mater.* **2015**, *289*, 219–234. [CrossRef]
3. Wang, X.Q.; He, M.C.; Xi, J.H.; Lu, X.F. Antimony distribution and mobility in rivers around the world's largest antimony mine of Xikuangshan, Hunan Province, China. *Microchem. J.* **2011**, *97*, 4–11. [CrossRef]
4. Li, X.D.; Thornton, I. Arsenic, Antimony and Bismuth in Soil and Pasture Herbage in Some Old Metalliferous Mining Areas in England. *Environ. Geochem. Health* **1993**, *15*, 135–144. [CrossRef]
5. Dietl, C.; Reifenhauser, W.; Peichl, L. Association of antimony with traffic-occurrence in airborne dust, deposition and accumulation in standardized grass cultures. *Sci. Total Environ.* **1997**, *205*, 235–244. [CrossRef]
6. Foldi, C.; Sauermann, S.; Dohrmann, R.; Mansfeldt, T. Traffic-related distribution of antimony in roadside soils. *Environ. Pollut.* **2018**, *237*, 704–712. [CrossRef]
7. Manta, D.S.; Angelone, M.; Bellanca, A.; Neri, R.; Sprovieri, M. Heavy metals in urban soils: A case study from the city of Palermo (Sicily), Italy. *Sci. Total Environ.* **2002**, *300*, 229–243. [CrossRef]
8. Biver, M.; Turner, A.; Filella, M. Antimony release from polyester textiles by artificial sweat solutions: A call for a standardized procedure. *Regul. Toxicol. Pharm.* **2021**, *119*, 104824. [CrossRef]
9. G.U. GU del 14 aprile 2006, Decreto Legislativo 3 aprile 2006, n. 152, Norme in Materia Ambientale. 2006. Available online: https://www.gazzettaufficiale.it/ (accessed on 30 September 2021).
10. United States Environmental Protection Agency (USEPA). *National Primary Drinking Water Regulations*; United States Environmental Protection Agency: Washington, DC, USA, 2009.
11. Filella, M.; Belzile, N.; Chen, Y.W. Antimony in the environment: A review focused on natural waters II. Relevant solution chemistry. *Earth Sci. Rev.* **2002**, *59*, 265–285. [CrossRef]
12. Filella, M.; Williams, P.A.; Belzile, N. Antimony in the environment: Knowns and unknowns. *Environ. Chem.* **2009**, *6*, 95–105. [CrossRef]
13. Filella, M.; Belzile, N.; Chen, Y.W. Antimony in the environment: A review focused on natural waters I. Occurrence. *Earth-Sci. Rev.* **2002**, *57*, 125–176. [CrossRef]
14. He, M.C.; Wang, N.N.; Long, X.J.; Zhang, C.J.; Ma, C.L.; Zhong, Q.Y.; Wang, A.H.; Wang, Y.; Pervaiz, A.; Shan, J. Antimony speciation in the environment: Recent advances in understanding the biogeochemical processes and ecological effects. *J. Environ. Sci.* **2019**, *75*, 14–39. [CrossRef]
15. Johnston, S.G.; Bennett, W.W.; Doriean, N.; Hockmann, K.; Karimian, N.; Burton, E.D. Antimony and arsenic speciation, redox-cycling and contrasting mobility in a mining-impacted river system. *Sci. Total Environ.* **2020**, *710*, 136354. [CrossRef]
16. Castaldi, P.; Silvetti, M.; Manzano, R.; Brundu, G.; Roggero, P.P.; Garau, G. Mutual effect of Phragmites australis, Arundo donax and immobilization agents on arsenic and trace metals phytostabilization in polluted soils. *Geoderma* **2018**, *314*, 63–72. [CrossRef]
17. Diquattro, S.; Garau, G.; Lauro, G.; Silvetti, M.; Deiana, S.; Castaldi, P. Municipal solid waste compost as a novel sorbent for antimony(V): Adsorption and release trials at acidic pH. *Environ. Sci. Pollut. Res.* **2018**, *25*, 5603–5615. [CrossRef] [PubMed]
18. Filella, M.; Williams, P.A. Antimony interactions with heterogeneous complexants in waters, sediments and soils: A review of binding data for homologous compounds. *Chem. Erde-Geochem.* **2012**, *72*, 49–65. [CrossRef]
19. Nakamaru, Y.M.; Altansuvd, J. Speciation and bioavailability of selenium and antimony in non-flooded and wetland soils: A review. *Chemosphere* **2014**, *111*, 366–371. [CrossRef] [PubMed]
20. Tighe, M.; Lockwood, P.; Wilson, S. Adsorption of antimony(v) by floodplain soils, amorphous iron(III) hydroxide and humic acid. *J. Environ. Monit.* **2005**, *7*, 1177–1185. [CrossRef]
21. Diquattro, S.; Garau, G.; Mangia, N.P.; Drigo, B.; Lombi, E.; Vasileiadis, S.; Castaldi, P. Mobility and potential bioavailability of antimony in contaminated soils: Short-term impact on microbial community and soil biochemical functioning. *Ecotoxicol. Environ. Saf.* **2020**, *196*, 110576. [CrossRef]
22. Garau, G.; Silvetti, M.; Vasileiadis, S.; Donner, E.; Diquattro, S.; Deiana, S.; Lombi, E.; Castaldi, P. Use of municipal solid wastes for chemical and microbiological recovery of soils contaminated with metal(loid)s. *Soil Biol. Biochem.* **2017**, *111*, 25–35. [CrossRef]
23. Yu, H.; Zheng, X.F.; Weng, W.L.; Yan, X.Z.; Chen, P.B.; Liu, X.Y.; Peng, T.; Zhong, Q.P.; Xu, K.; Wang, C.; et al. Synergistic effects of antimony and arsenic contaminations on bacterial, archaeal and fungal communities in the rhizosphere of Miscanthus sinensis: Insights for nitrification and carbon mineralization. *J. Hazard. Mater.* **2021**, *411*, 125094. [CrossRef] [PubMed]
24. Busby, R.R.; Barbato, R.A.; Jung, C.M.; Bednar, A.J.; Douglas, T.A.; Ringelberg, D.B.; Indest, K.J. Alaskan plants and their assembled rhizosphere communities vary in their responses to soil antimony. *Appl. Soil Ecol.* **2021**, *167*, 104031. [CrossRef]
25. Yadav, V.; Arif, N.; Kovac, J.; Singh, V.P.; Tripathi, D.K.; Chauhan, D.K.; Vaculik, M. Structural modifications of plant organs and tissues by metals and metalloids in the environment: A review. *Plant. Physiol. Biochem.* **2021**, *159*, 100–112. [CrossRef]
26. Zhu, Y.M.; Yang, J.G.; Wang, L.Z.; Lin, Z.T.; Dai, J.X.; Wang, R.J.; Yu, Y.S.; Liu, H.; Rensing, C.; Feng, R.W. Factors influencing the uptake and speciation transformation of antimony in the soil-plant system, and the redistribution and toxicity of antimony in plants. *Sci. Total Environ.* **2020**, *738*, 140232. [CrossRef]

27. Wang, X.Q.; Li, F.B.; Yuan, C.L.; Li, B.; Liu, T.X.; Liu, C.S.; Du, Y.H.; Liu, C.P. The translocation of antimony in soil-rice system with comparisons to arsenic: Alleviation of their accumulation in rice by simultaneous use of Fe(II) and NO3-. *Sci. Total Environ.* **2019**, *650*, 633–641. [CrossRef]

28. Teng, F.Y.; Zhang, Y.X.; Wang, D.Q.; Shen, M.C.; Hu, D.F. Iron-modified rice husk hydrochar and its immobilization effect for Pb and Sb in contaminated soil. *J. Hazard. Mater.* **2020**, *398*, 122977. [CrossRef]

29. Almas, A.R.; Pironin, E.; Okkenhaug, G. The partitioning of Sb in contaminated soils after being immobilization by Fe-based amendments is more dynamic compared to Pb. *Appl. Geochem.* **2019**, *108*, 104378. [CrossRef]

30. Nakamaru, Y.M.; Peinado, F.J.M. Effect of soil organic matter on antimony bioavailability after the remediation process. *Environ. Pollut.* **2017**, *228*, 425–432. [CrossRef]

31. Verbeeck, M.; Thiry, Y.; Smolders, E. Soil organic matter affects arsenic and antimony sorption in anaerobic soils. *Environ. Pollut.* **2020**, *257*, 113566. [CrossRef] [PubMed]

32. Lewinska, K.; Karczewska, A. A release of toxic elements from military shooting range soils as affected by pH and treatment with compost. *Geoderma* **2019**, *346*, 1–10. [CrossRef]

33. Clemente, R.; Hartley, W.; Riby, P.; Dickinson, N.M.; Lepp, N.W. Trace element mobility in a contaminated soil two years after field-amendment with a greenwaste compost mulch. *Environ. Pollut.* **2010**, *158*, 1644–1651. [CrossRef]

34. Abou Jaoude, L.; Castaldi, P.; Nassif, N.; Pinna, M.V.; Garau, G. Biochar and compost as gentle remediation options for the recovery of trace elements-contaminated soils. *Sci. Total Environ.* **2020**, *711*, 134511. [CrossRef]

35. Garau, M.; Garau, G.; Diquattro, S.; Roggero, P.P.; Castaldi, P. Mobility, bioaccessibility and toxicity of potentially toxic elements in a contaminated soil treated with municipal solid waste compost. *Ecotoxicol. Environ. Saf.* **2019**, *186*, 109766. [CrossRef]

36. G.U. GU del 21 ottobre 1999, Decreto Ministeriale 13 settembre 1999, Approvazione dei Metodi Ufficiali di Analisi Chimica del Suolo. 1999. Available online: https://www.gazzettaufficiale.it/ (accessed on 30 September 2021).

37. Brandstetter, A.; Sletten, R.S.; Mentler, A.; Wenzel, W.W. Estimating dissolved organic carbon in natural waters by UV absorbance (254 nm). *Z Pflanz Bodenkunde* **1996**, *159*, 605–607. [CrossRef]

38. Wenzel, W.W.; Kirchbaumer, N.; Prohaska, T.; Stingeder, G.; Lombi, E.; Adriano, D.C. Arsenic fractionation in soils using an improved sequential extraction procedure. *Anal. Chim. Acta* **2001**, *436*, 309–323. [CrossRef]

39. Insam, H. A New Set of Substrates Proposed for Community Characterization in Environmental Samples. In *Microbial Communities*; Insam, H., Rangger, A., Eds.; Springer: Berlin/Heidelberg, Germany, 1997.

40. Niklinska, M.; Chodak, M.; Laskowski, R. Pollution-induced community tolerance of microorganisms from forest soil organic layers polluted with Zn or Cu. *Appl. Soil Ecol.* **2006**, *32*, 265–272. [CrossRef]

41. Garland, J.L. Analytical approaches to the characterization of samples of microbial communities using patterns of potential C source utilization. *Soil Biol. Biochem.* **1996**, *28*, 213–221. [CrossRef]

42. Garau, G.; Castaldi, P.; Santona, L.; Deiana, P.; Melis, P. Influence of red mud, zeolite and lime on heavy metal immobilization, culturable heterotrophic microbial populations and enzyme activities in a contaminated soil. *Geoderma* **2007**, *142*, 47–57. [CrossRef]

43. Zak, J.C.; Willig, M.R.; Moorhead, D.L.; Wildman, H.G. Functional Diversity of Microbial Communities: A Quantitative Approach. *Soil Biol. Biochem.* **1994**, *26*, 1101–1108. [CrossRef]

44. Alef, K.; Nannipieri, P. *Methods in Applied Soil Microbiology and Biochemistry*; Academic Press: San Diego, CA, USA, 1995.

45. Garau, G.; Diquattro, S.; Lauro, G.P.; Deiana, S.; Castaldi, P. Influence of Pb(II) in the sorption of As(V) by a Ca-polygalacturonate network, a root mucilage model. *Soil Sci. Plant. Nutr.* **2019**, *65*, 305–315. [CrossRef]

46. Garau, G.; Mele, E.; Castaldi, P.; Lauro, G.P.; Deiana, S. Role of polygalacturonic acid and the cooperative effect of caffeic and malic acids on the toxicity of Cu(II) towards triticale plants (x Triticosecale Wittm). *Biol. Fertil. Soils* **2015**, *51*, 535–544. [CrossRef]

47. Garau, G.; Palma, A.; Lauro, G.P.; Mele, E.; Senette, C.; Manunza, B.; Deiana, S. Detoxification Processes from Vanadate at the Root Apoplasm Activated by Caffeic and Polygalacturonic Acids. *PLoS ONE* **2015**, *10*, e0141041. [CrossRef]

48. Silvetti, M.; Garau, G.; Demurtas, D.; Marceddu, S.; Deiana, S.; Castaldi, P. Influence of lead in the sorption of arsenate by municipal solid waste composts: Metal(loid) retention, desorption and phytotoxicity. *Bioresour. Technol.* **2017**, *225*, 90–98. [CrossRef]

49. Giannakis, G.V.; Kourgialas, N.N.; Paranychianakis, N.V.; Nikolaidis, N.P.; Kalogerakis, N. Effects of Municipal Solid Waste Compost on Soil Properties and Vegetables Growth. *Compost Sci. Util.* **2014**, *22*, 116–131. [CrossRef]

50. Manzano, R.; Silvetti, M.; Garau, G.; Deiana, S.; Castaldi, P. Influence of iron-rich water treatment residues and compost on the mobility of metal(loid)s in mine soils. *Geoderma* **2016**, *283*, 1–9. [CrossRef]

51. Reddy, N.; Crohn, D.M. Compost Induced Soil Salinity: A New Prediction Method and Its Effect on Plant Growth. *Compost Sci. Util.* **2012**, *20*, 133–140. [CrossRef]

52. Zhang, M.; Heaney, D.; Henriquez, B.; Solberg, E.; Bittner, E. A four-year study on influence of biosolids/MSW cocompost application in less productive soils in Alberta: Nutrient dynamics. *Compost Sci. Util.* **2006**, *14*, 68–80. [CrossRef]

53. Garau, G.; Roggero, P.P.; Diquattro, S.; Garau, M.; Pinna, M.V.; Castaldi, P. Innovative amendments derived from industrial and municipal wastes enhance plant growth and soil functions in PTE-polluted environments. *Ital. J. Agron.* **2021**, *4*, 159–170. [CrossRef]

54. Penalver-Alcala, A.; Alvarez-Rogel, J.; Conesa, H.M.; Gonzalez-Alcaraz, M.N. Biochar and urban solid refuse ameliorate the inhospitality of acidic mine tailings and foster effective spontaneous plant colonization under semiarid climate. *J. Environ. Manag.* **2021**, *292*, 112824. [CrossRef]

55. Slukovskaya, M.V.; Vasenev, V.I.; Ivashchenko, K.V.; Dolgikh, A.V.; Novikov, A.I.; Kremenetskaya, I.P.; Ivanova, L.A.; Gubin, S.V. Organic matter accumulation by alkaline-constructed soils in heavily metal-polluted area of Subarctic zone. *J. Soils Sediment* **2021**, *21*, 2071–2088. [CrossRef]

56. Diquattro, S.; Castaldi, P.; Ritch, S.; Juhasz, A.L.; Brunetti, G.; Scheckel, K.G.; Garau, G.; Lombi, E. Insights into the fate of antimony (Sb) in contaminated soils: Ageing influence on Sb mobility, bioavailability, bioaccessibility and speciation. *Sci. Total Environ.* **2021**, *770*, 145354. [CrossRef]

57. Dousova, B.; Lhotka, M.; Filip, J.; Kolousek, D. Removal of arsenate and antimonate by acid-treated Fe-rich clays. *J. Hazard. Mater.* **2018**, *357*, 440–448. [CrossRef] [PubMed]

58. Vandenbohede, A.; Wallis, I.; Alleman, T. Trace metal behavior during in-situ iron removal tests in Leuven, Belgium. *Sci. Total Environ.* **2019**, *648*, 367–376. [CrossRef] [PubMed]

59. Garau, G.; Porceddu, A.; Sanna, M.; Silvetti, M.; Castaldi, P. Municipal solid wastes as a resource for environmental recovery: Impact of water treatment residuals and compost on the microbial and biochemical features of As and trace metal-polluted soils. *Ecotoxicol. Environ. Saf.* **2019**, *174*, 445–454. [CrossRef]

60. Garau, G.; Morillas, L.; Roales, J.; Castaldi, P.; Mangia, N.P.; Spano, D.; Mereu, S. Effect of monospecific and mixed Mediterranean tree plantations on soil microbial community and biochemical functioning. *Appl. Soil Ecol.* **2019**, *140*, 78–88. [CrossRef]

61. Garcia, C.; Hernandez, T.; Costa, F. Potential use of dehydrogenase activity as an index of microbial activity in degraded soils. *Commun. Soil Sci. Plant Anal.* **1997**, *28*, 123–134. [CrossRef]

62. Nannipieri, P.; Ascher, J.; Ceccherini, M.T.; Landi, L.; Pietramellara, G.; Renella, G. Microbial diversity and soil functions. *Eur. J. Soil Sci.* **2003**, *54*, 655–670. [CrossRef]

63. Abad-Valle, P.; Iglesias-Jimenez, E.; Alvarez-Ayuso, E. A comparative study on the influence of different organic amendments on trace element mobility and microbial functionality of a polluted mine soil. *J. Environ. Manag.* **2017**, *188*, 287–296. [CrossRef]

64. Garau, G.; Silvetti, M.; Deiana, S.; Deiana, P.; Castaldi, P. Long-term influence of red mud on As mobility and soil physico-chemical and microbial parameters in a polluted sub-acidic soil. *J. Hazard. Mater.* **2011**, *185*, 1241–1248. [CrossRef]

65. Alvarenga, P.; Goncalves, A.P.; Fernandes, R.M.; de Varennes, A.; Vallini, G.; Duarte, E.; Cunha-Queda, A.C. Evaluation of composts and liming materials in the phytostabilization of a mine soil using perennial ryegrass. *Sci. Total Environ.* **2008**, *406*, 43–56. [CrossRef]

66. Bhattacharyya, K.G.; Sen Gupta, S. Adsorption of a few heavy metals on natural and modified kaolinite and montmorillonite: A review. *Adv. Colloid Interface Sci.* **2008**, *140*, 114–131. [CrossRef]

67. Raiesi, F.; Dayani, L. Compost application increases the ecological dose values in a non-calcareous agricultural soil contaminated with cadmium. *Ecotoxicology* **2021**, *30*, 17–30. [CrossRef] [PubMed]

68. Liu, X.; Zhang, M.; Li, Z.W.; Zhang, C.; Wan, C.L.; Zhang, Y.; Lee, D.J. Inhibition of urease activity by humic acid extracted from sludge fermentation liquid. *Bioresour. Technol.* **2019**, *290*, 121767. [CrossRef] [PubMed]

69. Castaldi, P.; Garau, G.; Melis, P. Maturity assessment of compost from municipal solid waste through the study of enzyme activities and water-soluble fractions. *Waste Manag.* **2008**, *28*, 534–540. [CrossRef] [PubMed]

70. Feigl, V.; Ujaczki, E.; Vaszita, E.; Molnar, M. Influence of red mud on soil microbial communities: Application and comprehensive evaluation of the Biolog EcoPlate approach as a tool in soil microbiological studies. *Sci. Total Environ.* **2017**, *595*, 903–911. [CrossRef] [PubMed]

71. Lebrun, M.; Miard, F.; Van Poucke, R.; Tack, F.M.G.; Scippa, G.S.; Bourgerie, S.; Morabito, D. Effect of fertilization, carbon-based material, and redmud amendments on the bacterial activity and diversity of a metal(loid) contaminated mining soil. *Land Degrad. Dev.* **2021**, *32*, 11. [CrossRef]

72. Lombi, E.; Zhao, F.J.; Wieshammer, G.; Zhang, G.Y.; McGrath, S.P. In situ fixation of metals in soils using bauxite residue: Biological effects. *Environ. Pollut.* **2002**, *118*, 445–452. [CrossRef]

73. Tu, C.; Guan, F.; Sun, Y.H.; Guo, P.P.; Liu, Y.; Li, L.Z.; Scheckel, K.G.; Luo, Y.M. Stabilizing effects on a Cd polluted coastal wetland soil using calcium polysulphide. *Geoderma* **2018**, *332*, 190–197. [CrossRef] [PubMed]

74. Garau, M.; Castaldi, P.; Patteri, G.; Roggero, P.P.; Garau, G. Evaluation of Cynara cardunculus L. and municipal solid waste compost for aided phytoremediation of multi potentially toxic element-contaminated soils. *Environ. Sci. Pollut. Res.* **2021**, *28*, 3253–3265. [CrossRef] [PubMed]

75. Kuzyakov, Y.; Friedel, J.K.; Stahr, K. Review of mechanisms and quantification of priming effects. *Soil Biol. Biochem.* **2000**, *32*, 1485–1498. [CrossRef]

76. Song, Y.Y.; Liu, C.; Song, C.C.; Wang, X.W.; Ma, X.Y.; Gao, J.L.; Gao, S.Q.; Wang, L.L. Linking soil organic carbon mineralization with soil microbial and substrate properties under warming in permafrost peatlands of Northeastern China. *Catena* **2021**, *203*, 105348. [CrossRef]

77. Zhu, P.F.; Zhu, J.R.; Pang, J.Y.; Xu, W.W.; Shu, L.Z.; Hu, H.Q.; Wu, Y.; Tang, C.P. Biochar Improves the Growth Performance of Maize Seedling in Response to Antimony Stress. *Water Air Soil Pollut.* **2020**, *231*, 154. [CrossRef]

78. Hiradate, S.; Ma, J.F.; Matsumoto, H. Strategies of plants to adapt to mineral stresses in problem soils. *Adv. Agron.* **2007**, *96*, 65–132. [CrossRef]

79. Kraemer, S.M.; Crowley, D.E.; Kretzschmar, R. Geochemical aspects of phytosiderophore-promoted iron acquisition by plants. *Adv. Agron.* **2006**, *91*, 1–46. [CrossRef]

80. Mitra, R.; Singh, S.B.; Singh, B. Radiochemical evidence validates the involvement of root released organic acid and phytosiderphore in regulating the uptake of phosphorus and certain metal micronutrients in wheat under phosphorus and iron deficiency. *J. Adioanalytical Nucl. Chem.* **2020**, *326*, 893–910. [CrossRef]
81. Nakib, D.; Slatni, T.; Di Foggia, M.; Rombola, A.D.; Abdelly, C. Changes in organic compounds secreted by roots in two Poaceae species (Hordeum vulgare and Polypogon monspenliensis) subjected to iron deficiency. *J. Plant Res.* **2021**, *134*, 151–163. [CrossRef] [PubMed]
82. Nedjimi, B. Phytoremediation: A sustainable environmental technology for heavy metals decontamination. *SN Appl. Sci.* **2021**, *3*, 286. [CrossRef]

soil systems

Article

Evaluating Potential Ecological Risks of Heavy Metals of Textile Effluents and Soil Samples in Vicinity of Textile Industries

Jaskaran Kaur [1], Sandip Singh Bhatti [1], Sartaj Ahmad Bhat [2,3], Avinash Kaur Nagpal [1], Varinder Kaur [4] and Jatinder Kaur Katnoria [1,*]

1   Department of Botanical and Environmental Sciences, Guru Nanak Dev University, Amritsar 143005, Punjab, India; jaskaranbot.rsh@gndu.ac.in or jaskarankaur0@gmail.com (J.K.); singh.sandip87@gmail.com (S.S.B.); avinash.botenv@gndu.ac.in or avnagpal@rediffmail.com (A.K.N.)
2   Department of Environmental Sciences, Government Degree College, Khanabal, Anantnag 192101, Jammu and Kashmir, India; sartajbhat88@gmail.com
3   River Basin Research Center, Gifu University, 1-1 Yanagido, Gifu 501-1193, Japan
4   Centre for Advanced Studies, Department of Chemistry, Guru Nanak Dev University, Amritsar 143005, Punjab, India; varinder_texchem@gndu.ac.in or varinder_gndu@yahoo.com
*   Correspondence: jatinder.botenv@gndu.ac.in or jatinkat@yahoo.co.in or jkat08@yahoo.com

Abstract: The present study pertains to assessing the heavy metal (Cd, Cr, Co, Cu, Pb, and Zn) contents of untreated and treated effluents of two textile industries and agricultural soil samples in the vicinity of these industries located in Ludhiana, Punjab (India). The genotoxicity of the effluents samples was estimated using *Allium cepa* root chromosomal aberration assay. The exposure of *Allium cepa* roots to untreated effluents from both industries resulted in the reduction of mitotic index (MI) and increase in chromosomal aberrations in the root tip meristematic cells when compared to those that were exposed to the treated effluents indicating the significant genotoxic potential of untreated effluents. Risk characterization of soil sample was carried out by calculating the potential ecological and human health risks of heavy metals. The hazard index was observed to be less than 1, indicating there was no potential health risk of heavy metals in soil samples. Furthermore, bioaccumulation potential studies on plant species grown in the vicinity of these industries have shown that bioaccumulation factor (BAF) varied as *Ricinus communis* L. > *Chenopodium album* L. > *Cannabis sativa* L. with Co and Pb having maximum and minimum values, respectively.

Keywords: health risk assessment; *Allium cepa* root chromosomal aberration assay; bioaccumulation; genotoxicity; heavy metals; industrial effluents

Citation: Kaur, J.; Bhatti, S.S.; Bhat, S.A.; Nagpal, A.K.; Kaur, V.; Katnoria, J.K. Evaluating Potential Ecological Risks of Heavy Metals of Textile Effluents and Soil Samples in Vicinity of Textile Industries. *Soil Syst.* **2021**, *5*, 63. https://doi.org/10.3390/soilsystems5040063

Academic Editors: Matteo Spagnuolo, Paola Adamo and Giovanni Garau

Received: 18 June 2021
Accepted: 27 September 2021
Published: 9 October 2021

Publisher's Note: MDPI stays neutral with regard to jurisdictional claims in published maps and institutional affiliations.

Copyright: © 2021 by the authors. Licensee MDPI, Basel, Switzerland. This article is an open access article distributed under the terms and conditions of the Creative Commons Attribution (CC BY) license (https://creativecommons.org/licenses/by/4.0/).

1. Introduction

Numerous obnoxious chemical agents continuously enter our environment due to various industrial, domestic, and other human activities. These chemicals have the tendency to pose threats to the survival of living beings, ultimately endangering the ecological balance [1–3]. The water pollution index on account of inorganic chemicals is considered to be one of the major indicators of environmental pollution, which has accelerated in past decades due to various anthropogenic activities, especially, agricultural practices and the discharges of effluents from industries into the natural water bodies. [4]. The release of huge quantities of treated, as well as untreated municipal wastes, to aquatic bodies has also become a problem in different developing countries [5]. Wastewater irrigation has been documented to cause the accumulation of heavy metals in agricultural soils and plants [6–9].

The contamination of water bodies due to genotoxic compounds like heavy metals and pesticides has been widely documented [10–12]. The presence of various unidentified and noxious toxicants possessing potential carcinogenicity has been widely demonstrated in various genotoxicity studies [13,14]. The reports on genotoxicity studies of various industrial wastewaters and other effluents have globally raised concern over the genotoxic and carcinogenic hazards of the contaminants present in the samples [1]. Since the chemical characterization alone cannot provide sufficient knowledge on their genotoxicity and potential hazard, different bioassays have been used to explore the same. Many bioassays have been effectively used to assess the genotoxicity of complex wastewaters and a number of bacterial and plant-based assays have been developed for the estimation of the genotoxic potential of water samples. Among these, the *Allium cepa* test takes a prominent position because it has a low chromosome number and large size of chromosomes [15].

Soil is an essential resource for sustaining two basic human necessities, that is, production of sufficient food and a clean environment by adsorbing different contaminants. However, certain plants grown on polluted land can uptake contaminants like heavy metals either as ions through their root system or by absorption through foliage, and they get accumulated in different plant parts such as in roots, stems, leaves, fruits, and grains [16]. Heavy metal contamination of soils is a very serious issue that has contributed significantly to the contamination of various food crops [3,17,18]. Although heavy metals exist in soils in natural concentrations (significantly low) deriving from parent rock materials, these trace amounts do not pose any harm to human health. However, anthropogenic inputs of wastewaters from various sources along with the dumped waste can significantly increase the heavy metal concentrations in soil [19,20]. Excessive levels of heavy metals in agricultural soils not only lead to the disorders of soil functions and crop growth but also, poses serious risks to human health by accumulating in food crops [21–24].

Potential human health risk (non-carcinogenic and carcinogenic) assessment has been recognized as an efficient tool for assessing risks of various pollutants in the environment and is essential for making decisions regarding regulations concerning pollution reduction in urban soil and minimizing human exposure to toxic pollutants [25–27]. Considering the ecological threats posed by contaminants in textile industry effluents, the present study was conducted to assess the effluents (treated and untreated) from two textile industries situated in Punjab, India for heavy metal contents, physico-chemical characteristics, and genotoxicity following the *Allium cepa* root chromosomal aberration assay. Heavy metal estimation and ecological risk assessment of the agricultural soil in the vicinity of these industries were also conducted. The study further focused on the evaluation of heavy metal bioaccumulation in three plant species viz., *Cannabis sativa* L., *Ricinus communis* L., and *Chenopodium album* growing in the vicinity of these industries, as well as the application of various pollution indices to determine the pollution level of analyzed heavy metals in the soil of study area.

2. Material and Methods

2.1. Collection of Samples

2.1.1. Textile Industrial Effluents

Untreated and treated effluents originating from two textile industries (Textile Industry A and Textile Industry B) being discharged into the Sutlej river, Ludhiana, Punjab, India, were selected for toxicity assessments. In the present study, the effluent samples from both textile industries were collected during March 2017. Effluent samples from the respective industries were collected in triplicate in clean bottles, brought to the laboratory, and stored at 4 °C until further analysis. The samples were coded as shown in Table 1. The physico-chemical analysis of the collected samples was carried out following standard protocols [28,29].

Table 1. Description of sample codes.

S. No.	Sample Code	Description of Sample
1.	AU	Untreated effluent sample collected from textile industry A
2.	AT	Treated effluent sample collected from textile industry A
3.	BU	Untreated effluent sample collected from textile industry B
4.	BT	Treated effluent sample collected from textile industry B
5.	SA	Soil sample collected from an agricultural field in the vicinity of industries A and B

2.1.2. Soil Sample

Soil samples in triplicate were taken from agricultural fields in the vicinity of the industries. For the collection of soil samples, the soil was dug to the depth of 20 cm [30]. Soil was collected from 4–5 parts of the field viz., east, west, north, south, and center and pooled to constitute the sample of the particular field. Approx. area of the agricultural field was 1000 sq. mts and situated 500 meters away from the main industrial units. Soil samples were stored in clean and airtight polyethylene bags. Soil samples were dried in the laboratory, cleaned by removing visible traces of leaves and other waste materials, homogenized, and sieved through a size 2 mm sieve for heavy metal analysis.

2.1.3. Plant Samples

Since there was no crop grown at the time of sampling in the agricultural field, the plant samples (leaves) of three wildly growing plant species viz. *Cannabis sativa* L. (Cannabaceae), *Chenopodium album* L. (Amaranthaceae), and *Ricinus communis* L. (Euphorbiaceae) on the boundaries of agricultural fields in the vicinity of industries were collected to explore their heavy metal bioaccumulation potential. The leaves were thoroughly washed using tap water followed by distilled water, oven-dried at 70 °C, grounded to a fine powder by pestle mortar, and stored in airtight polyethylene bags at 4 °C until further analysis.

2.2. Physico-Chemical Characteristics of Industrial Effluents and Soil

The soil extract (1:5 w/v) was prepared by adding 20 g of the collected soil sample in 100 mL of distilled water. This solution was kept in a mechanical shaker for 12 hours at room temperature and filtered through Whatman No. 1 filter paper [31]. The filtrate was termed soil extract and was used for further analysis of the physico-chemical parameters (pH, electrical conductivity, calcium, sodium, and magnesium). Total organic carbon of the soil was estimated using the dry combustion method [32]. A core measuring cylinder (100 ML) was used for bulk density (BD) estimation [33]. Soil texture was determined by the sieving and sedimentation method [34]. On the basis of size, different particles of soil were grouped as: sand: 0.5–2.00 mm; silt: 0.002–0.5 mm; clay: <0.002 mm. The analysis of the physico-chemical parameters (pH, temperature, total solids (TS), total dissolved solids (TDS), total suspended solids (TSS), total hardness, alkalinity, calcium, chloride, magnesium, sodium, and phosphate) of effluent samples was carried out following the standard protocols of the American Public Health Association [35,36]. The sodium content of both effluents and soil samples was measured using a Flame Photometer (Model-128; Make: Systronics). The pH of each effluent was measured using a pH meter (Model: μ pH system 361; Make: Systronics).

2.3. Heavy Metal Estimation

Heavy metal contents in collected samples were determined using the flame atomic absorption spectrophotometer (AAS) (Agilent 240 FS AA model), at variable/recommended wavelength of 228.80 nm for cadmium, 240.70 nm for chromium, 357.90 nm for cobalt, 324.80 nm for copper, 217.0 nm for lead, and 213.90 nm for zinc. Limits of detection (μg/L) for different metals were cadmium (1.5), cobalt (3), chromium (5), copper (1.2), lead (7), and zinc (1.6). The airflow rate was maintained at 13.50 L/min for all heavy metal determinations. The acetylene flow rate was set at 2.00 L/min for Cd, Co, Cu, Pb, and Zn, and at 2.90 L/min for Cr estimations while the lamp currents were set at 4.00 mA, 7.00 mA, 7.00 mA, 4.00 mA, 10.00 mA, and 5.00 mA for determination of Cd, Co, Cr, Cu, Pb, and Zn, respectively. All the glassware was thoroughly washed and oven-dried before use. Double distilled water and analytical grade reagents were used during the whole experiment. The standard solutions (1000 mg/L) of Agilent made for different metals were used to prepare solutions of varying concentrations as 0.5, 1, 1.5 (mg/L) for cadmium and zinc; 5, 10, 15 (mg/L) for chromium, lead, nickel, and cobalt; and 1, 3, 5 (mg/L) for copper using the serial dilution method. The accuracy (>95%) of the instrument was maintained throughout the experiment by thorough washing. For which, after every 10 sample readings, the standards were run to observe the accuracy of the instrument. Soil samples were digested using aqua regia, that is, a mixture of one part concentrated nitric (HNO_3) and three parts hydrochloric acid (HCl) following the method described by the authors of [37] with minor modifications. For this purpose, 1 g of finely ground soil sample was digested slowly with aqua regia on a hot plate in a fume hood till white fumes appeared, indicating the complete digestion of the soil sample. Plant sample digestion was carried out using a tri-acid mixture, that is, five parts of nitric acid (HNO_3) and one part of both perchloric ($HClO_4$) and sulfuric acid (H_2SO_4) as prescribed by Allen [38]. Only concentrated acids were used for both types of digestion. The digested soil and plant samples were filtered using Whatman No.1 filter paper and diluted with double distilled water up to a final volume of 50 mL.

2.4. Metal Bioaccumulation Factor (BAF)

In order to assess the accumulation of heavy metals from the soil in the agricultural fields in the vicinity of the industries into the three plant species (*Cannabis sativa* L., *Chenopodium album* L. and *Ricinus communis* L.), the bioaccumulation factor (BAF) was calculated. The bioaccumulation factor is commonly used to study the fate of different environmental contaminants in plants [39]. Ali et al. [40] documented that BAF is the ratio of the concentration of heavy metals in the crop to that in the soil. Accordingly, BAF was calculated using the following equation.

$$BAF = C_{plant}/C_{soil} \qquad (1)$$

where, C_{plant} stands for concentrations of heavy metal in plant leaves and C_{soil} stands for concentrations of heavy metal in soil.

2.5. Genotoxicity Assessment

The genotoxicity of both untreated and treated industrial effluents was determined using the *Allium cepa* root chromosomal aberration assay [41–43]. After the removal of primary roots of freshly purchased onion bulbs, bulbs were placed on Couplin jars containing distilled water (negative control) and industrial effluents for 48–72 h for rooting. The Couplin jars were kept in a BOD (Biochemical Oxygen Demand) incubator at $25 \pm 2\,^\circ C$ until roots grew. Care was taken to fill the coupling jars with exposure media on a daily basis so that the onion bulbs were emersed in solution and the root primordia were under continuous exposure to the treatment. Distilled water was used as a negative control during the study. The onion bulbs after treatment were thoroughly washed. The root tips were plucked with forceps and put in a solution of glacial acetic acid and Ethanol in the ratio of 1:3 (Farmer's fluid). The root tips were squashed in aceto-orcein stain to prepare slides. At least five slides consisting of approximately 500 dividing cells were examined under a light microscope to calculate mitotic index (MI) and to score different types of aberrations for each sample. The chromosomal aberrations were categorized into physiological (outcomes of spindle inhibition) and clastogenic (formed due to damage of DNA) based on the descriptions given earlier [42,44].

2.6. Pollution Assessment

The degree of pollution of the agricultural soil in the studied area was evaluated by using several indices, like the geoaccumulation index (Igeo), contamination factor (CF), degree of contamination (Cd_{eg}), modified degree of contamination (mCd_{eg}), Numerow's pollution index (PI), pollution load index (PLI), potential ecological risk factor (ER_i), and the potential ecological risk index (RI). A brief description of these soil contamination indices is given in Table S1.

2.7. Human Health Risk Assessment

The model for human health risk assessment given by the United States Environmental Protection Agency (USEPA) was used to assess the non-carcinogenic and carcinogenic effects of environmental toxicants like heavy metals on humans. Due to behavioral and physiological differences in this study area, people were divided into two groups, that is, adults and children. Soil contaminants, that is, heavy metals, pose health risks to the human body mainly by three exposure pathways, which include ingestion, inhalation, and dermal contact. So, the carcinogenic and non-carcinogenic threat of these exposure pathways was calculated in the present study. The methodology used in the present study for human health risk assessment was based on the guidelines given by the US Environmental Protection Agency [28,45–48].

2.7.1. Exposure Assessment

To calculate the human exposure dose, the average daily intake (*ADI*) of heavy metals in soil for three exposure pathways (ingestion, inhalation, and dermal contact) is calculated as follows:

Ingestion pathway:

$$ADI_{ingestion} = \frac{C \times IR_{ig} \times EF \times ED \times CF}{BW \times AT} \tag{2}$$

Inhalation pathway:

$$ADI_{inhalation} = \frac{C \times IR_{ih} \times EF \times ED}{BW \times AT \times PEF} \tag{3}$$

Dermal contact pathway:

$$ADI_{dermal\ contact} = \frac{C \times SA \times SAF \times DAF \times EF \times ED \times CF}{BW \times AT} \tag{4}$$

where $ADI_{ingestion}$, $ADI_{inhalation}$, $ADI_{dermal\ contact}$ is the average daily intake (mg/kg day) via ingestion, inhalation, and dermal contacts, respectively. *C* is the concentration of analyzed heavy metals in soil samples (mg/kg); IR_{ig} is the ingestion rate (100 and 200 mg/day for adults and children, respectively) [28,45,48]; IR_{ih} is the inhalation rate (12.8 m^3/day for adults and 7.63 m^3/day for children) [28]; *EF* is the exposure frequency (365 days/year) [28,45]; *ED* is the exposure duration (30 years for adults and 6 years for children) [28]; *CF* is the conversion factor for soil (10^{-6} kg/mg) [48]; *BW* is the body weight (70 and 20 kg for adults and children, respectively) [28]; *AT* is the average exposed time ($EF \times ED$) [28]; PEF is the particulate emission factor (1.36×10^9 m^3/kg) [28]; *SA* is the skin exposed area for soil (4350 and 1600 cm^2 for adults and children, respectively) [28]; *SAF* is the skin adherence factor (0.7 mg/cm^2 for adults and 0.2 mg/cm^2 for children) [28,48]; and *DAF* is the dermal absorption factor (0.001) [28].

2.7.2. Non-Carcinogenic Risk Assessment

The hazard quotient (*HQ*) is characterized for non-carcinogenic hazards and is defined as the average daily intake by the toxicity threshold value, which is referred to as the chronic reference dose (*RfD*) in mg/kg-day of the specific heavy metal. *HQ* is computed as the ratio of the average daily intake (*ADI*) and a reference dose (*RfD*). The equitation of *HQ* is given as follows [28,45]:

$$HQ = \frac{ADI}{RfD} \tag{5}$$

where *HQ* is the hazard quotient, *ADI* is the average daily intake (mg/kg day) and *RfD* is the reference dose (mg/kg day) of heavy metals via ingestion, inhalation, and dermal contact pathways. The reference dose (*RfD*) of studied heavy metals is shown in Table S2.

Hazard index (*HI*) is a cumulative non-cancer health risk that can be evaluated by the sum of the *HQ* (hazard quotient) values of various exposure pathways. It can be calculated as the sum of non-carcinogenic hazard quotients for all contaminants [45] as follows:

$$HI = \sum HQi \tag{6}$$

where *HQi* is the non-cancer hazard quotient for the ith contaminants.

HI < 1 indicated no non-carcinogenic health, whereas *HI* > 1 risk indicated adverse non-carcinogenic health risk [28,49].

2.7.3. Carcinogenic Risk Assessment

The carcinogenic risk assessment is the incremental probability of an individual developing cancer over a lifetime as a result of exposure to the potential carcinogen like heavy metals [27,50]. Carcinogenic risk and total carcinogenic risks are determined as follows:

$$CR = ADI \times SF \tag{7}$$

$$TCR = \sum CR \tag{8}$$

where *CR* is the carcinogenic risk; *ADI* is the average daily intake (mg/kg day); *SF* is the cancer slope factor over a lifetime (mg/kg day). The cancer slope factor (*SF*) of studied heavy metals is shown in Table S2.

The values of carcinogenic risk (*CR*) ranging from 1×10^{-6} to 1×10^{-4} are considered as safe limit for human health [28,45], whereas higher *CR* values than the limit of 1×10^{-4} cause lifetime cancer risks to the human body [45,49].

2.8. Statistical Analysis

Student's *t*-test ($p \leq 0.05$) was applied to find significant differences between values of heavy metals and genotoxicity parameters like mitotic inhibition (MI), physiological aberrations (PA), clastogenic aberrations (CA), and total aberration (TA) for untreated and treated effluents of the same industry. Chi-square test ($p \leq 0.05$) was used to calculate the statistically significant differences between the values of genotoxicity parameters (MI, PA, CA, and TA) for effluents and the negative control. Statistical analysis was performed using Minitab version 14.0 (State College, PA, USA).

3. Results and Discussion

3.1. Physico-Chemical Characteristics of Industrial Effluents and Soil

Physico-chemical parameters of studied soil and industrial effluent samples are shown in Table 2. The mean pH of soil, that is, 8.02 was observed to be within the permissible limit of 6.5–8.5 and is alkaline in nature. Electrical conductivity (EC), which indicates the soil salinity, was also found to be within the prescribed limit. The studied soil sample had a sand content of 33.49%, silt content of 26.05%, and clay of 40.45%. The Ca, Mg and Na contents (mg/kg) of the soil sample were observed to be 120.24, 176.64, and 343.08, respectively. pH, the most significant parameter for the assessment of water quality, ranged from 6.67 to 8.90 and remained within the prescribed limits. Bulk density (BD) plays a vital role in the growth of plants as high BD can decrease the root penetration in soil. The mean bulk density (BD) of the studied soil sample was found to be 1.08. Organic matter content plays a chief role in the fertility of agricultural soils. Total organic carbon (TOC) in the present study was observed to be 2.22%. The pH of treated effluent of textile industry A (AT) was observed to be acidic while all other effluent samples showed basic pH. Dissolved calcium and magnesium in water are the two most common minerals that determine water hardness. Total hardness (mg/L) for AU, AT, BU, and BT was found to be 111.33, 151.33, 191.33, and 104.67, respectively. The calcium content in effluent samples ranged between 17.90–60.65 mg/L and magnesium content was seen in the range of 17.58–70.33mg/L. The order of chloride content (mg/L) was observed to be: 232.41 (AU) > 142.47 (AT) > 114.07 (BU) > 66.74 (BT). The electrical conductivity varied from 511 μS/cm to 1908.67 μS/cm. The value of total suspended solids (mg/L) was found to be minimum for AU (106.67) and maximum for AT (585) while the content of total dissolved solids (mg/L) was found to be minimum for AT (201.67) and maximum for BU (3666.67). The value of alkalinity (mg/L) varied from 356.67 to 656.67; sodium content (mg/L) from 141.08 (BU) to 333.63 (AU); and phosphate content (mg/L) from 1.48 (AT) to 2.08 (BU). The value of total solids (mg/L) was found to be in the order of BU (3893.33) > AU (3473.33) > BT (2000) > AT (786.67).

According to Paul et al. [51], the basic nature of the pH of the industrial effluents was because of the usage of scouring and bleaching agents along with other various chemicals like caustic soda, hydrogen peroxide, and soap while pulping the waste. Similar results were also reported by Ramamurthy et al. [52]. Alkaline pH can have an adverse effect on soil permeability and soil microflora [53]. Total solid (TS) levels were higher in both AU and BU, which can lead to high turbidity of the water bodies into which these effluents are discharged. In the present study, higher levels of total dissolved solids (TDS) indicated high salt content in the effluents analyzed. Paul et al. [51] also reported higher values of TDS (2264–7072 mg/L) in textile effluents. The higher values of TDS are due to the addition of different chemicals during pulping and bleaching processes, which can have detrimental effects on aquatic flora and fauna.

The degree of hardness becomes higher as the calcium and magnesium content increases and is related to the concentration of multivalent cations dissolved in the water. The change in alkalinity depends on carbonates and bicarbonates, which in turn depends upon the release of CO_2. The amount of total alkalinity in the effluents during the present study ranged from 356.667 to 656.667 mg/L. The hardness of water is mainly due to the presence of calcium and magnesium ions, and it is an important indicator of the toxic effect of poisonous elements [54].

Table 2. Physico-chemical characteristics (Mean ± S.E.) of collected samples (textile industrial effluents and soil) from Ludhiana, Punjab (India).

Parameter	AU	AT	BU	BT	BIS Limits [a]	Soil	Soil Limits [b]
pH	6.67 ± 0.05	7.49 ± 0.00 *	7.43 ± 0.02	8.90 ± 0.02 *	6.5–8.5	8.02 ± 0.01	6.5–8.5
EC (μS/cm)	1908.67 ± 4.67	628.67 ± 1.86 *	1858.33 ± 1.67	511.00 ± 2.00 *	-	442.5 ± 4.79	450
TDS (mg/L)	3366.33 ± 6.67	201.67 ± 1.67 *	3666.67 ± 13.33	1813.33 ± 35.28 *	500–2000	-	-
TS (mg/L)	3473.33 ± 6.67	786.67 ± 13.34 *	3893.33 ± 13.33	2000.00 ± 40.00 *	-	-	-
TSS (mg/L)	106.67 ± 6.67	585 ± 12.58 *	226.67 ± 13.33	186.67 ± 13.33	-	-	-
Alkalinity (mg/L)	656.67 ± 33.34	456.67 ± 33.34 *	490.00 ± 57.74	356.67 ± 33.33	200–600	-	-
Hardness (mg/L)	111.33 ± 6.67	151.33 ± 6.67 *	191.33 ± 6.67	104.67 ± 6.67 *	200–600	-	-
Calcium (mg/L)	33.93 ± 2.67	17.90 ± 2.67 *	60.65 ± 2.67	28.59 ± 2.67 *	75–200	120.24 (mg/kg) ± 0.00	0–3500 mg/kg
Magnesium (mg/L)	17.58 ± 4.40	70.33 ± 4.40 *	26.37 ± 7.61	21.98 ± 4.40 *	30–100	176.64 (mg/kg) ± 6.09	0–500 mg/kg
Sodium (mg/L)	333.63 ± 1.62	308.20 ± 1.25 *	141.08 ± 0.58	262.42 ± 1.04 *	-	343.08 (mg/kg) ± 3.02	0–300 mg/kg
Chloride (mg/L)	232.41 ± 4.73	142.47 ± 4.73 *	114.07 ± 4.73	66.74 ± 0.58 *	250–1000	-	-
Phosphate (mg/L)	1.58 ± 0.03	1.48 ± 0.02	2.08 ± 0.13	1.50 ± 0.03 *	-	-	-
Bulk density (g/cc)	-	-	-	-	-	1.08 ± 0.01	-
Sand (%)	-	-	-	-	-	33.49 ± 0.72	-
Silt (%)	-	-	-	-	-	26.05 ± 0.19	-
Clay (%)	-	-	-	-	-	40.45 ± 0.68	-
TOC (%)	-	-	-	-	-	2.22 ± 0.15	-

(AU: Untreated effluent of textile industry A; AT: Treated effluent of textile industry A; BU: Untreated effluent of textile industry B; BT: Treated effluent of textile industry B; TOC: Total organic carbon). [a] BIS [55]; [b] Awashthi [56], * Indicates statistically significant difference between values of parameters in untreated and treated effluents of the same industry. (Independent Student's *t*-test, $p \leq 0.05$).

3.2. Heavy Metal Estimation

3.2.1. Heavy Metal Contents in Industrial Effluents

The results of a metal analysis of industrial effluents are given in Table 3. The contents (mg/L) of Cd, Cr, Co, Cu, Pb, and Zn observed for untreated effluents of textile industry A were 0.004, 0.06, 1.72, 0.02, 0.13, and 0.09, respectively, while the values (mg/L) of these metals for treated effluents were below detection limit (BDL), 0.05, 1.33, 0.02, 0.11, and 0.02, respectively. The heavy metal contents observed for untreated effluent of textile industry B were in the order Co (1.69) > Cd (1.33) > Pb (0.14) > Zn (0.13) > Cr (0.06) > Cu (0.03), while in the case of treated effluent, the order was Co (1.42) > Pb (0.11) > Zn (0.07) > Cr (0.06) > Cu (0.01) > Cd (0.001). Among the heavy metals analyzed, the contents of Co were found to be above the standard limits for the discharge of effluents from textile industries.

Among all metals, Co content was observed to be high in untreated and treated effluents of both textile industries A and B as compared to the prescribed limits. All other tested heavy metals were found to be within the permissible level. Metal contamination in textile effluents was reported to occur because of the wide usage of chemicals, colorants, mordants, and other additives like caustic soda, sodium carbonate, etc. during the manufacturing processes [57]. Adinew [58] also reported that different heavy metals such as cobalt, copper, and chromium in textile effluents were present within the dye chromophores. The presence of heavy metals in effluents produces several adverse effects on living organisms [59]. Metals like chromium, zinc, iron, mercury, and lead were reported to pose environmental challenges [60]. In the present study, there was a significant difference ($p > 0.05$) in Co, Pb, and Zn content between untreated and treated effluents of both textile industries.

Table 3. Heavy metal contents (Mean ± S.E.) of collected samples (effluent and soil) from Ludhiana, Punjab (India).

Heavy Metal	Content of Heavy Metals (mg/L) of Effluent				Normal Acceptable Range (USEPA)	FAO,1985	Content of Heavy Metals (mg/kg) in Soil	Indian Limits for Soil (mg/kg) [a]	European Union Standards (mg/kg) [b]
	AU	AT	BU	BT					
Cadmium	0.004 ± 0.00	N.D.	0.002 ± 0.00	0.001 ± 0.00	2	0.01	1.33 ± 0.05	3–6	1
Chromium	0.06 ± 0.00	0.05 ± 0.00 *	0.06 ± 0.00	0.06 ± 0.00	2	0.10	16.43 ± 0.60	-	100
Cobalt	1.72 ± 0.00	1.33 ± 0.00 *	1.69 ± 0.00	1.42 ± 0.00 *	-	0.05	214.60 ± 0.42	-	50
Copper	0.02 ± 0.00	0.02 ± 0.00	0.03 ± 0.00	0.01 ± 0.00 *	3	0.20	13.63 ± 1.88	135–270	100
Lead	0.13 ± 0.00	0.11 ± 0.00 *	0.14 ± 0.00	0.11 ± 0.00 *	0.1	5	57.33 ± 1.20	250–500	100
Zinc	0.09 ± 0.00	0.02 ± 0.00 *	0.13 ± 0.00	0.07 ± 0.00 *	5	2	92.52 ± 0.06	300–600	300

(AU: Untreated effluent of textile industry A; AT: Treated effluent of textile industry A; BU: Untreated effluent of textile industry B; BT: Treated effluent of textile industry B). * Indicates statistically significant difference between values of heavy metals in untreated and treated effluents of the same industry (Independent Student's t-test, $p \leq 0.05$). [a] Awashthi [56]; [b] European Union Standards (EU) [61].

3.2.2. Heavy Metal Contents in Soil

Table 3 shows the contents of various studied heavy metals in the soil of the agricultural field collected from the vicinity of textile industries. The contents (mg/kg) of Cd, Cr, Co, Cu, Pb, and Zn observed in samples were 1.33, 16.43, 214.60, 13.63, 57.33, and 92.52, respectively.

The chief sources of heavy metals in the roadside agricultural soils were documented to be the parent rock material, vehicular emissions, industrial activities, and agrochemicals like fertilizers and pesticides used for cultivation [18,62]. In the present study, soil samples were collected from agricultural fields in the vicinity of the textile industries. In the present study, three metals, that is, Cu, Cr, and Zn were observed to be low while Pb was higher in comparison to heavy metal contents reported from other parts of Punjab [63,64]. Also, Cd content was observed to be high in the present study in comparison to other parts of the world [65–70]. Cadmium is a toxic metal that causes serious health problems to humans, animals, and plants. Bhatti et al. [61] reported that the sources of the high levels of cadmium in the agricultural soils of Punjab were due to the usage of various agrochemicals like NPK (nitrogen, phosphate, potassium) fertilizers, pesticides, weedicides, etc. Industrial activities, lead mines, farmyard manure, and sewage sludge applications, etc. are reported to be the main sources of lead pollution in agriculture and plants [71]. Zinc pollution in roadside soils was caused by traffic-related activities such as vehicular emissions and weathering of crash barriers [18,72,73].

Among the different metals analyzed, the content of Co was observed to be maximum in the soil sample analyzed during the present study. Cobalt is documented to occur naturally in soils following two main pathways, that is, weathering of rocks comprising of minerals and breakdown of organic matter. The major mechanism involved in cobalt content in soil includes the anthropogenic usage of cobalt salts. Smaller amounts of cobalt can also enter the soil from the airborne transport of particulate emissions and application of sewage sludge onto fields. Heavy metal mobility in soil was reported to be inversely related to the strength of adsorption by soil constituents. However, the adsorption of cobalt to soils was reported to be rapid by Kim et al. [74]. Cobalt is reported to be one of the beneficial elements for the growth of higher plants although there is no report available regarding its direct role in plant metabolism [75]. However, some studies showed that cobalt was required for nitrogen fixation by bacteria in the root nodules of plants belonging to the leguminous family [76,77]. Cobalt has also been documented to be necessary for the processes of stem growth, elongating the coleoptiles, and expanding leaf discs. Moreover, cobalt reduced the peroxidase activity resulting in the breakdown of Indole acetic acid (IAA). Application of cobalt through seed treatments improved the germination of a seed, stand establishment, growth, yield, and quality [78]. Yet, a higher concentration of cobalt was found to be toxic, causing chlorosis and necrosis and inhibited root growth by retarding cell division, hindering the uptake and translocation of nutrients and water [78,79].

Copper and zinc are considered essential elements for plant nutrition, however, these can also cause toxic effects, if their concentrations exceed the required limits [80]. Plants mainly absorb zinc as a divalent cation, which acts either as a metal component or enzymes or as a functional, structural, or regulatory co-factor of many enzymes [81]. Despite being a non-essential element, cadmium can also get highly accumulated in plants as reported by Nadian [82]. Pb being a toxic element decreases the biomass growth and disrupts the total chlorophyll content of plants [83]. Naureen et al. [84] observed that the concentration of heavy metals in plants varied from species to species. Another study reported that the accumulation of selected metals varied greatly among plant species and uptake of an element by a plant was primarily dependent on the plant species and the soil characteristics [85]. Similar observations were also made by Rattan et al. [86]. On the other hand, Muchuweti et al. [87] reported that the excessive accumulation of heavy metals in soils was due to elevated levels of heavy metals in wastewater used for irrigation that led to increased uptake of metals in crops.

3.2.3. Heavy Metal Contents in Leaves of Plants

figfig:soilsystems-1285565-f001 shows the heavy metal contents in leaf samples of three plants *Cannabis sativa* L., *Chenopodium album* L., and *Ricinus communis* L. from the study area. The order of the heavy metals in the leaves of the three plants was observed to be Co > Zn > Pb > Cr > Cu > Cd. However, among three plant species, the order of heavy metal contents observed for Cd and Pb was *Cannabis sativa* L. > *Ricinus communis* L. > *Chenopodium album* L.; for Cr and Zn the order was *Chenopodium album* L. > *Ricinus communis* L. > *Cannabis sativa* L.; for Co the order was *Ricinus communis* L. > *Chenopodium album* L. > *Cannabis sativa* L.; and for Cu it was *Ricinus communis* L. > *Cannabis sativa* L. > *Chenopodium album* L.

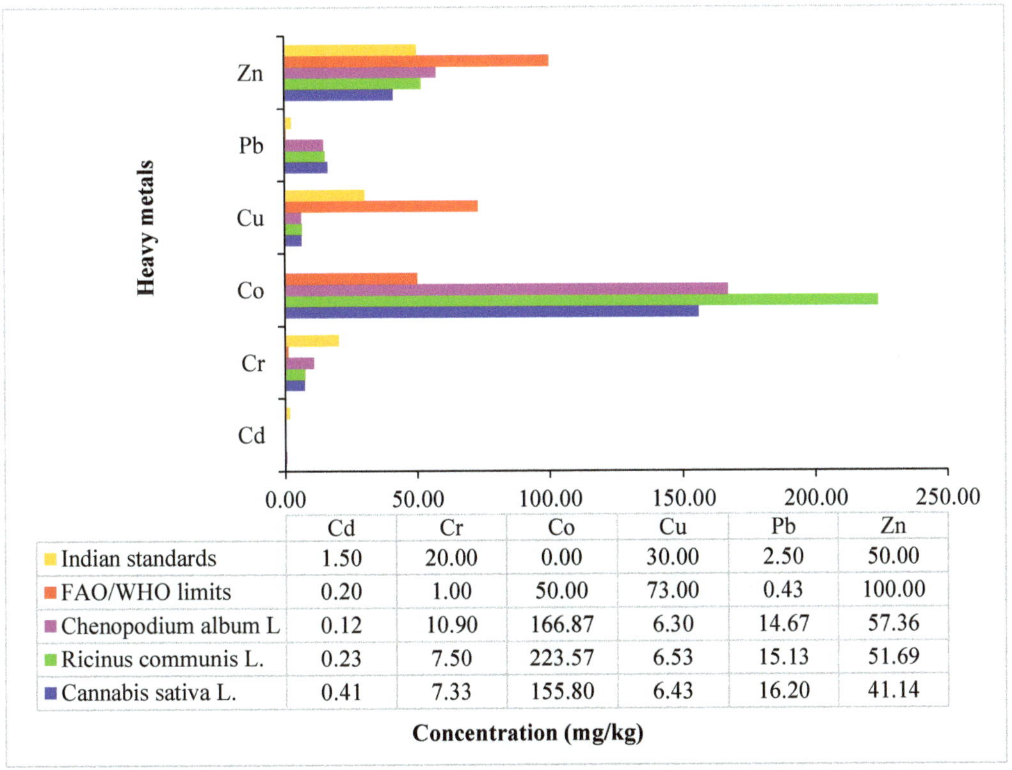

	Cd	Cr	Co	Cu	Pb	Zn
Indian standards	1.50	20.00	0.00	30.00	2.50	50.00
FAO/WHO limits	0.20	1.00	50.00	73.00	0.43	100.00
Chenopodium album L	0.12	10.90	166.87	6.30	14.67	57.36
Ricinus communis L.	0.23	7.50	223.57	6.53	15.13	51.69
Cannabis sativa L.	0.41	7.33	155.80	6.43	16.20	41.14

Figure 1. Heavy metal contents in three plant species viz., *Cannabis sativa* L., *Ricinus communis* L., and *Chenopodium album* L. FAO/WHO, Adapted with permission from [56]. 2001, Indian standards: Awashthi.

In all the collected plant samples, the concentrations of different metals were observed to be above the European permissible limits like 50 mg/kg for cobalt, 0.2 mg/kg for cadmium, and 1 mg/kg for chromium. However, two metals and contents were determined to be lower than the Indian standards like Cd < 1.5 mg/kg and Cr < 20 mg/kg [88,89]. The concentration of copper in all three plant species was recorded to be less than both Indian standards of 30 mg/kg and European permissible limits of 73 mg/kg. Lead content was found to be higher than both Indian standards (2.5 mg/kg) and European permissible limits (0.43 mg/kg). The leaves of *Ricinus communis* L. and *Chenopodium album* L. had a higher zinc content than the Indian standards of 50 mg/kg and lower than that of European limits of 100 mg/kg. The content of zinc in *Cannabis sativa* L. was recorded to be less than both Indian standards and European permissible limits.

The variations of heavy metal contents in soil have a direct influence on the accumulation of heavy metals in plants. However, the heavy metal accumulation in plants also depends on the type of plant species, plant organelles, and traffic density. The plant species included in the present study were preferred because these were wildly grown along the boundaries of the agricultural field at the time of sampling. The contribution of human beings to metal concentrations in the terrestrial environment has arisen mainly from mining, smelting, and industrial activities [90].

3.3. Metal Bioaccumulation Factor (BAF)

BAF is one of the main indices that provide insight into the heavy metal uptake capacity of plant species. In the present study, BAF values were used to estimate and compare the extent of accumulation of various metals such as Cd, Co, Cr, Cu, Pb, and Zn in leaves of the three plants from the soil. Figure 2 presents the BAF values for different plant samples collected during the study. A BAF value above 1 was observed only for cobalt (1.042) in the leaves of *Ricinus communis* L, which indicates a high level of metal bioaccumulation in *Ricinus communis* L. The order of accumulation of heavy metals in the leaves of *Ricinus communis* L. was Co (1.04) > Zn (0.56) > Cu (0.48) > Cr (0.46) > Pb (0.26) > Cd (0.18); for *Chenopodium album* L. the order was Co (0.78) > Cr (0.66) > Zn (0.62) > Cu (0.46) > Pb (0.26) > Cd (0.09); and for *Cannabis sativa* L. it was Co (0.73) > Cu (0.47) > Cr (0.45) > Cd (0.31) > Pb (0.28).

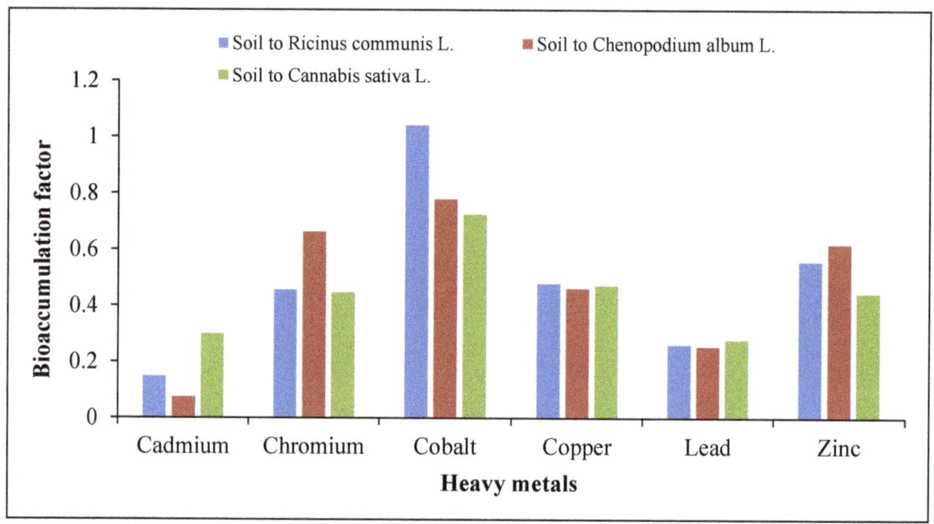

Figure 2. Bioaccumulation factor of Soil to plant.

The heavy metal accumulation examination provides very important information about the phytoremediation potential of the plant species. Transport and accumulation of heavy metals from soil to edible parts act as the major pathway for the entry of heavy metals into the food chain, which ultimately leads to various harmful effects [91]. The exposure of heavy metals to human beings leads to various health problems such as nervous system disorder, skin ailments, stomach problems, kidney damage, bone, and lung diseases [92–94]. The BAF of heavy metals depends upon the bioavailability of metals, which in turn depends upon the concentration of metal in soil, its chemical forms, the difference in uptake capability for different metals, and the growth rate of different plant species [53,95]. Different metals are accumulated in plants at variable rates depending on various factors such as physiology, requirements, and the metal uptake mechanism of plants, the physico-chemical characteristics of soil such as soil texture, soil pH and soil organic matter, as well as quantity of heavy metals present in the soil [94,96–98]. According to the guidelines given by Baker [99], BAF values greater than 1 indicated that the plant was an accumulator for the metal being analyzed and considered as harmful for plant health [53,100]. It is documented that *Ricinus communis* has good tolerance and phytoremediation potential for the removal of nickel (Ni) from contaminated land areas [101,102]. In the present study also, maximum BAF values for Co were observed in *Ricinus communis* L., which showed that this plant had higher metal bioaccumulation capacity than *Chenopodium album* L. and *Cannabis sativa* L. The mechanisms of cobalt accumulation are still not properly defined. However, there are some Cu accumulators which have the potential for Co accumulation as well due to similar mechanisms in the accumulation of different heavy metals [103]. Considering the Co toxicity on the targeted cellular system of plants that can hyperaccumulate, Co has been found to be evolved in regulating Fe homeostasis thus avoiding the accumulation of free ions that can induce oxidative stress [104]. Wong et al. [105] reported that heavy metals in carbonate-bounded form were more bioavailable than the presence of metal in any other fractions. Tamoutisidis et al. [106] revealed that heavy metals were transported passively from the root system to the shoot system through xylem vessels and were accumulated in the zones of high transpiration rates.

3.4. Genotoxicity of Industrial Effluent

The genotoxic potential of textile industrial effluents, before and after treatment was evaluated on the basis of percent aberrant cells. The results of genotoxic potential analysis of effluents and distilled water (negative control) using the *Allium cepa* test system are shown in Table 4. Among all physiological aberrations, delayed anaphases and stickiness were the most frequent type of aberrations while chromatin bridges dominated among clastogenic aberrations. Total chromosomal abnormalities in the meristematic cells of root tips of *Allium cepa* exposed to untreated effluents were significantly higher as compared to those exposed to treated effluents. The reduced mitotic index clearly indicates the cell division reduction in the root meristematic cells, which may be due to the collaborating effects of a complex mixture of cytotoxic chemicals like metals present in the textile industrial effluents. The total percentage aberration including both physiological (laggards, vagrants, stickiness, delayed anaphases, and c-mitosis) and clastogenic (chromatin bridges and chromosomal breaks) aberrations are shown in Figure 3. The total percent aberrant cells were observed to be 29.36% for AU, 27.48% for BU, 19.69% for AT, 16.52% for BT, and 3.84% for the negative control. The results obtained indicate the less toxic nature of the treated effluents of both textile industries (A and B) as compared to the untreated effluents, which can be due to the decrease in heavy metal contents.

Table 4. Genotoxic potential of industrial effluents collected from textile industries of Ludhiana (Punjab), India.

Parameter	NC	AU	AT	BU	BT
Average TDC	494	431	608	500	668
MI (%)	44.37 ± 1.01	13.83 ± 0.13 [#]	22.17 ± 0.40 [#,*]	17.46 ± 0.21 [#]	24.32 ± 0.26 [#,*]
PA (%)	3.43 ± 0.19	26.67 ± 0.30 [#]	17.50 ± 0.22 [#,*]	26.08 ± 0.48 [#]	15.13 ± 0.55 [#,*]
CA (%)	0.41 ± 0.01	2.70 ± 0.17 [#]	2.19 ± 0.15 [#]	1.40 ± 0.20 [#]	1.40 ± 0.06 [#]
TA (%)	3.84 ± 0.19	29.37 ± 0.40 [#]	19.69 ± 0.36 [#,*]	27.48 ± 0.44 [#]	16.52 ± 0.59 [#,*]

(NC: Negative Control; (AU: Untreated effluent of textile industry A; AT: Treated effluent of textile industry A; BU: Untreated effluent of textile industry B; BT: Treated effluent of textile industry B); TDC: Total Dividing Cells; MI: Mitotic Index; PA: Physiological Aberration; CA: Clastogenic Aberration; TA: Total Aberration). * Indicates statistically significant difference between values of genotoxicity parameters (MI, PA, CA, and TA) for untreated and treated effluents of same industry (Independent Student's *t*-test, $p \leq 0.05$). # Indicates statistically significant difference between values of genotoxicity parameters (MI, PA, CA, and TA) for effluents and negative control (Chi square test, $p \leq 0.05$).

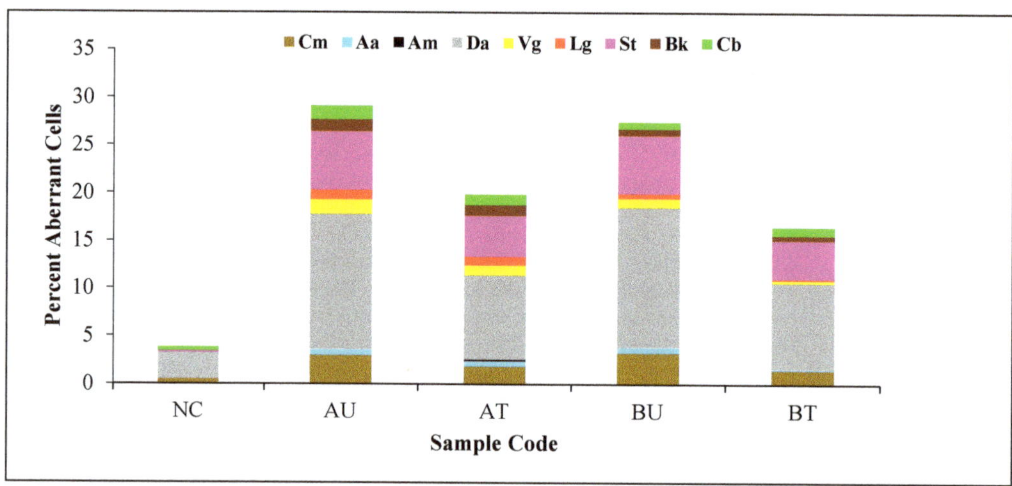

Figure 3. Induction of physiological and clastogenic chromosomal aberrations in root tip cells of *Allium cepa* under exposure to industrial effluents and distilled water (Negative Control). Physiological aberrations (Cm: C-mitosis; Da: Delayed anaphase; Lg: Laggards; St: Stickiness; Vg: Vagrants; Aa: Abnormal anaphases; Am: Abnormal metaphases); Clastogenic aberrations (Cb: Chromatin bridges; Bk: Chromosomal breaks), Sample codes (AU: Untreated effluent of textile industry A; AT: Treated effluent of textile industry A; BU: Untreated effluent of textile industry B; BT: Treated effluent of textile industry B); NC: Negative control.

The results observed in the present study indicate the mitogenic, as well as the clastogenic effects of the textile effluents, which were evident from the low value of the mitotic index (MI) and higher values of the chromosomal aberrations assay when compared to the results obtained from treatment with negative control (distilled water). The statistical analysis (Chi square test) also revealed that there is a significant difference between the values of the genotoxicity parameters viz., physiological aberration (PA), clastogenic aberration (CA), and total aberration (TA) along with mitotic index (MI) for effluents of both textile industries (A and B) and negative control at $p \leq 0.5$. Chromosomal aberration is an important indicator for assessing the genotoxicity of textile effluents [107]. The *Allium cepa* root chromosomal aberration assay has been widely used for cytotoxic as well as genotoxic mitotic studies [41,108–110]. The reduction in the values of mitotic index, in the present study, indicated the cytotoxic effects, whereas the induction of chromosomal and nuclear abnormalities showed genotoxic effects. Both cytotoxic and genotoxic effects were endorsed by various environmental pollutants [44].

The higher genotoxic response of root tip cells under exposure to untreated effluent of textile industry A as compared to untreated effluent of B industry can be attributed to the presence of high content of Cd, Co, Na, and Cl. Some authors have reported the cytotoxic and genotoxic effects using the *Allium cepa* test system following exposure to heavy metals [111,112]. Cd has been shown to reduce the mitotic index (MI) and enhance the induction of chromosomal aberrations, as well as micronuclei, in various studies [113–116]. Grover and Kaur [117] studied the genotoxic potential of textile and paper mill effluents and sewage water following the *Allium cepa* chromosomal aberration assay and reported that the industrial effluents induced the formation of micronuclei and chromosomal abnormalities in the root tip cells of *Allium cepa*. Genotoxicity of both untreated and treated textile industrial effluents was evaluated using the *Allium cepa* test system by Vijayalakshmidevi and Muthukumar, [118] and they observed a reduction in the mitotic index, as well as induction of various types of chromosomal aberrations in the root tips exposed to effluent. It is possible that some chemicals in the complex chemical mixtures could have stimulatory effects on the mitotic process while some others might have mito-depressive effects [43]. Similar results were also reported for genotoxicity of textile industrial effluents using the *Allium cepa* test system by other authors [42,108,119,120] and they demonstrated the induction of chromosomal abnormalities and decrease in the mitotic index in root tips cells treated with the effluent. Therefore, mitotic responses observed in this study could be due to the overall collaborative effects such as additive, antagonistic, and synergistic of the complex chemical mixtures in the effluents on the root meristematic cells.

3.5. Pollution Assessment

The Igeo, *CF*, Cd_{eg}, mCd_{eg}, PI, PLI, ER_i and RI of the agricultural soil in the present study were calculated based on the heavy metals content in the studied soil sample. Table 5 indicates that the *CF* value of heavy metals in studied area was ranked as: Co (21.46) > Cd (13.57) > Pb (2.87) > Zn (1.30) > Cu (0.55) > Cr (0.47) and the Igeo value ranked as Co (3.84) > Cd (3.18) > Pb (0.93). Pollution levels of contamination factor (*CF*) were classified as: low contamination (*CF* < 1), moderate contamination (*CF* value in the range of 1–3), considerable contamination (*CF* value in the range of 3–6), and very high contamination (*CF* > 6) by Taylor and Mclennan [121] and Hakanson [122]. The *CF* value indicated that the soil is extremely polluted by Cd and Co, moderately polluted by Pb and Zn whereas unpolluted by Cu and Cr. The result of Igeo showed that soil is heavily contaminated by Cd and Co whereas uncontaminated by Cr, Cu, and Zn on the basis of the classification given by Muller [123] and Taylor and Mclennan [121]. Considering the Cd_{eg} values > 32, PI values > 3, and PLI > 3, the studied soil was found to be extremely polluted with heavy metals whereas mCd_{eg} (6.70) value in the present study indicated that soil has a high degree of contamination. ER_i values for Cr, Cu, Pb, and Zn were observed to be below 40, indicating low potential ecological risk from these metals whereas Cd showed considerable potential ecological risk and Co exhibited very high potential ecological risk. In the present work, ER_i values showed that Co is the major pollutant in the area which indicates that agricultural management is a probable cause of heavy metals accretion. The potential ecological risk index (RI) demonstrated that the study area had considerable ecological risk considering the RI value in the range of 300–600.

Table 5. Metal pollution indices for collected soil samples from Ludhiana, Punjab (India).

Metal	Igeo	CF	Cd_{eg}	mCd_{eg}	PI	PLI	ER_i	RI
Cd	3.18	13.57	40.22	6.70	15.90	26.41	407.14	533.74
Cr	−1.68	0.47					0.94	
Co	3.84	21.46					107.3	
Cu	−1.46	0.55					2.73	
Pb	0.93	2.87					14.33	
Zn	−0.203	1.30					1.30	

Igeo: geoaccumulation index; CF: contamination factor; Cd_{eg}: degree of contamination; mCd_{eg}: modified degree of contamination; PI: Numerow's pollution index; PLI: pollution load index; ER_i: potential ecological risk factor; RI: potential ecological risk index.

3.6. Human Health Risk Assessment

The non-carcinogenic, hazard quotient (*HQ*), and hazard index (*HI*) of analyzed heavy metals (Cd, Cr, Co, Cu, Pb, and Zn) through three exposure pathways, that is, ingestion, dermal contact, and inhalation for adults and children were calculated and results were shown in Table 6. The values of $HQ_{ingestion}$, HQ_{dermal}, and $HQ_{inhalation}$ for all studied heavy metals were found to be lower than 1 for both adults and children and thus indicated that there is no obvious risk to the population. The total carcinogenic risks (TCR) were calculated only for Cr as cancer slope factors (*SF*) for all three exposure pathways (ingestion, dermal contact, and inhalation) are not available for other heavy metals. The total carcinogenic risk value was found to be in the range of the permissible limit of 1×10^{-6} to 1×10^{-4}, as provided by USEPA [28]. The results of cancer risk were shown in Figure 4.

Table 6. Exposure values for non-carcinogenic risks for adults and children from different exposure pathways in the study area.

Receptor	Exposure Pathway	Cd	Cr	Co	Cu	Pb	Zn
Adult	ADI ingestion	1.9×10^{-6}	2.347×10^{-5}	3.066×10^{-4}	1.947×10^{-5}	8.19×10^{-5}	1.322×10^{-4}
	ADI dermal	5.786×10^{-8}	7.147×10^{-7}	9.335×10^{-6}	5.929×10^{-7}	2.494×10^{-6}	4.0246×10^{-6}
	AID inhalation	1.788×10^{-10}	2.209×10^{-9}	2.885×10^{-8}	1.833×10^{-9}	7.708×10^{-9}	1.244×10^{-8}
	Total	1.958×10^{-6}	2.42×10^{-5}	3.159×10^{-4}	2.01×10^{-5}	8.440×10^{-5}	1.36×10^{-4}
	HQ ingestion	1.9×10^{-3}	7.824×10^{-3}	1.533×10^{-2}	4.868×10^{-4}	5.85×10^{-2}	4.406×10^{-4}
	HQ dermal	5.786×10^{-3}	2.382×10^{-4}	5.834×10^{-4}	4.941×10^{-5}	4.759×10^{-3}	6.708×10^{-5}
	HQ inhalation	1.788×10^{-7}	7.724×10^{-5}	5.053×10^{-3}	4.582×10^{-8}	2.190×10^{-6}	4.147×10^{-8}
	HI	7.686×10^{-3}	8.139×10^{-3}	2.097×10^{-2}	5.362×10^{-4}	6.326×10^{-2}	5.077×10^{-4}
Children	ADI ingestion	1.33×10^{-5}	1.643×10^{-4}	2.146×10^{-3}	1.363×10^{-4}	5.733×10^{-4}	9.252×10^{-4}
	ADI dermal	2.128×10^{-8}	2.629×10^{-7}	3.434×10^{-6}	2.181×10^{-7}	9.173×10^{-7}	1.480×10^{-6}
	ADI inhalation	3.731×10^{-10}	4.609×10^{-9}	6.020×10^{-8}	3.823×10^{-9}	1.608×10^{-8}	2.595×10^{-8}
	Total	1.332×10^{-5}	1.64567×10^{-4}	2.149×10^{-3}	1.365×10^{-4}	5.742×10^{-4}	9.267×10^{-4}
	HQ ingestion	1.33×10^{-2}	5.477×10^{-2}	0.1073	3.408×10^{-3}	0.410	3.084×10^{-3}
	HQ dermal	2.128×10^{-3}	8.763×10^{-5}	2.146×10^{-4}	1.817×10^{-5}	1.751×10^{-3}	2.467×10^{-5}
	HQ inhalation	3.731×10^{-7}	1.611×10^{-4}	1.054×10^{-2}	9.559×10^{-8}	4.569×10^{-6}	8.651×10^{-8}
	HI	1.543×10^{-2}	5.502×10^{-2}	0.118	3.426×10^{-3}	0.411	3.109×10^{-4}

(ADI: Average daily intake; HQ: Hazard quotient; HI: Hazard index).

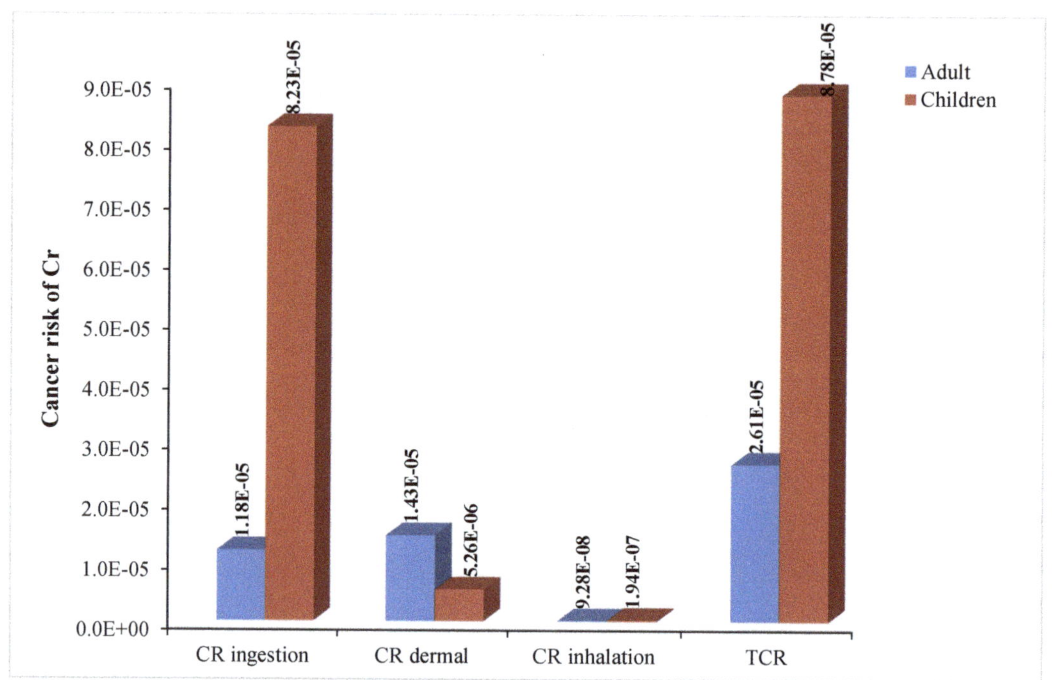

Figure 4. Distribution of carcinogenic risk of chromium (Cr) for adults and children in the study area. (CR; carcinogenic risk; TCR: total carcinogenic risk).

4. Conclusions

The present study pertained to exploring the potential ecological risks of heavy metals of textile effluents in soil samples in the vicinity of textile industries Ludhiana, Punjab (India). The metal bioaccumulation potential of some plant species grown in its environs was also explored. The Co content in untreated and treated effluent samples indicated the possibility of accumulation of cobalt in agricultural soil samples and plant samples in the vicinity of textile industries, which is a serious matter of concern. The genotoxicity assay showed that treated as well as untreated effluents of both industries induced chromosomal aberrations and the percent aberrations in treated samples were significantly lower than untreated samples. The heavy metal bioaccumulation factor analysis showed that phytoremediation using wildly grown plants like *Ricinus communis* L., *Chenopodium album* L. and *Cannabis sativa* L., can be one of the environmentally friendly techniques for cleaning contaminated soil environs. Furthermore, Igeo and *CF* revealed that heavy metals showed no contamination to extreme contamination in the studied soil whereas Cd_{eg}, PI, and PLI indicated extreme pollution. The results of ER_i studies indicated that Co is the prime metal responsible for ecological threats in the study area. It is also emphasized that bioanalytical tools such as the *Allium cepa* root chromosomal aberration assay should be incorporated along with chemical analysis for evaluating the efficacy of industrial effluent treatment plants so as to indicate the harmful consequences in the biological systems.

Supplementary Materials: The following are available online at https://www.mdpi.com/2571-8789/5/4/63/soilsystems5040063/s1, Table S1: Descriptions of the soil contamination indices used in the study, Table S2. Summary of reference doses (*RfD*) and slope factors (*SF*) of heavy metals.

Author Contributions: J.K.: She has carried out all experimental and has drafted/written the manuscript; S.S.B.: Helped in statistical analysis of data; S.A.B.: review and editing the manuscript; A.K.N.: Contributed for assessment of health risk; V.K.: Co-supervisor of the first author and helped in reviewing the whole manuscript; J.K.K.: Supervisor of the first author and helped in the drafting of manuscript/guidance for result compilation and finalizing the manuscript. All authors have read and agreed to the published version of the manuscript.

Funding: This research was funded by University Potential for Excellence (UPE) as the first author is recipient of fellowship under this program.

Institutional Review Board Statement: Not applicable.

Informed Consent Statement: Not applicable.

Data Availability Statement: The data presented in this study are available on request from the corresponding author.

Acknowledgments: The authors are thankful to the University Grants Commission (UGC), New Delhi for providing financial assistance through various schemes like University Potential for Excellence (UPE), Centre with Potential for Excellence in Particular Area (CPEPA), Department of Science and Technology—Promotion of University Research and Scientific Excellence (DST-PURSE), Departmental Research Support-Special Assistance Program (DRS-SAP). Thanks are due to the Head of the Department of Botanical and Environmental Sciences for providing laboratory facilities and the Central Facility, Emerging Life Sciences, Guru Nanak Dev University, Amritsar for providing sophisticated instrumentation to carry out the present work.

Conflicts of Interest: The authors declare no conflict of interest.

References

1. Tabrez, S.; Ahmad, M. Oxidative stress-mediated genotoxicity of wastewaters collected from two different stations in northern India. *Mutat. Res.* **2011**, *726*, 15–20. [CrossRef] [PubMed]
2. Dey, S.; Islam, A. A Review on Textile Wastewater Characterization in Bangladesh. *Resour. Environ.* **2015**, *5*, 15–44.
3. Brevik, E.C.; Slaughter, L.; Singh, B.R.; Steffan, J.J.; Collier, D.; Barnhart, P.; Pereira, P. Soil and human health: Current status and future needs. *Air Soil Water Res.* **2020**, *13*, 1–23. [CrossRef]
4. Mandour, R.A.; Azab, Y.A. The Prospective Toxic Effects of Some Heavy Metals Overload in Surface Drinking Water of Dakahlia Governorate, Egypt. *J. Occup. Environ. Med.* **2011**, *2*, 245–253.
5. Huang, D.; Liu, X.; Jiang, S.; Wang, H.; Wang, J.; Zhang, Y. Current state and future perspectives of sewer networks in urban China. *Front. Environ. Sci. Eng.* **2018**, *12*, 2. [CrossRef]
6. Singh, S.; Kumar, M. Heavy metal load of soil, water and vegetables in periurban Delhi. *Environ. Monit. Assess.* **2006**, *120*, 79–91. [CrossRef]
7. Sharma, R.K.; Agarwal, M.; Marshall, F.M. Heavy metals contamination of soil and vegetables in suburban areas of Varanasi, India. *Ecotoxicol. Environ. Saf.* **2007**, *66*, 258–266. [CrossRef]

8.   Zehra, S.; Arshad, M.; Mahmood, T.; Waheed, A. Assessment of heavy metal accumulation and their translocation in plant species. *Afr. J. Biotechnol.* **2009**, *8*, 2802–2810.

9.   Doherty, V.F.; Sogbanmu, T.O.; Kanife, U.C.; Wright, O. Heavy metals in vegetables collected from selected farm and market sites in Lagos, Nigeria. *Glob. Adv. Res. J. GJAR* **2012**, *1*, 137–142.

10.   Egito, L.C.M.; Medeiros, M.D.G.; Medeiros, S.R.B.D.; Agnez-Lima, L.F. Cytotoxic and genotoxic potential of surface water from the Pitimbu river, Northeastern/RN Brazil. *Gen. Mol. Biol.* **2007**, *30*, 435–441. [CrossRef]

11.   Alam, Z.M.; Ahmad, S.; Malik, A.; Ahmad, M. Genotoxic and mutagenic potential of agricultural soil irrigated with tannery effluents at Jajmau (Kanpur), India. *Arch. Environ. Contam. Toxicol.* **2009**, *57*, 463–476. [CrossRef]

12.   Alam, Z.M.; Ahmad, S.; Malik, A.; Ahmad, M. Mutagenicity and genotoxicity of tannery effluents used for irrigation at Kanpur, India. *Ecotoxicol. Environ. Saf.* **2010**, *73*, 1620–1628. [CrossRef]

13.   Magdaleno, A.; Puig, A.; de Cabo, L.; Salinas, C.; Arreghini, S.; Korol, S.; Bevilacqua, S.; Liopez, L.; Moretton, J. Water pollution in an urban Argentine river. *Bull. Environ. Contam. Toxicol.* **2001**, *67*, 408–415. [CrossRef] [PubMed]

14.   Ohe, T.; Watanabe, T.; Wakabayashi, K. Mutagens in surface waters: A review. *Mutat. Res.* **2004**, *567*, 109–149. [CrossRef]

15.   Siddiqui, A.H.; Ahmad, M. The *Salmonella* mutagenicity of industrial, surface and ground water sample of Aligarh region of India. *Mutat. Res.* **2003**, *541*, 21–29. [CrossRef]

16.   Fatoki, O.S. Trace Zinc and Copper Concentrations in Roadside Vegetation and Surface Soils: A Measurement of Local Atmospheric Pollution in Alice, South Africa. *Int. J. Environ. Stud.* **2000**, *57*, 501–513. [CrossRef]

17.   Keshavarzi, A.; Kumar, V.; Ertunc, G.; Brevik, E.C. Ecological risk assessment and source apportionment of heavy metals contamination: An appraisal based on the Tellus soil survey. *Environ. Geochem. Health* **2021**, *43*, 2121–2142. [CrossRef] [PubMed]

18.   Bhatti, S.S.; Bhatt, S.A.; Kumar, V.; Kaur, M.; Sambyal, V.; Singh, J.; Vig, A.P.; Nagpal, A.K. Ecological risk assessment of metals in roadside agricultural soils: A modified approach. *Hum. Ecol. Risk Assess.* **2018**, *24*, 186–201. [CrossRef]

19.   Kachenko, A.G.; Singh, B. Heavy metals contamination in vegetables grown in urban and metal smelter contaminated sites in Australia. *Water Air Soil Poll.* **2006**, *169*, 101–123. [CrossRef]

20.   Atafar, Z.; Mesdaghinia, A.; Nouri, J.; Homaee, M.; Yunesian, M.; Ahmadimoghaddam, M.; Mahvi, A.H. Effect of fertilizer applications on soil heavy metal concentration. *Environ. Monit. Assess.* **2010**, *160*, 83–89. [CrossRef]

21.   Cui, Y.J.; Zhu, Y.G.; Zhai, R.H.; Chen, D.Y.; Huang, Y.Z.; Qiu, Y.; Liang, J.Z. Transfer of metals from soil to vegetables in an area near a smelter in Nanning, China. *Environ. Int.* **2004**, *30*, 785–791. [CrossRef]

22.   Lee, C.S.; Li, X.D.; Shi, W.Z.; Cheung, S.C.; Thornton, L. Metal contamination in urban, suburban and country park soils of Hong Kong: A study based on GIS and multivariate statistics. *Sci. Total Environ.* **2006**, *356*, 45–61. [CrossRef]

23.   Tepanosyan, G.; Sahakyan, L.; Belyaeva, O.; Maghakyan, N.; Saghatelyan, A. Human health risk assessment and riskiest heavy metal origin identification in urban soils of Yerevan, Armenia. *Chemosphere* **2017**, *184*, 1230–1240. [CrossRef] [PubMed]

24.   Adimalla, N.; Wang, H. Distribution, contamination, and health risk assessment of heavy metals in surface soils from northern Telangana, India. *Arab. J. Geosci.* **2018**, *11*, 684. [CrossRef]

25.   Chen, H.Y.; Teng, Y.G.; Lu, S.J.; Wang, Y.; Wang, J. Contamination features and health risk of soil heavy metals in China. *Sci. Total Environ.* **2015**, *512*, 143–153. [CrossRef]

26.   Eziz, M.; Mohammad, A.; Mamut, A.; Hini, G. A human health risk assessment of heavy metals in agricultural soils of Yanqi Basin, Silk Road Economic Belt, China. *Hum. Ecol. Risk Assess.* **2018**, *24*, 1352–1366. [CrossRef]

27.   Adimalla, N. Heavy metals pollution assessment and its associated human health risk evaluation of urban soils from Indian cities: A review. *Environ. Geochem. Health* **2020**, *42*, 173–190. [CrossRef]

28.   US Environmental Protection Agency. *Supplemental Guidance for Developing Soil Screening Levels for Superfund Sites*; OSWER 9355; Office of Emergency and Remedial Response, US Environmental Protection Agency: Washington, DC, USA, 2002; pp. 4–24.

29.   Eaton, A.D.; Clesceri, L.S.; Rice, E.W.; Greenberg, A.E. *Standard Methods for The Examination of Water and Wastewater*, Centennial ed.; American Public Health Association, American Water Works Association, Water Environment Federation: Washington, DC, USA, 2005.

30.   Kumar, V.; Bhatti, S.S.; Nagpal, A.K. Seasonal analysis of physico-chemical parameters of agricultural soil samples collected from banks of rivers Beas and Sutlej, Punjab, India. *J. Chem. Pharm. Res.* **2016**, *8*, 439–449.

31.   Trivedi, R.K.; Goel, P.K.; Trisal, C.L. (Eds.) Aquatic ecosystem. In *Practical Methods in Ecology and Environmental Sciences*; Enviro Media Pub: Bhopal, India, 1987; pp. 57–113.

32.   Nelson, D.W.; Sommer, L.E. Total Carbon, Organic Carbon and Organic Matter. Methods of Soil Analysis, Part 2. In *Chemical and Microbiological Properties*, 2nd ed.; ASA-SSSA: Madison, WI, USA, 1982; pp. 579–595.

33.   Jacob, H.; Clarke, G. Part 4. Physical method. In *Methods of Soil Analysis*; Soil Science Society of America: Madison, WI, USA, 2002; p. 1692.

34.   ISO 11277. *Soil Quality-Determination of Particle Size Distribution in Mineral Soil Material-Method by Sieving and Sedimentation*; ISO: Geneva, Switzerland, 2009.

35.   American Public Health Association. *Standard Methods for the Examination of Water and Wastewater*, 20th ed.; American Public Health Association: Washington, DC, USA, 1998.

36.   Sharma, D.; Katnoria, J.K.; Vig, A.P. Chemical changes of spinach waste during composting and vermicomposting. *Afr. J. Biotechnol.* **2011**, *10*, 3124–3127.

37. Ehi-Eromosole, C.O.; Adaramodu, A.A.; Anake, W.U.; Ajanaku, C.O.; Edobor-Osoh, A. Comparison of three methods of digestion for trace metal analysis in suface dust collected from an Ewaste recycling site. *Nat. Sci.* **2012**, *10*, 1–6.
38. Allen, S.E.; Grimshaw, H.M.; Rowland, A.P. Chemical Analysis. In *Methods in Plant Ecology*; Moore, P.D., Chapman, S.B., Eds.; Blackwell Scientific Publications: Oxford, UK; London, UK, 1986; pp. 285–344.
39. Huang, M.; Zhou, S.; Sun, B.; Zhao, Q. Heavy metals in wheat grain: Assessment of potential health risk for inhabitants in Kunshun, China. *Sci. Total Environ.* **2008**, *405*, 54–61. [CrossRef] [PubMed]
40. Ali, H.; Khan, E.; Sajad, M.A. Phytoremediation of heavy metals-concepts and applications. *Chemosphere* **2013**, *91*, 869–881. [CrossRef] [PubMed]
41. Katnoria, J.K.; Arora, S.; Bhardwaj, R.; Nagpal, A.K. Evaluation of genotoxic potential of industrial waste contaminated soil extracts of Amritsar, India. *J. Environ. Biol.* **2011**, *32*, 363–367. [PubMed]
42. Pathiratne, A.; Hemachandra., C.K.; De Silva, N. Efficacy of *Allium cepa* test system for screening cytotoxicity and genotoxicity of industrial effluents originated from different industrial activities. *Environ. Monit. Assess.* **2015**, *187*, 1–12. [CrossRef]
43. Hemachandra, C.K.; Pathiratne, A. Combination of physico-chemical analysis, *Allium cepa* test system and *Oreochromis niloticus* erythrocyte based comet assay/nuclear abnormalities tests for cyto-genotoxicty assessments of treated effluents discharged from textile industries. *Ecotoxicol. Environ. Saf.* **2016**, *131*, 54–64. [CrossRef]
44. Leme, D.M.; Marin-Morales, M.A. *Allium cepa* test in environmental monitoring: A review on its application. *Mutat. Res.* **2009**, *682*, 71–81. [CrossRef]
45. US Environmental Protection Agency. Risk Assessment Guidance for Superfund. In *Human Health Evaluation Manual, Part A*; EPA/540/1–89/002; Office of Emergency and Remedial Response, US Environmental Protection Agency: Washington, DC, USA, 1989; Volume 1.
46. US Environmental Protection Agency. Guidelines for carcinogen risk assessment, EPA/630/P-03/001F. In *Risk Assessment Forum*; US Environmental Protection Agency: Washington, DC, USA, 2005.
47. US Environmental Protection Agency. *Risk Assessment Guidance for Superfund Volume I: Human Health Evaluation Manual (part F, Supplemental Guidance for Inhalation Risk Assessment)*; OSWER 9285; Office of Superfund Remediation and Technology Innovation, US Environmental Protection Agency: Washington, DC, USA, 2009; pp. 7–82.
48. US Environmental Protection Agency. *Exposure Factors Handbook*, 2011 ed.; EPA/600/R-09/052F; National Center for Environmental Assessment: Washington, DC, USA, 2011.
49. US Environmental Protection Agency. *Exposure Factors Handbook*; Volume 1: General Factors; US Environmental Protection Agency, Office of Research and Development: Washington, DC, USA, 1997.
50. Li, Z.; Ma, Z.; van der Kuijp, T.J.; Yuan, Z.; Huang, L. A review of soil heavy metal pollution from mines in China: Pollution and health risk assessment. *Sci. Total Environ.* **2014**, *468*, 843–853. [CrossRef]
51. Paul, S.A.; Chavan, S.K.; Khambe, S.D. Studies on characterization of textile industrial waste water in Solapur city. *Int. J. Chem. Sci.* **2012**, *10*, 632–642.
52. Ramamurthy, N.; Balasaraswathy, S.; Sivasakthivelan, P. Biodegradation and Physico-chemical changes of textile effluent by various fungal species. *Rom. J. Biophys.* **2011**, *21*, 113–123.
53. Singh, A.; Sharma, R.K.; Agrawal, M.; Marshall, F.M. Health risk assessment of heavy metals via dietary intake of foodstuffs from the wastewater irrigated site of a dry tropical area of India. *Food Chem. Toxicol.* **2010**, *48*, 611–619. [CrossRef]
54. Vyas, P.B. An aerobic treatment for Dudh Sagar dairy, Mahesana. *J. Ind. Pollut. Control.* **2011**, *27*, 29–32.
55. Bureau of Indian Standards. *Indian Standard Drinking Water Specification*; Bureau of Indian Standards: New Delhi, India, 2012; p. 10500.
56. Awashthi, S.K. (Ed.) *Central and State Rules as Amended for 1999: Prevention of Food Adulteration Act No. 37 of 1954*; Ashoka Law House: New Delhi, India, 2000.
57. Yaseen, D.A.; Scholz, M. Textile dye wastewater characteristics and constituents of synthetic effluents: A critical review. *Int. J. Environ. Sci. Technol.* **2018**, *16*, 1193–1226. [CrossRef]
58. Adinew, B. Textile effluent treatment and decolorization techniques—A review. *Chem. Bulg. J. Sci. Educ.* **2012**, *21*, 434–456.
59. Chukwu, L.O. Physico-chemical characterization of pollutant load of treated industrial effluents in Lagos metropolis, Nigeria. *J. Indus. Poll. Control* **2006**, *22*, 17–22.
60. Hussein, F.H. Chemical properties of treated textile dyeing wastewater. *Asian J. Chem.* **2013**, *25*, 9393–9400. [CrossRef]
61. European Union. Heavy Metals in Wastes, European Commission on Environment 2000. Available online: http://ec.europa.eu/environment/waste/studies/pdf/heavy_metalsreport.pdf (accessed on 21 July 2014).
62. Zhao, Z.; Hazelton, P. Evaluation of accumulation and concentration of heavy metals in different urban roadside soil types in Miranda Park, Sydney. *J. Soils Sediment.* **2016**, *16*, 2548–2556. [CrossRef]
63. Bhatti, S.S.; Sambyal, V.; Nagpal, A.K. Heavy metals bioaccumulation in Berseem (*Trifoliumalexandrium*) cultivated in areas under intensive agriculture, Punjab, India. *SpringerPlus* **2016**, *5*, 1–11. [CrossRef]
64. Bhatti, S.S.; Sambyal, V.; Singh, J.; Nagpal, A.K. Analysis of soil characteristics of different land uses and metal bioaccumulation in wheat grown around rivers: Possible human health risk assessment. *Environ. Dev. Sustain.* **2017**, *19*, 571–588. [CrossRef]
65. Aschale, M.; Sileshi, Y.; Kelly-Quinn, M.; Hailu, D. Pollution Assessment of Toxic and Potentially Toxic Elements in Agricultural Soils of the City Addis Ababa, Ethiopia. *Bull. Environ. Contam. Toxicol.* **2017**, *98*, 234–243. [CrossRef]

66. Chabukdhara, M.; Munjal, A.; Nema, A.K.; Gupta, S.K.; Kaushal, R.K. Heavy metal contamination in vegetables grown around peri-urban and urban-industrial clusters in Ghaziabad, India. *Hum. Ecol. Risk Assess.* **2016**, *22*, 736–752. [CrossRef]

67. Tian, K.; Huang, B.; Xing, Z.; Hu, W. Geochemical baseline establishment and ecological risk evaluation of heavy metals in greenhouse soils from Dongtoi, China. *Ecol. Indic.* **2017**, *72*, 510–520. [CrossRef]

68. Guan, Z.H.; Lil, X.G.; Wang, L. Heavy metal enrichment in roadside soils in the eastern Tibetan Plateau. *Environ. Sci. Pollut. Res.* **2018**, *25*, 7625–7637. [CrossRef] [PubMed]

69. Krailertrattanachai, N.; Ketrot, D.; Wisawapipat, W. The Distribution of Trace Metals in Roadside Agricultural Soils, Thailand. *Int. J. Environ. Res. Public Health* **2019**, *16*, 714. [CrossRef] [PubMed]

70. Szwalec, A.; Mundała, P.; Kędzior, R.; Pawlik, J. Monitoring and assessment of cadmium, lead, zinc and copper concentrations in arable roadside soils in terms of different traffic conditions. *Environ. Monit. Assess.* **2020**, *192*, 155. [CrossRef] [PubMed]

71. Khan, M.A.; Ahmad, I.; Rahman, I. Effect of Environmental Pollution on Heavy Metals Content of *Withamnia somnifera*. *J. Chin. Chem. Soc.* **2007**, *54*, 339–343. [CrossRef]

72. Ahmed, F.; Fakhruddin, A.N.M.; Imam, M.D.T.; Khan, N.; Khan, T.A.; Rahman, M.M.; Abdullah, A.T.M. Spatial distribution and source identification of heavy metal pollution in roadside surface soil: A study of Dhaka Aricha highway, Bangladesh. *Ecol. Process.* **2016**, *5*, 1–16. [CrossRef]

73. Aryal, R.; Beecham, S.; Sarkar, B.; Chong, M.N.; Kinsela, A.; Kandasamy, J.; Vigneswaran, S. Readily wash-off road dust and associated heavy metals on motorways. *Water Air Soil Pollut.* **2017**, *228*, 1–12. [CrossRef]

74. Kim, J.H.; Gibb, H.J.; Howe, P.D. Cobalt and Inorganic Cobalt Compounds. In *Concise International Chemical Assessment Document, No. 69*; World Health Organization: Geneva, Switzerland, 2006; pp. 1–82.

75. Gad, N. Role and importance of cobalt nutrition on groundnut (*Arachis hypogaea*) production. *World Appl. Sci. J.* **2012**, *20*, 359–367.

76. Witte, C.P.; Tiller, S.A.; Taylor, M.A.; Davies, H.V. Addition of Nickel to Murashiga and Skoog medium in plant tissue culture activates urease and may reduce metabolic stress. *Plant Cell Tiss. Org. Cult.* **2002**, *86*, 103–104. [CrossRef]

77. Nagpal, N.K. *Water Quality Guidelines for Cobalt*; Ministry of Water, Land and Air Protection, Water Protection Section, Water, Air and Climate Change Branch: Victoria, TX, USA, 2004.

78. Caselles, J.M.; Pérez-Espinosa, A.; Pérez Murcia, M.D.; Moral, R.; Gomez, I. Effect of Increased Cobalt Treatments on Cobalt Concentration and Growth of Tomato Plants. *J. Plant Nutr.* **1997**, *20*, 805–811. [CrossRef]

79. Jayakumar, K.; Jaleel, C.A. Uptake and accumulation of cobalt in plants: A study based on exogenous cobalt in soybean. *Bot. Res. Int.* **2009**, *2*, 310–314.

80. Tchounwou, P.B.; Yedjou, C.G.; Patlolla, A.K.; Sutton, D.J. Heavy Metal Toxicity nad the Environment. *Mol. Clin. Environ. Toxicol. Exp. Suppl.* **2012**, *101*, 133–164. [CrossRef]

81. Alloway, B.J. *Heavy Metals in Soils*; Blackie Academic and Professional: London, UK, 1990; pp. 1–330.

82. Nadian, H. Cd and Mn uptake and bioaccumulation in *Trifolium alexandrinum* L.: Interaction with mycorrhizal colonization. In Proceedings of the Fourth International Iran and Russia Conference, Shahrekord, Iran, 8–10 September 2004; pp. 595–601.

83. McDermott, S.; Wu, J.; Cai, B.; Lawson, A.; Marjorie Aelion, C. Probability of intellectual disability is associated with soil concentrations of arsenic and lead. *Chemosphere* **2011**, *84*, 31–38. [CrossRef] [PubMed]

84. Naureen, A.; Irshad, M.; Hussain, F.; Mahmood, Q. Comparing heavy metals accumulation potential in natural vegetation and soil adjoining wastewater canal. *J. Chem. Soc. Pak.* **2011**, *33*, 661–665.

85. Chunilall, V.; Kindness, A.; Johnalagada, S.B. Heavy metal uptake by two edible Amaranthus herbs grown on soils contaminated with lead, mercury, cadmium and nickel. *J. Environ. Sci. Health* **2005**, *40*, 375–385. [CrossRef] [PubMed]

86. Rattan, R.K.; Datta, S.P.; Chhonkar, P.K.; Suribabu, K.; Singh, A.K. Long-term impact of irrigation with sewage effluents on heavy metal content in soils, crops and groundwater—A case study. *Agric. Ecosyst. Environ.* **2005**, *109*, 310–322. [CrossRef]

87. Muchuweti, M.; Birkett, J.W.; Chinyanga, E.; Zvauya, R.; Scrimshaw, M.D.; Lester, J.N. Heavy metal content of vegetables irrigated with mixtures of wastewater and sewage sludge in Zimbabwe: Implications for human health. *Agric. Ecosyst. Environ.* **2006**, *112*, 41–48. [CrossRef]

88. Lokeshwari, H.; Chandrappa, G. Impact of heavy metal contamination of Bellandur Lake on soil and cultivated vegetation. *Curr. Sci.* **2006**, *91*, 622–627.

89. Zango, M.S.; Anim-Gyampo, M.; Ampadu, B. Health risks of heavy metals in selected food crops cultivated in small-scale gold-mining areas in Wassa-Amenfi-West district of Ghana. *J. Nat. Sci.* **2013**, *3*, 96–105.

90. Collins, R.N.; Kinsela, A. The aqueous phase speciation and chemistry of cobalt in terrestrial environments. *Chemosphere* **2010**, *79*, 763–771. [CrossRef] [PubMed]

91. Nasar, H.M.; Sultan, S.; Gomes, R.; Noor, S. Heavy metal pollution of soil and vegetable grown near roadside at Gazipur. *Bangladesh J. Agri. Res.* **2012**, *37*, 9–17. [CrossRef]

92. Duruibe, J.O.; Ogwuegbu, M.D.C.; Egwurugwu, J.N. Heavy metal pollution and human biotoxic effects. *Int. J. Phys. Sci.* **2007**, *2*, 112–118.

93. Kim, H.; Song, B.; Kim, H.; Park, J. Distribution of trace metals at two abandoned mine sites in Korea and arsenic-associated health risk for the residents. *Toxicol. Environ. Health Sci.* **2009**, *1*, 83–90. [CrossRef]

94. Khan, A.; Khan, S.; Khan, M.A.; Qamar, Z.; Waqas, M. The uptake and bioaccumulation of heavy metals by food plants, their effects on plants nutrients, and associated health risk: A review. *Environ. Sci. Pollut. Res.* **2015**, *22*, 13772–13799. [CrossRef]

95. Bi, X.; Feng, X.; Yang, Y.; Li, X.; Shin, G.P.Y.; Li, F. Allocation and source attribution of lead and cadmium in maize (*Zea mays* L.) impacted by smelting emissions. *Environ. Pollut.* **2009**, *157*, 834–839. [CrossRef]
96. Ahmad, J.U.; Goni, M.A. Heavy metal contamination in water, soil, and vegetables of the industrial areas in Dhaka, Bangladesh. *Environ. Monit. Assess.* **2010**, *166*, 347–357. [CrossRef]
97. Ding, C.; Zhang, T.; Wang, X.; Zhou, F.; Yang, Y.; Yin, Y. Effects of soil type and genotype on lead concentration in rootstalk vegetables and the selection of cultivars for food safety. *J. Environ. Manag.* **2013**, *122*, 8–14. [CrossRef]
98. Hu, J.; Wu, F.; Wu, S.; Sun, X.; Lin, X.; Wong, M.H. Phytoavailability and phytovariety codetermine the bioaccumulation risk of heavy metal from soils, focusing on Cd-contaminated vegetable farms around the Pearl River Delta, China. *Ecotoxicol. Environ. Saf.* **2013**, *91*, 18–24. [CrossRef] [PubMed]
99. Baker, A.J. Accumulators and excluders-strategies in the response of plants to heavy metals. *J. Plant Nutr.* **1981**, *3*, 643–654. [CrossRef]
100. Bashir, M.; Khalid, S.; Rashid, U.; Adrees, M.; Ibrahim, M.; Islam, M.S. Assessment of selected heavy metals uptake from soil by vegetation of two areas of district Attock, Pakistan. *Asian J. Chem.* **2014**, *26*, 1063–1068. [CrossRef]
101. Ma, Y.; Rajkumar, M.; Freitas, H. Isolation and characterization of Ni mobilizing PGPB from serpentine soils and their potential in promoting plant growth and Ni accumulation by *Brassica* spp. *Chemosphere* **2009**, *75*, 719–725. [CrossRef]
102. Adhikari, T.; Kumar, A. Phytoaccumulation and tolerance of *Ricinus communis* L. to nickel. *Int. J. Phytoremediation* **2012**, *14*, 481–492. [CrossRef]
103. Faucon, M.P.; Shutcha, M.N.; Meerts, P. Revisiting copper and cobalt concentrations in supposed hyperaccumulators from SC Africa: Influence of washing and metal concentrations in soil. *Plant Soil* **2007**, *301*, 29–36. [CrossRef]
104. Oven, M.; Grill, E.; Golan-Goldhirsh, A.; Kutchan, T.M.; Zenk, M.H. Increase of free cysteine and citric acid in plant cells exposed to cobalt ions. *Phytochemistry* **2002**, *60*, 467–474. [CrossRef]
105. Wong, S.C.; Li, X.D.; Zhang, G.; Qi, S.H.; Min, Y.S. Heavy metals in agricultural soils of the Pearl River Delta, South China. *Environ. Pollut.* **2002**, *119*, 33–44. [CrossRef]
106. Tamoutsidis, E.; Lazaridou, M.; Papadopoulos, I.; Spanos, T.; Papathanasiou, F.; Tamoutsidou, M.; Mitlianga, P.; Vasiliou, G. The effect of treated urban wastewater on soil properties, plant tissue composition and biomass productivity in berseem clover and corn. *J. Food Agric. Environ.* **2009**, *7*, 782–786.
107. Jadhav, S.; Phugare, S.; Patil, P.S.; Jadhav, J.P. Biochemical degradation pathway of textile dye Remazol red and subsequent toxicological evaluation by cytotoxicity, genotoxicity and oxidative stress studies. *Int. Biodeterior. Biodegrad.* **2011**, *65*, 733–743. [CrossRef]
108. Carita, R.; Marin-Morales, M.A. Induction of chromosome aberrations in the *Allium cepa* test system caused by exposure of seeds to industrial effluents in contaminated with azo dyes. *Chemosphere* **2008**, *72*, 722–725. [CrossRef] [PubMed]
109. Panneerselman, N.; Palanikumar, L.; Gopinathan, S. Chromosomal aberration by Glycidol in *Allium cepa* L. root meristem cell. *Int. J. Pharm. Sci. Res.* **2012**, *3*, 300–304.
110. Kaur, J.; Kaur, V.; Pakade, Y.B.; Katnoria, J.K. A study on water quality monitoring of Buddha Nullah, Ludhiana, Punjab (India). *Environ. Geochem. Health* **2020**, *43*, 2699–2722. [CrossRef] [PubMed]
111. Barbosa, J.S.; Cabral, T.M.; Ferreira, D.N.; Agnez-Lima, L.F.; Batistuzzo de Medeiros, S.R. Genotoxicity assessment in aquatic environment impacted by the presence of heavy metals. *Ecotoxicol. Environ. Saf.* **2010**, *73*, 320–325. [CrossRef]
112. Hemachandra, C.K.; Pathiratne, A. Assessing toxicity of copper, cadmium and chromium levels relevant to discharge limits of industrial effluents into inland surface waters using common onion, *Allium cepa* bioassay. *Bull. Environ. Contam. Toxicol.* **2015**, *94*, 199–203. [CrossRef]
113. Zhang, Y.X.; Yang, X.L. The toxic effects of cadmium on cell division and chromosomal morphology of *Hordeum vulgare*. *Mutat. Res.* **1994**, *312*, 121–126. [CrossRef]
114. Zhang, Y.; Xiao, H. Antagonistic effect of calcium, zinc and selenium against cadmium induced chromosomal aberrations and micro-nuclei in root cells of *Hordeum vulgare*. *Mutat. Res.* **1998**, *420*, 1–6. [CrossRef]
115. Fojtova, M.; Kovarik, A. Genotoxic effect of cadmium is associated with apoptotic changes in tobacco cells. *Plant Cell Environ.* **2000**, *23*, 531–537. [CrossRef]
116. Unyayar, S.; Celik, A.; Cekic, F.O.; Gozel, A. Cadmium-induced genotoxicity, cytotoxicity and lipid peroxidation in *Allium sativum* and *Vicia faba*. *Mutagenesis* **2006**, *21*, 77–81. [CrossRef] [PubMed]
117. Grover, I.S.; Kaur, S. Genotoxicity of wastewater samples from sewage and industrial effluent detected by the *Allium* root anaphase aberration and micronucleus assays. *Mutat. Res. Fund. Mol. Mech. Mut.* **1999**, *426*, 183–188. [CrossRef]
118. Vijayalakshmidevi, S.R.; Muthukumar, K. Improved biodegradation of textile dye effluent by coculture. *Ecotoxicol. Environ. Saf.* **2015**, *114*, 23–30. [CrossRef]
119. Sudhakar, R.; Gowda, K.N.; Venu, G. Mitotic abnormalities induced by silk dyeing industry effluents in the cells of *Allium cepa*. *Cytologia* **2001**, *66*, 235–239. [CrossRef]
120. Khan, S.; Anas, M.; Malik, A. Mutagenicity and genotoxicity evaluation of textile industry wastewater using bacterial and plant bioassays. *Toxicol. Rep.* **2019**, *6*, 193–201. [CrossRef]
121. Taylor, S.R.; Mclennan, S.M. The geochemical evolution of the continental crust. *Rev. Geophys.* **1995**, *33*, 241–265. [CrossRef]
122. Hakanson, L. An ecological risk index for aquatic pollution control. A sedimentological approach. *Water Res.* **1980**, *14*, 975–1001. [CrossRef]

123. Muller, G. Index of geoaccumulation in sediments of the Rhine River. *J. Geol.* **1969**, *2*, 108–118.

 *soil systems*

Article

Investigating Lead Bioavailability in a Former Shooting Range by Soil Microanalyses and Earthworms Tests

Carlo Porfido [1,*], Concetta Eliana Gattullo [1], Ignazio Allegretta [1], Nunzio Fiorentino [2], Roberto Terzano [1], Massimo Fagnano [2] and Matteo Spagnuolo [1]

[1] Dipartimento di Scienze del Suolo, Della Pianta e Degli Alimenti, Università Degli Studi di Bari "Aldo Moro", Via Amendola 165/A, 70126 Bari, Italy; concettaeliana.gattullo@uniba.it (C.E.G.); ignazio.allegretta@uniba.it (I.A.); roberto.terzano@uniba.it (R.T.); matteo.spagnuolo@uniba.it (M.S.)

[2] Dipartimento di Agraria, Università Degli Studi di Napoli "Federico II", Via Università 100, 80055 Portici, Italy; nunzio.fiorentino@unina.it (N.F.); massimo.fagnano@unina.it (M.F.)

* Correspondence: carlo.porfido@uniba.it

Abstract: Shooting ranges are among the major anthropogenic sources of Pb contamination in soils worldwide. Once they have reached the soil, bullet residues can have different fates according to the characteristics of the soil environment, leading to the formation of different Pb weathering products whose stability is crucial for Pb accessibility to soil biota. In this study, Pb availability in a former polluted shooting range was investigated with a combination of conventional soil analyses, X-ray microanalyses and assays with the bio-indicator earthworm *Eisenia andrei*. Chemical extractions evidenced a rather low mobility of soil Pb, while micro-X-ray fluorescence spectroscopy (μXRF) and scanning electron microscopy coupled with microanalysis (SEM-EDX) showed the formation of a weathering crust around Pb-containing bullet slivers dispersed within the soil. Such crusts consisted of a mixture of orthophosphates, including the highly insoluble Cl-pyromorphite. Furthermore, no acute toxicity effects and low Pb concentration values were measured in earthworm tissues (94.9 mg kg^{-1}) and coelom fluids (794 μg L^{-1}) after 28 days of exposure to the polluted soil. These results allow us to assume that most of the Pb in the shooting range soil underwent stabilization processes promoted by phosphatic fertilization. The soil was in fact used for agriculture after being dismissed for firing activities. Such a combined approach can be applied to study Pb bioavailability in other shooting ranges or, more generally, in soils heavily polluted with Pb.

Keywords: firing ranges; *Eisenia andrei*; micro-X-ray fluorescence; SEM-EDX; Pb phosphates; pyromorphite

Citation: Porfido, C.; Gattullo, C.E.; Allegretta, I.; Fiorentino, N.; Terzano, R.; Fagnano, M.; Spagnuolo, M. Investigating Lead Bioavailability in a Former Shooting Range by Soil Microanalyses and Earthworms Tests. *Soil Syst.* 2022, 6, 25. https://doi.org/10.3390/soilsystems6010025

Academic Editor: Mallavarapu Megharaj

Received: 11 February 2022
Accepted: 10 March 2022
Published: 13 March 2022

Publisher's Note: MDPI stays neutral with regard to jurisdictional claims in published maps and institutional affiliations.

Copyright: © 2022 by the authors. Licensee MDPI, Basel, Switzerland. This article is an open access article distributed under the terms and conditions of the Creative Commons Attribution (CC BY) license (https://creativecommons.org/licenses/by/4.0/).

1. Introduction

Lead (Pb) is a non-essential and non-beneficial element whose average content in uncontaminated soils worldwide is estimated to equal 17 mg kg^{-1} [1]. However, anthropogenic sources, such as mining and smelting, shooting ranges, automobile exhausts and sewage sludges, can cause Pb contamination in quantities up to several thousand mg kg^{-1} with huge risks for the environment and human beings. For instance, exposure to Pb-contaminated soil and ingestion of soil Pb-bearing particles and dusts has been indicated as the major pathway for Pb poisoning in children, which can pose risks for cognitive development and various diseases [2]. Despite Pb being quite stable in soils due to strong complexation or adsorption by humic substances, clay minerals and Fe oxides [1], plants and other soil inhabitants can suffer from Pb exposure, with lethal effects [3–5]. Pb can accumulate in their tissues and thereby be transferred to higher levels of the trophic chain [6].

In recent years, various studies have demonstrated the advantages of laboratory and synchrotron micro-X-ray florescence (μXRF) and hyperspectral elaboration of μXRF data in investigating the distribution and speciation of potentially toxic elements (PTEs) in polluted soils [7–14]. Indeed, through the assessment of element distributions in soil thin sections and their mutual correlations it is possible to infer, for instance, chemical speciation and mineralogical information, which are crucial especially for those (minor) phases otherwise not detectable with other techniques (e.g., X-ray diffraction) due to their weak abundance and/or crystallinity. Such outcomes can be further combined with more detailed observations at the sub-micrometric scale by means of scanning electron microscopy coupled with microanalysis (SEM-EDX), which additionally allows the detection of light elements not observable by μXRF. However, if, on the one hand, soil microanalysis offers an effective tool for predicting PTE behaviour and mobility based on chemical speciation or associations with soil phases, on the other hand, the use of bio-indicator species allows the appraisal of the actual impact of the pollutant on soil organisms. For decades, earthworms have been elected as the favoured bio-indicators in soils because of their peculiar living and feeding mode. Indeed, they are intimately exposed to soil-bound contaminants through integument contact and ingestion [15]. Moreover, earthworms accumulate pollutants in their bodies at higher rates than other soil organisms [16,17]. The combination of soil microanalysis and earthworm bioassays can therefore be used to evaluate the occurrence and speciation of PTEs in soil as well as their bioaccessibility.

In the present study, Pb bioavailability in a polluted soil (1575 mg Pb kg^{-1}) from a former shooting range area (Acerra, Italy) was studied using a synergistic approach combining: (a) physicochemical characterization through conventional analytical methods; (b) soil microanalysis by μXRF and SEM-EDX observations; (c) toxicological tests using earthworms *Eisenia andrei* Bouché. The outcomes obtained from this research could be useful in assessing the actual environmental risks of Pb contamination in the area under investigation as well as devising possible remediation strategies.

2. Materials and Methods

2.1. Soil Sampling and Physicochemical Characterization

The polluted soil (S2) was collected at the former shooting range of Acerra (40°59′37.74″ N; 14°24′7.27″ E), district of Naples, Southern Italy. The shooting range covered an area of approximately 6 ha and was dismissed in 2003 after about 10 years of activity. Once abandoned, the site has been colonized by spontaneous vegetation and then used for agricultural production from 2011 to 2014 and finally confiscated by Regional Authorities. In 2019 the site was assigned to the University of Naples for developing a phytoremediation project as part of the activities of the Rizobiorem project. A huge amount of bullet residues, mainly in the form of metallic slivers, was dispersed in the ground, featuring levels of Pb, Sb and polycyclic aromatic hydrocarbons (PAH) exceeding the national regulatory thresholds for agricultural sites (Ministerial Decree n. 46/2019) [18]. An agricultural uncontaminated soil (S1) in the close vicinity of the shooting area was sampled and used as a reference control soil. Both S1 and S2 were collected (2019) in three replicates at 0–20 cm soil layer after removing the very superficial undecomposed litter. Soil samples were air-dried and sieved through a 2 mm mesh.

The soil physicochemical characterization included the following parameters: particle size distribution (pipette and sieving method) [19]; electrical conductivity (1:5 soil:water solution ratio) [20]; pH H$_2$O (1:2.5 soil:water solution ratio, pH meter GLP 22, Crison^{TM}); calcium carbonate (titrimetric method [21]); total nitrogen (Kjeldahl); available phosphorous (Olsen); and organic carbon (Walkley and Black) [22]. Furthermore, the promptly bioavailable fraction of PTEs was estimated by ammonium nitrate extraction [23,24] and the pseudototal PTE contents were quantified by aqua regia digestion (ISO 12914) and ICP-MS (PerkinElmer Nexion 300, Waltham, MA, USA) [25,26]. Pseudototal PTE content was compared to screening values (SV) reported in Italian M.D. 46/2019.

Soil main minerals were determined by means of X-ray powder diffraction (XRPD) using a Miniflex II X-ray diffractometer (Rigaku Corporation, Tokyo, Japan) equipped with a Cu tube (Cu Kα, 30 kV, 15 mA). Data were acquired between 3 and 70° 2θ with a step width of 0.02° 2θ and a counting time of 3 s/step.

2.2. Soil µXRF and SEM EDX Analyses

Petrographic soil thin sections (30 µm thickness) of both control and polluted soils were prepared after embedding the soils in epoxy resin (L.R. White Resin, Polyscience Europe GmbH, Hirschberg an der Bergstrasse, Germany) [7] and were analysed with a micro-X-ray fluorescence spectrometer (µXRF, M4 Tornado, Bruker Nano GmbH, Berlin, Germany). Elemental distribution maps were acquired under vacuum (20 mbar) using a Rh tube X-ray source (50 kV, 600 µA, 30 W, spot size of 25 µm) with policapillary optics and two 30 mm^2 XFlash® silicon drift detectors. Micro-X-ray maps were collected with a step size of 25 µm and an acquisition time of 10 ms per pixel. Details of Pb slivers were acquired with a smaller step size (5 µm) and by repeating the scanning 5 times to increase the signal-to-noise (S/N) ratio. X-ray fluorescence hyperspectral data were processed using both Bruker M4 software and a combination of PyMca 5.1.3 (Copyright (c) 2004–2019 European Synchrotron Radiation Facility (ESRF), Grenoble, France) [27] and Datamuncher software [28] in order to evaluate both element distribution and correlations. Furthermore, a field emission gun (FEG) SEM Zeiss Σigma 300 VP (Zeiss Oberkochen, Oberkochen, Germany) was used to observe the samples with a sub-micrometric resolution. The instrument, working at 15 kV, was equipped with an energy-dispersive spectrometer (EDX) C-MaxN SDD with an active area of 20 mm^2 (Oxford Instruments, Oxford, UK).

2.3. Earthworm Testing and Analysis

Soil mesocosms of S1 and S2 were prepared in triplicate using plastic boxes containing approximately 500 g of soil (dry weight, sieved at 2 mm), following the OECD guidelines [29]. Ten mature (clitellate) earthworms of *E. andrei* (provided by our laboratory stockbreeding) of about 350−400 mg individual weight were introduced into each mesocosm after 24 h of gut depuration in Petri dishes (in the dark at 22 ± 2 °C). The mesocosms were then placed in an incubator chamber (21 ± 1 °C, $65 \pm 5\%$ RH) for 28 days, adjusting the moisture content when required. Finally, the earthworms were recovered, cleaned, purged for 48 h (in Petri dishes on a wet filter paper), rinsed with deionized water and weighed. Five specimens per mesocosm were used for tissue bioaccumulation analyses, while the other five individuals were used for coelom fluid extrusion and analysis. Lead concentration in earthworm tissues and in the coelom fluid were determined using a total reflection X-ray fluorescence spectrometer (S2 Picofox—Bruker Nano GmbH, Berlin, Germany) equipped with a Mo microfocus tube (30 W, 50 kV, 600 µA), a multilayer monochromator and a XFlash SDD with a 30 mm^2 active area. For Pb determination in tissues, samples were dried (48 h, 50 °C) and finely ground using a mixer mill (MM 400, Retsch GmbH, Haan, Germany). Then, suspensions were prepared, with 50 mg of each specimen placed in a 15 mL polypropylene centrifuge tube to which 2.5 mL of Triton X-100 (Sigma-Aldrich, Darmstadt, Germany) was added along with 10 µL of a 1000 µg L^{-1} yttrium (Y) standard solution (Sigma-Aldrich) as internal standard (IS). Coelom fluids were extruded, applying a voltage of 5 V for 3 s to each earthworm. Then, 10 µL of the extruded fluid was mixed with 80 µL of polyvinyl alcohol (PVA) and 10 µL of Y standard solution, as IS. All samples were analysed for 1000 s (live time) following the method described in [30,31]. The average weight of the earthworms before and after the testing period was calculated by dividing the mass of all earthworms by their number for both S1 and S2 mesocosms. The bioconcentration factor (BF) was calculated by dividing tissue mean Pb concentration by total Pb concentration in the soil.

3. Results and Discussion

3.1. Soil Main Properties

Soil physicochemical properties for both S1 and S2 are reported in Table 1. Both S1 and S2 were sandy loam soils with slightly alkaline pH according to USDA classification [32]. The organic matter (OM) content was higher in S2 (53 g kg^{-1}) than in S1 (31 g kg^{-1}). In addition, the available P was higher in S2 (P_2O_5 = 0.60 g kg^{-1}), almost double that of S1. As for the total carbonate content, S1 was highly calcareous and S2 was moderately calcareous. Both soils were non-saline according to their EC values.

Table 1. Physicochemical properties of the unpolluted (S1) and polluted (S2) soils.

	Sand	Silt $g\ kg^{-1}$	Clay	pH (H_2O)	EC $dS\ m^{-1}$	OC [a] $g\ kg^{-1}$	OM [b] $g\ kg^{-1}$	N $g\ kg^{-1}$	C/N	$CaCO_3$ $g\ kg^{-1}$	P_2O_5 $g\ kg^{-1}$
S1	600	240	160	7.8	0.21	18	31	2.2	8	155	0.29
S2	590	265	145	7.4	0.21	30	53	2.5	12	70	0.60

	Cd $mg\ kg^{-1}$	Sb $mg\ kg^{-1}$	Pb $mg\ kg^{-1}$	Zn $mg\ kg^{-1}$	Cd * $mg\ kg^{-1}$	Sb * $mg\ kg^{-1}$	Pb * $mg\ kg^{-1}$	Zn * $mg\ kg^{-1}$
S1	0.3	0.3	17	209	0.03	0.61	0.01	0.3
S2	3.3	22.7	1575	328	0.15	6.36	0.02	1.3
SV	5	10	100	300	n.a.	n.a.	n.a.	n.a.

[a] Organic carbon. [b] Organic matter. * NH_4NO_3 extractable. SV screening value M.D. n. 46/2019; n.a.: not available.

The soils showed similar mineralogy, the main minerals detected being silicates and aluminium silicates, such as quartz, feldspars, albite and illite/muscovite. The presence of calcite and apatite group minerals was also observed. No Pb-bearing phases were detected by XRPD. The concentrations of Sb, Pb and Zn in S2 were higher than the national screening values (SV) for agricultural soil [18]; in particular, Pb concentration was 1575 mg kg^{-1} compared to a SV of 100 mg kg^{-1}.

3.2. Soil Microanalysis

Micro-XRF was used to map and compare elemental distributions in soil thin sections of both the control and the polluted soil (Figure 1). The major elements, e.g., Al, Si, P, S, K, Ca and Fe, occurred with similar abundances and correlations in the two soils and could be attributed to the soil mineral constituents. Based on XRPD analysis, the overlapping distribution of Al, Si and K testifies to the presence of aluminium silicates, while the dark blue Si-containing particles in Figure 1 (middle panels) are instead related to quartz grains. Calcium is mainly associated with P (purple particles, Figure 1, right panels), presumably forming calcium phosphate phases, such as apatite; in other particles, Ca occurs uncorrelated with other elements, hence it is likely attributable to calcite (pure red particles, Figure 1, right panels), since C and O cannot be detected by XRF.

As expected, Pb was detected only in the polluted soil (S2), concentrated in particles ranging in size from hundreds to thousands of microns, visible as green areas (Figure 1b, right). Except for P, no other element was correlated with Pb. Several Pb-containing particles were then selected on the PTE-polluted soil thin sections and analysed again with increased resolution and accuracy in order to better investigate the nature of these Pb-containing particles as well as Pb vs. P correlations.

In Figure 2, the results of the µXRF hyperspectral data analysis performed on three Pb-containing particles, selected as the most representative, are shown. For each particle, the optical image is coupled with its false-colour map, which depicts the correlation between Pb and P fluorescent signals detected at the particle section surface.

From the scatterplot of the intensities (expressed as counts) of the Pb L-lines and P Kα-lines recorded in each pixel of the three micrographs shown in Figure 2, five main P/Pb correlations were identified and grouped with different colours (Figure 2a). The green region of the scatterplot contains those pixels containing almost only Pb, while the yellow, magenta, red and blue regions contain those pixels with an increasingly higher content of P, beside Pb. By applying such colour-based segmentation to the Pb-particle micrographs (Figure 2b–d, right), it can be generally observed that, starting from the inner only-Pb-containing core (green areas of the particles), the P amount gradually increases towards the particle boundaries, with the higher P/Pb signal ratio (blue) observed at the border of the particle. Such a pattern is clearly visible for the particle shown in Figure 2c (right), while such distribution is slightly more structured in the other particles (Figure 2b,d, right).

This depends on the particles shape and on the cutting section, which was reasonably diametrical for particle (c) and more external for the others. By looking at the composition of the particles and at their morphology with a light microscope, they appear as metallic Pb slivers deriving from bullet residues. Indeed, many bullet residues are still present in the investigated soils. Furthermore, the comparison between the correlation maps and their corresponding light micrographs (Figure 2b–d, left) shows other additional interesting features: in the case of Figure 2c, for instance, the sliver shows a Pb-rich nucleus (sliver core) which appears pale in visible light, becoming darker as the P/Pb ratio increases moving toward the borders of the particle.

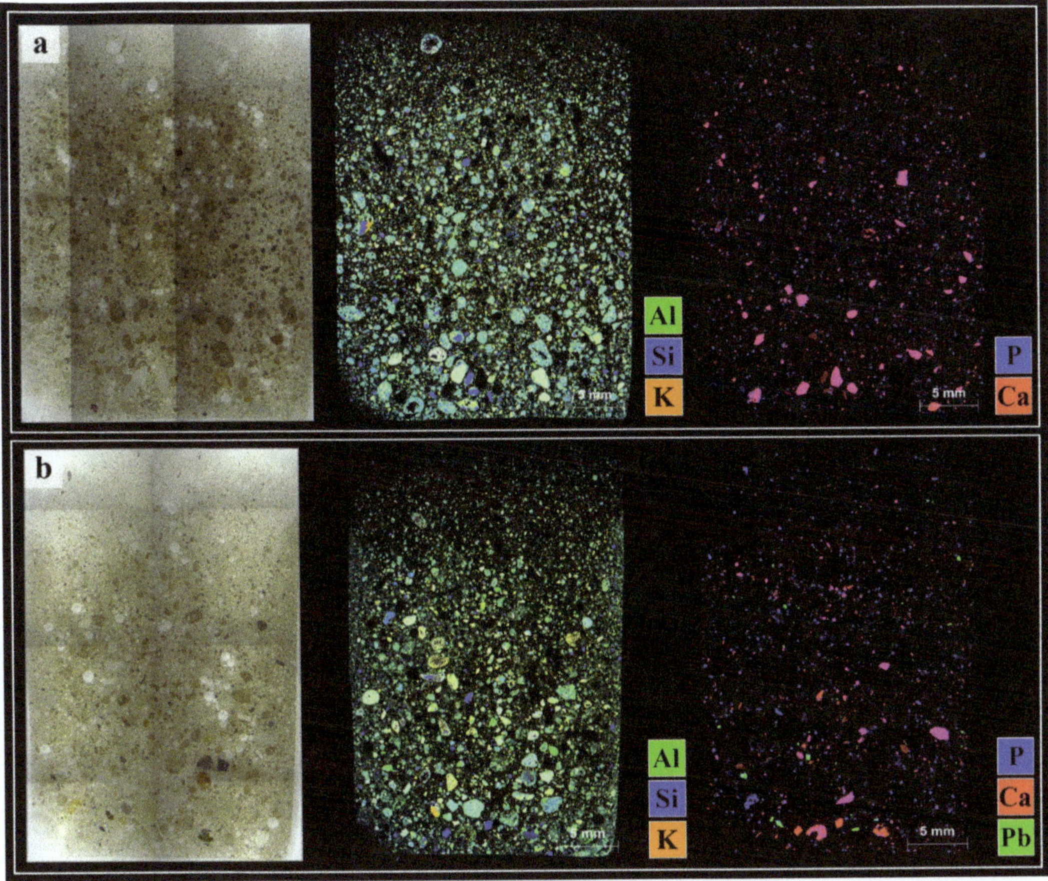

Figure 1. Thin section images (left) and corresponding µXRF maps (middle, right) of S1 (**a**) and S2 (**b**) soils. ((**a,b**) middle panels) Overlay of Al (green), Si (blue) and K (orange) distribution maps. ((**a,b**) right panels) Overlay of P (blue), Ca (red) and Pb (green) distribution maps. Brighter pixels correspond to higher element concentrations. The co-presence of two or more elements in the same pixels gives rise to different degrees of colour combinations.

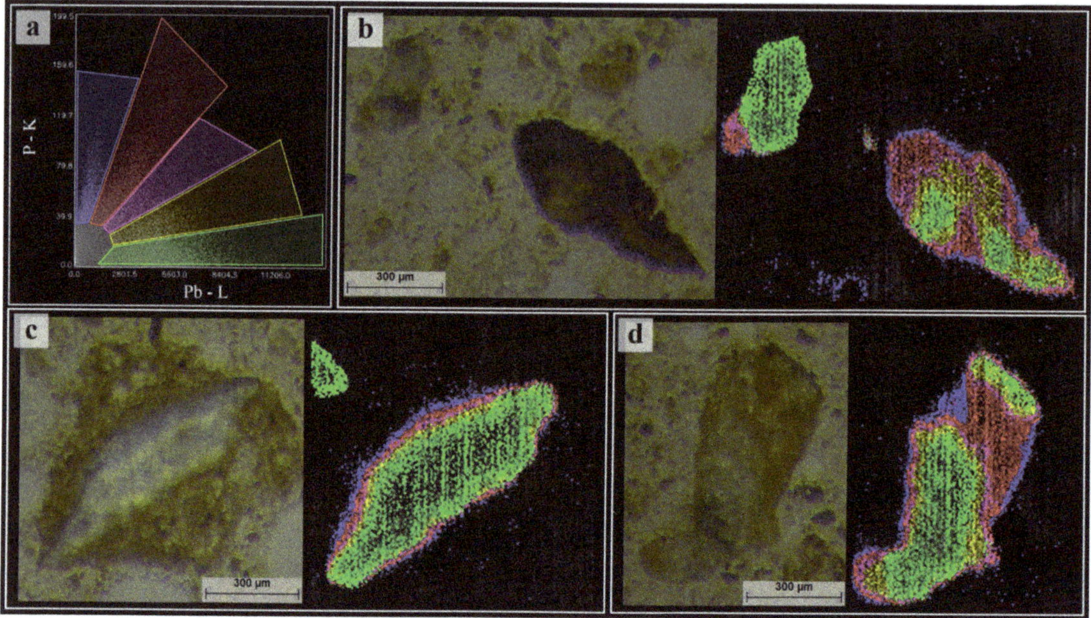

Figure 2. (**a**) P (K fluorescent signal) vs. Pb (L fluorescent signal) scatterplot obtained using spectral μXRF deconvolution data of (**b–d**) particles. (**b–d**) particles micrographs (**left**) and related correlation maps (**right**) coloured according to the different P/Pb correlations identified through the scatterplot (**a**)).

Other interesting correlations observed after spectral deconvolution are those between Cl and P and between Ca and P, shown in the scatterplots in Figure 3a,c. Although with low intensity, the Cl–K signal is correlated with that of P (Figure 3a) and is not detected where only the Pb signal is present (without P; grey regions in Figure 3b). Chlorine is located only in the portions of the particles where P was also detected. The scatterplot in Figure 3c instead shows three main Ca–P correlations grouping pixels whose Ca content is yellow < green < red. By looking at the corresponding false-color micrographs (Figure 3d), the primary evidence is that such Ca–P correlations occur in the same pixels where also P–Pb and Cl–P correlations were observed. Moreover, it can be noticed that the higher Ca intensities are located at the very external borders of the Pb-containing particles whilst they decrease as they move towards the inner part (green > yellow).

Beside P, Ca and Cl, no other chemical element has been detected within these Pb-containing fragments. The ubiquitous occurrence of P and Ca should indicate the presence of Ca phosphates (or even Cl apatite), while the presence of P and Cl, together with Pb should be attributed to the formation of Pb phosphates over the metallic slivers, such as secondary and tertiary Pb phosphates (Pb_2HPO_4 and $Pb_3(PO_4)_2$, respectively) and/or pyromorphite. The latter refers to three species, depending on the anion X in the structure $Pb_5(PO_4)_3X$ (namely, chloropyromorphite, where $X = Cl^-$; hydroxypyromorphite, where $X = OH^-$; and fluoropyromophite, where $X = F^-$). Unfortunately, neither F nor O and H can be detected by μXRF, therefore none of these three forms can be excluded, chloropyromorphite most likely being present.

Lead orthophosphates, above all pyromorphite, are highly insoluble phases. In 1932, Jowett and Price [33] determined an extremely low solubility product (K_{sp}) of $10^{-79.115}$ for $Pb_5(PO_4)_3Cl$, which is the most insoluble Pb orthophosphate known. In the same study, the authors demonstrated that secondary and tertiary orthophosphates transform to pyromorphite even in the presence of low chloride concentrations, thus explaining the fact that, while the first two phases are not found in nature, pyromorphite is rather common.

Subsequent studies [34,35] proposed for $Pb_5(PO_4)_3Cl$ the K_{sp} of $10^{-25.05}$ for soil with a pH in the usual range. However, even considering the higher value of $10^{-25.05}$, pyromorphite can be still considered several orders of magnitude less soluble than the most common Pb minerals in soils [2]. Indeed, the use of phosphates for amending Pb-polluted soils is a well-known and common remediation practice. By promoting the formation of Pb insoluble species, such as pyromorphite, phosphate addition leads to the reduction of Pb bioavailability and toxicity [2,36].

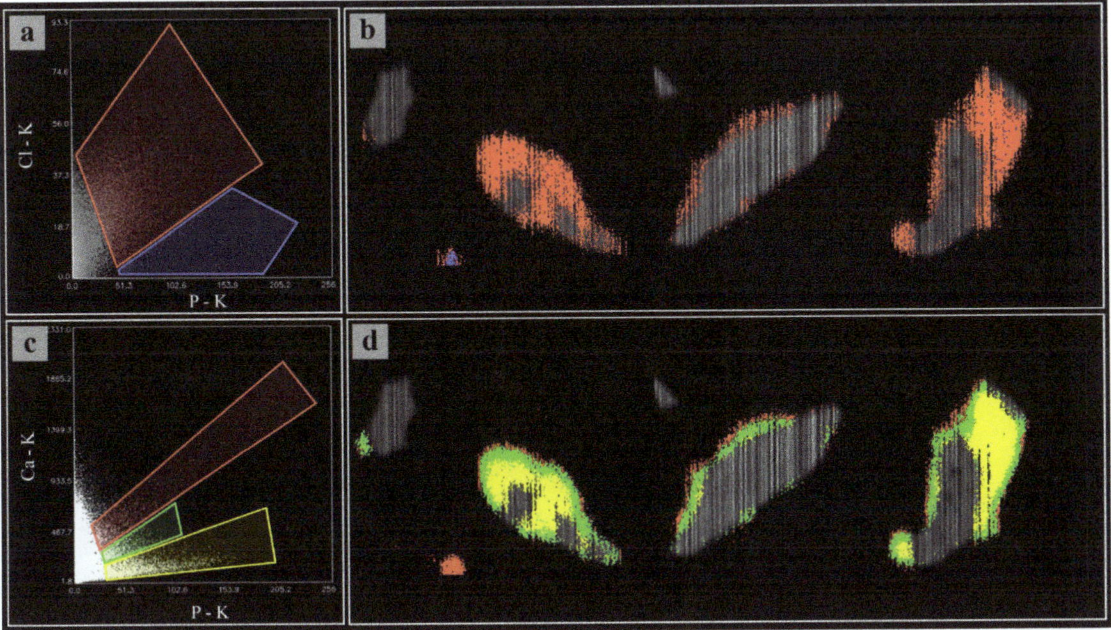

Figure 3. (**a**) Cl (K fluorescent signal) vs. P (K fluorescent signal) and (**c**) Ca (K fluorescent signal) vs. P (K fluorescent signal) scatterplots obtained using spectral μXRF deconvolution data of the three particles of Figure 2. (**b,d**) The three particles colored according to the different Cl–P and Ca–P correlations identified through the scatterplots (the background grey micrographs corresponding to Pb–L signals).

SEM-EDX results (Figure 4) also confirmed μXRF observations and allowed us to investigate additional features of the Pb particles and P-bearing encrustations. In the backscattered electron (BSE) image in Figure 4a (left), the darker grey regions, mainly present at the particle boundaries, indicate an average lower Z, thus the presence of lighter elements in addition to Pb. As shown by the corresponding EDX layered image (Figure 4a, right) and spectrum (Figure 4d1), this latter being relative to the white point analyses reported in the EDX image (Figure 4a, right), these low Z elements are P, Ca and Cl. In Figure 4b, which depicts a detail of the P-bearing regions, a cluster of darker grains (identified mostly as K-feldspars) ranging from a sub-micrometric scale to several micrometres in size is dispersed in a binding light grey matrix. Such matrices contain Pb, P, Ca and Cl together with Si and Al, these latter attributable to silicate fragments (region 2 and related spectrum). In Figure 4c (right), the ubiquitous occurrence of Pb, P, Ca and Cl is noticed in a compact region (spectrum d3) and within an aggregate-like structure, as well as in Figure 4b. In both Figure 4b,c, inner portions show almost only the presence of Pb (spectrum d4).

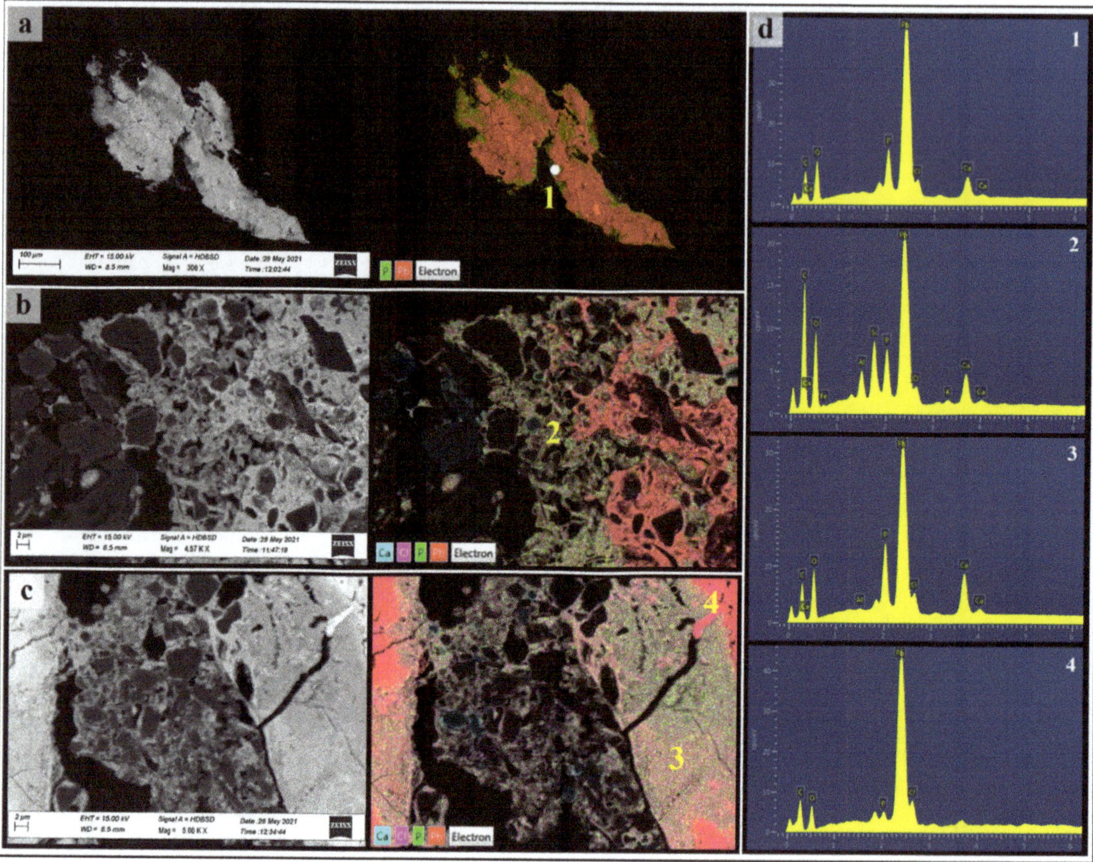

Figure 4. (Left panel) BSE micrographs of a Pb-containing particle. **(b,c)** Higher magnification details of the weathering crust of the particle shown in **(a)**. **(Middle panel)** EDX layered images. **(d, right panel)** EDX spectra of the points (white in **(a)**) and/or regions (yellow numbers in **(b,c)**) indicated within the middle panel: (1) typical EDX spectrum of the P-bearing coating at particle boundaries; (2) Pb–Ca phosphates matrix in a silicate environment forming an aggregate-like structure; (3) compact Pb and Ca phosphate solid solution; (4) metallic or weakly weathered Pb.

The μXRF and SEM-EDX results described so far suggest that a weathering shell formed over time around the metallic slivers arising from the past firing range activities. Such shells consist, perhaps, of a mixture of Ca phosphate and Pb orthophosphate species, among which it has been possible to distinguish chloropyromorphite. Previous studies demonstrated that Pb stabilization as pyromorphite occurs through an intermediate Pb and Ca solid solution upon hydroxyapatite (HA) dissolution. Such an intermediate phase $(Pb_{(10-x)}Ca_x(PO_4)_6(OH)_2–PbCaHA–)$ transforms into pyromorphite over time through Ca^{2+} exchange by Pb^{2+} [37–39]. In the studied slivers (Figures 2 and 3), the different correlations found between P–Pb and Ca–P seem to suggest a layering within the weathering crust, the lower Ca signal intensity indicating higher rates of Pb^{2+} substitution in the inner layers and hence a higher presence of pyromorphite. The external layers, which feature the highest Ca–P and P–Pb correlations, suggest early stages of PbCaHA formation. These sliver coatings can be further considered rather stable under the soil pH conditions. Indeed, the pH value of 7.4 measured in the polluted soil (Table 1) falls well within the pH range 3–11 in which chloropyromorphite has been found to be the most stable Pb species [36].

It could be argued that neutral to sub-alkaline pHs are yet inhibitory for pyromorphite formation due to the reduced solubility of P, which lowers both the kinetics and rate of Pb stabilization [40]. Conversely, more acidic conditions and hence soluble orthophosphate ions are required to react with Pb and precipitate pyromorphite [41]. However, a high amount of available P was measured in the soil (Table 1), and soil micro-XRF mapping also confirmed a high occurrence of P, either correlated with Ca or not: such P could have contributed to pyromorphite formation and, consequently, to Pb stabilization. Among the most common P forms in soil, apatite and phosphate rock are quite insoluble at neutral pHs. By contrast, P fertilizers, such as monocalcium and dicalcium phosphate (Ca $(H_2PO_4)_2$ and $CaHPO_4$, respectively), decrease soil pH upon dissolution, increasing the solubility of both P and Pb [40]. Based on these observations, it can be hypothesized that in the studied soil, which was used for agriculture after being dismissed as a shooting range, stabilization processes were triggered by soil phosphatic fertilization. The temporary soil acidification would have favoured pyromorphite precipitation at the fragment surfaces, thus protecting metallic Pb from further weathering processes and, at the same time, strongly limiting the possible mobilization of this hazardous element. Indeed, despite in alkaline and sub-alkaline conditions, the most commonly weathering phases of Pb bullet residues found in shooting ranges are lead carbonates and oxides (e.g., cerussite and litharge) [40], other previous studies showed the natural formation of pyromorphite in Pb-rich soils with high concentrations of available phosphates [42,43].

3.3. Bioassays with E. andrei

Considering the similarity of the physicochemical characteristics of S1 and S2 soils (Table 1), it may be hypothesized that the earthworm testing results could be only minimally influenced by the different soil properties (mainly the OM and carbonate content). In addition, the pH values of S1 and S2 (7.8 and 7.4, respectively) allow us to exclude pH-shock effects for earthworms, since they prefer circumneutral soils, even if *E. andrei* could also tolerate acidic conditions [44].

All earthworms survived after the 28-day exposure period, evidencing no acute toxicity effects, as previously experienced by other authors for similar background concentrations [5]. Moreover, the average weight of the earthworms increased after 28 days to a similar extent in both S1 and S2 mesocosms (*t*-test, $p < 0.05$). On the other hand, earthworms introduced into the polluted mesocosm accumulated significantly higher amounts of Pb in tissues and coelom fluid than the control ($p < 0.01$). Morgan and Morgan [45] found a positive correlation between Pb tissues and soil concentrations, different to other potentially toxic metals, such as Cu, Zn and Cd, whose tissue concentrations decreased proportionally as their soil concentrations increased. However, differently from these elements, Pb concentrations in earthworm tissues were generally below the soil Pb concentration, due to the higher retention of Pb by organic and inorganic soil components [46]. Even if the Pb tissue concentration of earthworms grown in S2 exceeds by ten times that of animals grown in the control soil, the bioconcentration factor (BF) obtained for S2 is very low (Table 2). Indeed, in a comprehensive study on Pb uptake in earthworms exposed to Pb-polluted soil, Langdon et al. reported a Pb tissue concentration of 266, 406 and 637 mg kg^{-1} (with corresponding BFs of 0.27, 0.14 and 0.16) for *E. andrei* exposed to 1000, 3000 and 4000 mg Pb kg^{-1}, respectively [5]. Based on these findings, an average uptake of approximately 300 mg Pb kg^{-1} in earthworm tissues exposed to S2 (1575 mg Pb kg^{-1}) should be expected, but it was only 95 mg kg^{-1} (Table 2). However, in the OECD-style experiments carried out by Langdon et al., Pb mesocosm contamination was obtained by artificially spiking the testing soils with aqueous solutions of $Pb(NO_3)_2$: this would imply higher Pb availability (and hence possibly higher Pb accumulation rates) than in field-collected soils, in which Pb can be already at least partly stabilized through complexation or adsorption to soil components (therefore less bioaccesible). Nonetheless, even by applying the regression equation proposed by Morgan and Morgan (e.g., $\log_{10}Pb_{worm} = -1.073 + 1.042 \log_{10}Pb_{soil}$), who instead worked with field-collected soils [45], the Pb uptake should be at least two times higher than our experimental data. The value found for the control soil (11 mg kg^{-1}) is instead consistent with Langdon's results and explainable as the normal trace amount of Pb in the organisms.

Table 2. Results of bioassays with *E. andrei* in the unpolluted (S1) and polluted (S2) soils. Ecotoxicology data (mean value ± standard deviation, n = 5) with different letters in the column are statistically different according to *t*-testing ($p < 0.05$).

	Pb Concentration in Soil mg kg^{-1}	Mortality	Mean Earthworm Weight g		Pb Mean Earthworm Tissue Concentration mg kg^{-1}	Bioconcentration Factor	Pb Mean Earthworm Fluid Concentration µg L^{-1}
			Day 0	Day 28			
S1	17	0	0.36 ± 0.04 ns	0.43 ± 0.04 ns	11.1 ± 2.2 b	-	251.8 ± 55.4 b
S2	1575	0	0.35 ± 0.06 ns	0.44 ± 0.07 ns	94.9 ± 19.8 a	0.06	794.4 ± 232.2 a

ns: Not significant.

Total PTE concentration in coelom fluid has been proposed in previous studies as a bioavailability indicator [30,31,47]. Differently from the determination of total concentration in tissues, PTE quantification in fluids is not affected by possible soil residues entrapped in the gut [48]; furthermore, it is a less laborious and time-consuming procedure, and it is unnecessary to kill the animals. Looking at our results (Table 2), the increase of Pb in coelom fluid would suggest some metal sequestration mechanisms occurring within the chloragogenous tissue, which surrounds the earthworm gut epithelium, by means of metal-binding metal-inducible cysteine-rich metalloproteins, namely, metallothioneins (MTs) [49]. Albeit Pb is not recognized as a MT-inducer [50], it was detected together with Cd and Zn in the chloragogenous tissue of earthworms collected from an abandoned Zn–Pb mine in Draethen (UK), by means of proton-induced X-ray emission analysis (PIXE) [51]. In such work, Morgan et al. evidenced the primary role of Cd (and much less of Zn) in inducing MT expression in earthworms living in heavily polluted soil. This would match the possible scenario of the firing soil studied in this work and could probably explain the Pb accumulation in coelom fluids. Indeed, besides Pb, high levels of Cd, Sb and Zn (which are elements related to the past firing activity) have been detected in S2 soil (Table 1) and could have worked as MT inducers, thus favouring Pb binding, also.

Overall, the results of earthworm tests seem to depict a very limited bioavailability for soil Pb. On the one hand, neither mortality nor weight decrease was observed; rather earthworms grew over the 28-day experimental period. On the other hand, Pb bioconcentration in tissues was lower than reported in the literature for similar soil Pb concentrations, whilst a relevant amount of Pb in the coelom fluids suggested a metal-induced metal-sequestration mechanism. Such findings appear in accordance with the results of the soil characterization and microanalyses, which showed a negligible NH_4NO_3-extractable Pb fraction (Table 1) and predominant Pb stabilization by phosphates. Indeed, significant correlations between promptly bioavailable metal fractions in soils (including Pb) and their bioconcentration in earthworms have been indicated in previous works [52]. In addition, the high amount of OM in S2 (Table 1) may have contributed to limiting Pb bioavailability. At the same time, the formation of pyromorphite makes Pb inaccessible in the gastrointestinal tract, since these phases remain insoluble even after ingestion [2]. Formation of pyromorphite may also occur inside the intestine in the presence of phosphate, thanks to gastrointestinal acidic conditions [40]. These latter considerations would further confirm a low Pb accessibility in the studied case on the base of P availability.

4. Conclusions

The combination of soil microanalyses and earthworm bioavailability testing proposed in this study allowed us to comprehensively assess Pb occurrence and distribution in the investigated shooting range soil, suggesting a rather limited mobility and bioaccessibility of Pb. In these soils, Pb occurred as metallic slivers, which over time underwent weathering processes resulting in the formation of a phosphate superficial crust, in some cases encapsulating other soil materials within, such as small silicate minerals. The coexistence of Pb and Ca, combined with different ratios to P, suggested Pb^{2+} to Ca^{2+} exchange mechanisms, hence the stabilization of Pb into Pb orthophosphates. These latter were attributed to insoluble Cl pyromorphite formation due to the presence of Cl in the weathered regions of the particles. The occurrence of these processes is likely to strongly reduce Pb mobility and leachability in the studied soil. Earthworm bioassays confirmed this prevision, since exposure to the contaminated soil did not show acute toxicity effects (although not in the aims of the present study, these toxicity results can be extended to the other PTEs found in the investigated soils, i.e., Cd and Sb) nor high levels of Pb bioconcentration. Additional tests would be required to investigate chronic exposure effects and/or plant availability. Based on the overall results of this study, we believe that the approach pursued in this work could be successfully used for other polluted soils.

Author Contributions: Conceptualization, C.P., R.T. and M.S.; methodology, C.P. and I.A.; software, C.P. and I.A.; investigation, C.P., I.A., C.E.G. and N.F.; data curation, C.P.; writing—original draft preparation, C.P.; writing—review and editing, C.E.G., I.A., R.T., N.F., M.F. and M.S.; visualization, C.P.; supervision, M.S.; project administration, M.S.; funding acquisition, M.S., M.F. and R.T. All authors have read and agreed to the published version of the manuscript.

Funding: This research was supported by PRIN 2017 (Progetti di Ricerca di Rilevante Interesse Nazionale)—2017BHH84R—"Role of Soil-Plant-Microbial Interactions at Rhizosphere Level on the Biogeochemical Cycle and Fate of Contaminants in Agricultural Soils Under Phytoremediation with Biomass Crops (RIZOBIOREM)." Ignazio Allegretta was supported by a research grant for the project PON R&I "Studio del sistema suolo-pianta mediante tecniche analitiche innovative che impiegano raggi X"—Progetto AIM1809249—attività 1, linea 1.

Institutional Review Board Statement: Not applicable.

Informed Consent Statement: Not applicable.

Data Availability Statement: Data are available upon request by contacting Dr. Carlo Porfido (e-mail: carlo.porfido@uniba.it).

Acknowledgments: Micro-XRF and TXRF analyses were performed at the Micro-X-ray Lab of the University of Bari Aldo Moro (Italy).

Conflicts of Interest: The authors declare no conflict of interest.

References

1. Alloway, B.J. *Heavy Metals in Soil*; Blackie and Son Ltd.: London, UK, 1990; p. 339. [CrossRef]
2. Scheckel, K.G.; Diamond, G.L.; Burgess, M.F.; Klotzbach, J.M.; Maddaloni, M.; Miller, B.W.; Partridge, C.R.; Serda, S.M. Amending soils with phosphate as means to mitigate soil lead hazard: A critical review of the state of the science. *J. Toxicol. Environ. Health Part B* **2013**, *16*, 337–380. [CrossRef] [PubMed]
3. Antonovics, J.; Bradshaw, A.; Turner, R. Heavy metal tolerance in plants. In *Advances in Ecological Research*; Academic Press: Cambridge, MA, USA, 1971; Volume 7, pp. 1–85. [CrossRef]
4. Česynaitė, J.; Praspaliauskas, M.; Pedišius, N.; Sujetovienė, G. Biological assessment of contaminated shooting range soil using earthworm biomarkers. *Ecotoxicology* **2021**, *30*, 2024–2035. [CrossRef] [PubMed]
5. Langdon, C.J.; Hodson, M.E.; Arnold, R.E.; Black, S. Survival, Pb-uptake and behaviour of three species of earthworm in Pb treated soils determined using an OECD-style toxicity test and a soil avoidance test. *Environ. Pollut.* **2005**, *138*, 368–375. [CrossRef] [PubMed]
6. Roodbergen, M.; Klok, C.; van der Hout, A. Transfer of heavy metals in the food chain earthworm Black-tailed godwit (*Limosa limosa*): Comparison of a polluted and a reference site in The Netherlands. *Sci. Total Environ.* **2008**, *406*, 407–412. [CrossRef] [PubMed]
7. Allegretta, I.; Porfido, C.; Martin, M.; Barberis, E.; Terzano, R.; Spagnuolo, M. Characterization of As-polluted soils by laboratory X-ray-based techniques coupled with sequential extractions and electron microscopy: The case of Crocette gold mine in the Monte Rosa mining district. *Environ. Sci. Pollut. Res.* **2018**, *25*, 25080–25090. [CrossRef] [PubMed]
8. Gattullo, C.E.; D'Alessandro, C.; Allegretta, I.; Porfido, C.; Spagnuolo, M.; Terzano, R. Alkaline hydrothermal stabilization of Cr(VI) in soil using glass and aluminum from recycled municipal solid wastes. *J. Hazard. Mater.* **2018**, *344*, 381–389. [CrossRef] [PubMed]

9.   Sosa, N.N.; Kulkarni, H.; Datta, S.; Beilinson, E.; Porfido, C.; Spagnuolo, M.; Zárate, M.A.; Surber, J. Occurrence and distribution of high arsenic in sediments and groundwater of the Claromecó fluvial basin, southern Pampean plain. *Sci. Total Environ.* **2019**, *695*, 133673. [CrossRef]

10. Gattullo, C.E.; Allegretta, I.; Porfido, C.; Rascio, I.; Spagnuolo, M.; Terzano, R. Assessing chromium pollution and natural stabilization processes in agricultural soils by bulk and micro X-ray analyses. *Environ. Sci. Pollut. Res.* **2020**, *27*, 22967–22979. [CrossRef] [PubMed]

11. MacLean, L.C.W.; Beauchemin, S.; Rasmussen, P.E. Lead speciation in house dust from canadian urban homes using EXAFS, Micro-XRF, and Micro-XRD. *Environ. Sci. Technol.* **2011**, *45*, 5491–5497. [CrossRef]

12. McIntosh, K.G.; Cordes, N.L.; Patterson, B.M.; Havrilla, G.J. Laboratory-based characterization of plutonium in soil particles using micro-XRF and 3D confocal XRF. *J. Anal. At. Spectrom.* **2015**, *30*, 1511–1517. [CrossRef]

13. Landrot, G.; Khaokaew, S. Determining the fate of lead (Pb) and phosphorus (P) in alkaline Pb-polluted soils amended with P and acidified using multiple synchrotron-based techniques. *J. Hazard. Mater.* **2020**, *399*, 123037. [CrossRef] [PubMed]

14. Li, Q.; Hu, X.; Hao, J.; Chen, W.; Cai, P.; Huang, Q. Characterization of Cu distribution in clay-sized soil aggregates by NanoSIMS and micro-XRF. *Chemosphere* **2020**, *249*, 126143. [CrossRef] [PubMed]

15. Weltje, L. Mixture toxicity and tissue interactions of Cd, Cu, Pb and Zn in earthworms (Oligochaeta) in laboratory and field soils: A critical evaluation of data. *Chemosphere* **1998**, *36*, 2643–2660. [CrossRef]

16. Martin, M.H.; Coughtrey, P.J. The use of terrestrial animals as monitors and indicators of environmental contamination by heavy metals. In *Biological Monitoring of Heavy Metal Pollution*; Springer: Dordrecht, The Netherlands, 1982; pp. 221–310. [CrossRef]

17. Ireland, M.P. *Heavy Metal Uptake and Tissue Distribution in Earthworms*; Springer Science and Business Media LLC: Berlin/Heidelberg, Germany, 1983; pp. 247–265.

18. Ministerial Decree. n. 46/2019. Available online: https://www.gazzettaufficiale.it/eli/id/2019/06/07/19G00052/sg (accessed on 29 September 2021).

19. Gee, G.W.; Bauder, J.W. Particle-size analysis. In *Methods of Soil Analysis. Part 1: Physical and Mineralogical Methods*, 2nd ed.; Klute, A., Ed.; American Society of Agronomy: Madison, WI, USA, 1986; pp. 383–411.

20. Slavich, P.; Petterson, G. Estimating the electrical conductivity of saturated paste extracts from 1:5 soil, water suspensions and texture. *Soil Res.* **1993**, *31*, 73–81. [CrossRef]

21. Bloom, P.R.; Meter, K.; Crum, J.R. Titration method for determination of clay-sized carbonates. *Soil Sci. Soc. Am. J.* **1985**, *49*, 1070–1073. [CrossRef]

22. Sparks, D.L.; Page, A.L.; Helmke, P.A.; Loeppert, R.H.; Soltanpour, P.N.; Tabatabai, M.A.; Johnston, C.T.; Sumner, M.E. *Methods of Soil Analysis: Part 3 Chemical Methods 5.3*, 1st ed.; Soil Science Society of America Book Series 5; Soil Science Society of America, Inc.; American Society of Agronomy, Inc.: Madison, WI, USA, 1996; p. 1390. [CrossRef]

23. Duri, L.G.; Visconti, D.; Fiorentino, N.; Adamo, P.; Fagnano, M.; Caporale, A.G. Health risk assessment in agricultural soil potentially contaminated by geogenic Thallium: Influence of plant species on metal mobility in soil-plant system. *Agronomy* **2020**, *10*, 890. [CrossRef]

24. BBodSchV. Bundes-Bodenschutz- und Altlastenverordnung (BBodSchV) vom 12. Juli 1999. Bundesgesetzblatt I 1999, 1554 (English Version) [Federal Soil Protection and Contaminated Sites Ordinance, Dated 12 July 1999]. 1999. Available online: https://www.elaw.org/es/content/germany-federal-soil-protection-and-contaminated-sites-ordinance-bbodschv-12-july-1999 (accessed on 29 September 2021).

25. Gupta, S.; Vollmer, M.; Krebs, R. The importance of mobile, mobilisable and pseudo total heavy metal fractions in soil for three-level risk assessment and risk management. *Sci. Total Environ.* **1996**, *178*, 11–20. [CrossRef]

26. Kingston, H.M.; Haswell, S.J. *Microwave-Enhanced Chemistry: Fundamentals, Sample Preparation, and Applications*; American Chemical Society: Washington, DC, USA, 1997.

27. Solé, V.; Papillon, E.; Cotte, M.; Walter, P.; Susini, J. A multiplatform code for the analysis of energy-dispersive X-ray fluorescence spectra. *Spectrochim. Acta Part B At. Spectrosc.* **2007**, *62*, 63–68. [CrossRef]

28. Alfeld, M.; Janssens, K. Strategies for processing mega-pixel X-ray fluorescence hyperspectral data: A case study on a version of Caravaggio's painting Supper at Emmaus. *J. Anal. At. Spectrom.* **2015**, *30*, 777–789. [CrossRef]

29. OECD. Guideline for Testing of Chemicals no. 207. Earthworm, Acute Toxicity Test. 1984. Available online: https://www.oecd.org/chemicalsafety/risk-assessment/1948293.pdf (accessed on 3 January 2021).

30. Allegretta, I.; Porfido, C.; Panzarino, O.; Fontanella, M.C.; Beone, G.M.; Spagnuolo, M.; Terzano, R. Determination of as concentration in earthworm coelomic fluid extracts by total-reflection X-ray fluorescence spectrometry. *Spectrochim. Acta Part B At. Spectrosc.* **2017**, *130*, 21–25. [CrossRef]

31. Porfido, C.; Allegretta, I.; Panzarino, O.; Laforce, B.; Vekemans, B.; Vincze, L.; de Lillo, E.; Terzano, R.; Spagnuolo, M. Correlations between as in earthworms' coelomic fluid and as bioavailability in highly polluted soils as revealed by combined laboratory X-ray techniques. *Environ. Sci. Technol.* **2019**, *53*, 10961–10968. [CrossRef] [PubMed]

32. USDA Soil Survey Staff, Natural Resources Conservation Service, United States Department of Agriculture. Official Soil Series Descriptions. Available online: https://www.nrcs.usda.gov/wps/portal/nrcs/detail/soils/survey/geo/?cid=nrcs142p2_053587 (accessed on 3 January 2021).

33. Jowett, M.; Price, H.I. Solubilities of the phosphates of lead. *Trans. Faraday Soc.* **1932**, *28*, 668–681. [CrossRef]

34. Lindsay, W.L. *Chemical Equilibria in Soils*; John Wiley and Sons: Hoboken, NJ, USA, 1979.

35. Scheckel, K.G.; Ryan, J.A. Effects of aging and pH on dissolution Kinetics and stability of Chloropyromorphite. *Environ. Sci. Technol.* **2002**, *36*, 2198–2204. [CrossRef] [PubMed]

36. Nriagu, J.O. Lead orthophosphates—IV Formation and stability in the environment. *Geochim. Cosmochim. Acta* **1974**, *38*, 887–898. [CrossRef]

37. Mavropoulos, E.; Rossi, A.M.; Costa, A.M.; Perez, C.A.C.; Moreira, J.C.; Saldanha, M. Studies on the mechanisms of lead immobilization by hydroxyapatite. *Environ. Sci. Technol.* **2002**, *36*, 1625–1629. [CrossRef]

38. Mavropoulos, E.; Rocha, N.C.; Moreira, J.C.; Rossi, A.M.; Soares, G.A. Characterization of phase evolution during lead immobilization by synthetic hydroxyapatite. *Mater. Charact.* **2004**, *53*, 71–78. [CrossRef]

39. Miretzky, P.; Fernandez-Cirelli, A. Phosphates for Pb immobilization in soils: A review. *Environ. Chem. Lett.* **2008**, *6*, 121–133. [CrossRef]

40. Chrysochoou, M.; Dermatas, D.; Grubb, D.G. Phosphate application to firing range soils for Pb immobilization: The unclear role of phosphate. *J. Hazard. Mater.* **2007**, *144*, 1–14. [CrossRef] [PubMed]

41. Porter, S.K.; Scheckel, K.G.; Impellitteri, C.A.; Ryan, J.A. Toxic metals in the environment: Thermodynamic considerations for possible immobilization strategies for Pb, Cd, As, and Hg. *Crit. Rev. Environ. Sci. Technol.* **2004**, *34*, 495–604. [CrossRef]

42. Cao, X.; Ma, L.Q.; Chen, M.; Hardison, D.W.; Harris, W.G. Lead transformation and distribution in the soils of shooting ranges in Florida, USA. *Sci. Total Environ.* **2003**, *307*, 179–189. [CrossRef]

43. Cotter-Howells, J. Lead phosphate formation in soils. *Environ. Pollut.* **1996**, *93*, 9–16. [CrossRef]

44. Edwards, C.A.; Bohlen, P.J. *Biology and Ecology of Earthworms*, 3rd ed.; Chapman & Hall: London, UK, 1996.

45. Morgan, J.; Morgan, A. Earthworms as biological monitors of cadmium, copper, lead and zinc in metalliferous soils. *Environ. Pollut.* **1988**, *54*, 123–138. [CrossRef]

46. Zimdahl, R.L.; Skogerboe, R.K. Behavior of lead in soil. *Environ. Sci. Technol.* **1977**, *11*, 1202–1207. [CrossRef]

47. Tan, Q.-G.; Ke, C.; Wang, W.-X. Rapid Assessments of metal bioavailability in marine sediments using coelomic fluid of sipunculan worms. *Environ. Sci. Technol.* **2013**, *47*, 7499–7505. [CrossRef]

48. Chapman, P.M. Effects of gut sediment contents on measurements of metal levels in benthic invertebrates—A Cautionary Note. *Bull. Environ. Contam. Toxicol.* **1985**, *35*, 345–347. [CrossRef]

49. Adamowicz, A. Morphology and ultrastructure of the earthworm *Dendrobaena veneta* (Lumbricidae) coelomocytes. *Tissue Cell* **2005**, *37*, 125–133. [CrossRef]

50. Kägi, J.H. Overview of metallothionein. *Methods Enzymol.* **1991**, *205*, 613–626. [CrossRef]

51. Morgan, A.; Stürzenbaum, S.; Winters, C.; Grime, G.; Aziz, N.A.A.; Kille, P. Differential metallothionein expression in earthworm (*Lumbricus rubellus*) tissues. *Ecotoxicol. Environ. Saf.* **2004**, *57*, 11–19. [CrossRef]

52. Dai, J.; Becquer, T.; Rouiller, J.H.; Reversat, G.; Bernhard-Reversat, F.; Nahmani, J.; Lavelle, P. Heavy metal accumulation by two earthworm species and its relationship to total and DTPA-extractable metals in soils. *Soil Biol. Biochem.* **2003**, *36*, 91–98. [CrossRef]

soil systems

Article

Prospects for the Use of *Echinochloa frumentacea* for Phytoremediation of Soils with Multielement Anomalies

Svetlana V. Gorelova [1], Anna Yu. Muratova [2], Inga Zinicovscaia [3,4,*] Olga I. Okina [5] and Aliaksandr Kolbas [6,7]

[1] Natural Science Institute, Tula State University, 300012 Tula, Russia; salix35@gmail.com
[2] Laboratory of Environmental Biotechnology, Institute of Biochemistry and Physiology of Plants and Microorganisms, Russian Academy of Sciences, 410049 Saratov, Russia; amuratova@yahoo.com
[3] Department of Nuclear Physics, Joint Institute for Nuclear Research, 141980 Dubna, Moscow Region, Russia
[4] Department of Nuclear Physics, Horia Hulubei National Institute for R&D in Physics and Nuclear Engineering, Str. Reactorului No. 30, P.O. Box MG-6, 077125 Bucharest-Magurele, Romania
[5] Laboratory of Chemico-Analytical Investigations, Geological Institute of the Russian Academy of Sciences, 119017 Moscow, Russia; okina@bk.ru
[6] Ecology Center, Brest State University Named after A.S. Pushkin, 224016 Brest, Belarus; kolbas77@mail.ru
[7] Polessky Agrarian and Ecological Institute at the National Academy of Sciences of Belarus, 224020 Brest, Belarus
*   Correspondence: zinikovskaia@mail.ru; Tel.: +7-916-867-73-26

Citation: Gorelova, S.V.; Muratova, A.Y.; Zinicovscaia, I.; Okina, O.I.; Kolbas, A. Prospects for the Use of *Echinochloa frumentacea* for Phytoremediation of Soils with Multielement Anomalies. *Soil Syst.* **2022**, *6*, 27. https://doi.org/10.3390/soilsystems6010027

Academic Editors: Matteo Spagnuolo, Paola Adamo and Giovanni Garau

Received: 31 January 2022
Accepted: 14 March 2022
Published: 16 March 2022

Publisher's Note: MDPI stays neutral with regard to jurisdictional claims in published maps and institutional affiliations.

Copyright: © 2022 by the authors. Licensee MDPI, Basel, Switzerland. This article is an open access article distributed under the terms and conditions of the Creative Commons Attribution (CC BY) license (https:// creativecommons.org/licenses/by/ 4.0/).

Abstract: In a model experiment, some adaptive characteristics, the bioaccumulation of toxic elements from technogenically-contaminated soils with polyelement anomalies, and rhizosphere microflora of Japanese millet, *Echinochloa frumentacea*, were studied using biochemical, microbiological, physico-chemical (AAS, ICP-MS, INAA), and metagenomic (16S rRNA) methods of analysis. Good adaptive characteristics (the content of photosynthetic pigments, low molecular weight antioxidants) of *E. frumentacea* grown on the soils of metallurgical enterprises were revealed. The toxic effect of soils with strong polyelement anomalies (multiple excesses of MPC for Cr, Ni, Zn, As, petroleum products) on biometric parameters and adaptive characteristics of Japanese millet were shown. The rhizosphere populations of *E. frumentacea* grown in the background soil were characterized by the lowest taxonomic diversity compared to the rhizobiomes of plants grown in contaminated urban soils. The minimal number of all groups of microorganisms studied was noted in the soils, which contain the highest concentrations of both inorganic (heavy metals) and organic (oil products) pollutants. The taxonomic structure of the rhizospheric microbiomes of *E. frumentacea* was characterized. It has been established that *E. frumentacea* accumulated Mn, Co, As, and Cd from soils with polyelement pollution within the average values. V was accumulated mainly in the root system (transfer factor from roots to shoots 0.01–0.05) and its absorption mechanism is rhizofiltration. The removal of Zn by shoots of *E. frumentacea* increased on soils where the content of the element exceeded the MPC and was 100–454 mg/kg of dry weight (168–508 g/ha). Analysis of the obtained data makes it possible to recommend *E. frumentacea* for phytoremediation of soil from Cu and Zn at a low level of soil polyelement contamination using grass mixtures.

Keywords: *Echinochloa frumentacea*; contaminated soils; toxic elements; heavy metals; adaptation; AOS; ascorbic acid; GSH; phytoremediation; bioaccumulation; rhizosphere microflora

1. Introduction

Global climate changes, pollution levels, and high-impact farming have induced a strong decline in soil quality, making the sustainable use of land more challenging than in the past [1]. For soils contaminated with trace element (TE) there are various phytomanagement options for reducing the environmental risks. Phytoextraction in conjunction with the application of conditioners can be considered as the way to minimize the dispersion and biological action of TE on soil and to increase vegetation cover on polluted soils [2–4].

Recently, the cultivation of non-food crops for the production of plant-based feedstock, the bioremediation of metal-contaminated soils, and risk management have been developing at the field scale [5–9].

Chemically-induced hyperaccumulation is impaired by various environmental risks, e.g., metal leaching from the root zone and toxic effects on the microbiome [7,10]. Regarding secondary TE accumulators, the desirable characteristics for these plant species are (1) relatively fast growth and high biomass; (2) extended root system for exploring large soil volumes; (3) good tolerance to high concentrations of TE in plant tissues; (4) high translocation factor; (5) adaptability to specific environments/sites; (6) easy agricultural management [11], (7) good interaction with associated bacteria [12,13].

Global climate change makes it possible to introduce relatively thermophilic new biomass crops for phytoremediation, especially species of the *Poaceae* family [14]. These plants, due to C_4 assimilation pathways [15], are of particular interest for phytoremediation due to the production of significant biomass and, as a consequence, the accumulation of significant amounts of carbon and heavy metals.

The selective accumulation of Pb, Cu, Zn, and Cd in the roots and the possibility to remove the root remains makes technical sorghum, sugar sorghum, and *Sudan grass* extremely suitable for phytoremediation purposes [16]. Some sorghum cultivars show a high antioxidant status during the bioaccumulation of Pb and Cu on anthropogenically disturbed substrates with polyelement contamination [17].

The capacity of *Miscanthus* species to accumulate inorganic contaminants into the root system and to reduce the dissipation of persistent organic contaminants makes them good candidates for soil phytostabilization and phytodegradation. The noninvasive hybrid *Miscanthus giganteus*, with high lignocellulosic content, has a high potential in the biorefinery and bioenergy industries [18]. The applicability of *Arundo donax* for cadmium removal from contaminated soil and water was shown by Sabeen et al. [19].

Echinochloa frumentacea is a cost-effective crop through seeds and biomass production. It can be applied for sustainable crop rotation for the enhancement of soil development processes, nutrient cycles, and microbial community.

Echinochloa frumentacea was previously described as an effective extractor (As) ([20], responsive to fertilizers, hymexazole, and rhizobacteria inoculation [21]. There is the experience of using this species in the phytoremediation of polyelement-contaminated substrates such as the stabilization of municipal wastewater sludge [22] and phytoremediation of soil contaminated with cadmium, copper, and polychlorinated biphenyls [23]. *Echinochloa* species can also be used efficiently for chromium and cadmium extraction [24].

Other advantages are the sufficient diversity of varieties, their high seed germination (up to 10 years), that they develop well on soils poor in mineral nutrition, give from two to eight mows, and provide grain and green mass with average yields of 1.5–3 t/ha and 30–50 t/ha, respectively. The latter fact is very important for effective phytoextraction since TE removal by plants arises from two factors: (1) TE concentration in dry plant tissue [25] and (2) the amount of harvested biomass [9].

The acceptable concentration of TEs in plant raw material significantly expands the area of its valorization. Potential plant-based products are (1) non-food products such as biofuels [26], fibers, wood, essential oils, etc., and (2) animal feed, depending on the contamination level. Plant parts generally accumulate metals to different degrees [27], and some parts are usually less contaminated and may often be used for consumption or at least for non-food purposes even if the plant has been grown on contaminated or marginal soil [28]. Seeds generally accumulate the lowest TEs concentrations in a plant [29].

Other harvestable parts can be used in various processes, i.e., composting to fertilize TE-deficient soil, incineration, ashing, vacuum and oxidative pyrolysis, liquid extraction, the synthesis of hydrogen fuel, biofuels such as bioethanol, biogas, and activated carbons, hydrothermal oxidation, gasification, etc. In addition, metals can be recovered and reused for economic purposes by phytomining [30].

The objective of this study is to determine the phytoremediation potential and prospects for the use of *Echinochloa frumentacea* on urban soils with polyelement anomalies (especially for large urbanized ecosystems with high levels of industrial and vehicular pollution). During the study, the tasks were:

1. to assess the adaptive potential of *Echinochloa frumentacea* on HM-contaminated soils;
2. to assess bioaccumulation and the transfer factor of heavy metals (HMs) from polyelement-contaminated soils to *Echinochloa frumentacea*;
3. to study the rhizosphere microflora of *Echinochloa frumentacea* on HM-contaminated soils;
4. on the basis of the data obtained, to assess the possibility of using *Echinochloa frumentacea* for the phytoremediation of HM-contaminated chernozems.

2. Materials and Methods

2.1. Soil Sampling

Model laboratory experiments with urban ecosystem soils characterized by polyelement anomalies [17] were performed. To carry out the experiments, the soils were collected in the following sites: the sanitary protection zones (SPZ) of large metallurgical enterprises: Kosogorsky Metallurgical Plant (KMP), Tulachermet PJSC (TCh), the embankment opposite the arms and machine-building plants (Embankment), as well as a major highway and a sanitary protection zone of the central avenue of the city (Lenin Ave.). The gray forest soils of the state museum–estate of Leo N. Tolstoy "Yasnaya Polyana" served as background (Background). Sampling and soil preparation for determination of toxic elements was carried out in accordance with RNS R 53123-2008 [31]. The sampling depth was 0–25 cm.

2.2. Soil Characterization

To characterize the soil used for analysis of the pH, content of petroleum products, total carbon, mobile phosphorus, available forms of nitrogen–nitrates from water soluble ammonia were determined.

Preparation of soil extract and determination of its acidity were performed according to Russian National Standard (RNS) 26483-85 [32] using a Metler Toledo Delta 320 pH meter. Determination of the total organic carbon in soil was performed according to [33]. The content of mobile (available) phosphorus P_2O_5 in mg/100 g was determined by the photocolorimetric method according to Kirsanov [34]. The determination of nitrates and water-soluble ammonium in the soil was carried out by standard photocolorimetric methods according to RNS 26489-85 [35] and RNS 26488-85 [36]. Petroleum products were determined by the gravimetric method [37]. Elemental analysis of soil was determined using X-ray fluoresce in the chemical analytical laboratory of the Geological Institute of the Russian Academy of Sciences.

2.3. Experiment Design

The object of the study was the Japanese millet, *Echinochloa frumentacea* Link.

The soils of sampling points for the model experiment were placed in plastic containers with drainage that did not have a drain for water. Sowing of seeds and observation of plants was carried out in laboratory conditions at a temperature of 21–23 °C using natural light. Watering was carried out with distilled water as the topsoil dried up. The seed germination was determined on the 7th day.

2.4. Pigments Determination

A spectrophotometric method was used to determine the quantitative content of pigments in plants. The determination of pigments was carried out in an ethanol extract using 1 cm thick cuvettes in triplicate biological and analytical repetition. Optical density was determined at wavelengths of 665 nm, 649 nm, and 470 nm. Calculation of the quantitative content of chlorophylls and carotenoids was carried out according to the formulas [38], followed by conversion to g of fresh weight [39].

2.5. Plants Analysis

Several analytical techniques were applied to determine the elemental composition of plants materials. Content of As, Br, K, La, Na, Mo, Sm, U, W, Ba, Ce, Co, Cr, Cs, Hf, Ni, Rb, Sb, Sc, Sr, Ta, Tb, Th, Yb, Zn, Al, Ca, Cl, I, Mn, and V was determined using instrumental neutron activation analysis (INAA) at the fast-pulsed reactor IBR-2 in FLNP JINR. For Al, Ca, Cl, I, Mn, and V determination, samples of about 0.3 g were irradiated for 3 min and measured for 15 min. To determine As, Br, K, La, Na, Mo, Sm, U, W, Ba, Ce, Co, Cr, Cs, Hf, Ni, Rb, Sb, Sc, Sr, Ta, Tb, Th, Yb, and Zn, samples were irradiated for 3 days and measured after 4 and 20 days of irradiation. Gamma spectra of induced activity were measured using three spectrometers based on HPGe detectors with an efficiency of 40–55% and resolution of 1.8–2.0 keV for total absorption peak 1332 keV of the isotope ^{60}Co and Canberra spectrometric electronics. The analysis of the spectra was performed using the Genie2000 software from Canberra, while the calculation of concentration was carried out using software "Concentration" developed in FLNP. The quality control of measurement was assured by the use of the following reference materials: IAEA-336 (Lichen), NIST SRM 1572 (*Citrus* Leaves), NIST SRM 1575 (*Pine* Needles).

The content of Cu, Pb, and Cd in samples was determined by using an iCE 3400 Atomic Absorption Spectrometer (AAS) with electrothermal (graphite furnace) atomization (Thermo Fisher Scientific, Waltham, MA, USA). Details of samples preparation for analysis and measurement can be found in [40].

Along with INAA, elemental composition of plant samples grown on contaminated was determined using an Element 2 mass spectrometer (Thermo Fisher Scientific of GmbH, Dreieich, Germany) after adding indium as an internal standard. Before the measurement, the instrument was adjusted in such a way that the sensitivity was at least 1,000,000 cps when analyzing a solution of indium with a concentration of 1 µg/L. The instrument was calibrated using multielement standard solutions ICP-MS-68A Solution B, ICP-MS-E, and ICP-MS-B (High-Purity Standards, Charleston, SC, USA). The good correlation of the concentration of elements obtained by two technique was attained.

To determine Mg and Fe content, a KVANT-2 spectrometer (KORTEK, Moscow, Russia) was used. The measurement was carried out in an air-acetylene flame using absorption lines of 248.3 nm for Fe and 285.2 for Mg. The instrument was calibrated using a multielement standard solution ICP-MS-68A Solution A (High-Purity Standards, USA).

For ICP-MS and AAS analysis, plant samples were digested using a microwave system (MARS5, CEM Corporation, Charlotte, NC, USA) in XP-1500 Teflon liners. The liners were preliminarily kept with 10.0 mL of an aqueous solution of nitric acid (1:1) at 160 °C, then cooled and rinsed with deionized water (18.2 MΩ.cm, Milli-Q, ADVANTAGE A10, Millipore Corporation, Molsheim, France). A 0.5 g sample was placed in a liner and nitric acid was added, then the mixture was kept at room temperature for 48 h. Then, hydrogen peroxide was added, and after the end of the vigorous reaction, decomposition was carried out in a microwave system at 180 °C. The resulting solution was analyzed after dilution with deionized water. Each average plant sample was analyzed twice, the content of the element in the sample was calculated as the average of two independent values.

Simultaneously with the plants, the analysis of "blank" samples and standard samples of plants certified for microelement composition was carried out. Elodea canadensis EK-1, birch leaf LB-1, and grass mixture Tr-1 (Institute of Geochemistry SB RAS, Irkutsk, Russia) were used as standard samples. For every 10 routine samples, one blank sample and one standard sample were analyzed.

2.6. Determination of Ascorbic Acid (AA) and Glutathione (GSH)

Determination of ascorbic acid and glutathione was carried out by titrimetric method using fresh shoots of *Echinochloa frumentacea*. To determine the content of ascorbic acid, an aliquot of the centrifuged supernatant of plant shoots was titrated in a solution of metaphosphoric acid (5 mL) with 0.001 N solution of 2,6-dichlorophenolindophenol (2,6-DCPIP) to a slightly pink color.

To determine the amount of glutathione, 2–3 drops of a 15% KI solution and 5 drops of a 1% starch solution were added to an aliquot of the supernatant (5 mL) and titrated with a 0.001 N KIO_3 solution until a faint blue color was obtained.

The content of AA and GSH was calculated using the following formulas and expressed in mg/g of fresh weight

$$c\,(AA) = [(a \times K) \times 0.88 \times M] \div (m \times n)$$

$$c\,(GSH) = [(b - a) \times K \times 0.307 \times M] \div (m \times n)$$

where c(AA), c(GSH) is content of ascorbic acid and glutathione in plant material (mg/g fresh weight), a is the volume of titrated 2,6-dichlorophenolindophenol (mL), b is the volume of titrated KIO_3 (mL), K is the ratio of volumes of titrated KIO_3 and 2,6-dichlorophenolindophenol (in the present study K = 1.5 ± 0.02), n is the sample weight (mg), M is the total volume of extract (mL), m is the volume of aliquot (mL), 0.88 is the volume of AA (mL), equivalent to 1 mL of 0.001 N solution of 2,6-dichlorophenolindophenol, and 0.307 is the volume of GSH (mL), (equivalent to 1 mL of 0.001 N solution of KIO_3) [41].

2.7. Microbiological Analysis of Rhizosphere Soil

To analyze the rhizosphere soil, the plants were taken out from the pot, the bulk soil was shaken off from the plant roots, and the roots with the remaining adhering soil (thickness, no more than 2–3 mm) were used for analysis. The sampling was made in three replicates per pot. The sample of root with attached rhizosphere soil (1.5–2.0 g) was placed into 0.25-mL Erlenmeyer flask with 100 mL of sterile tap water and was shaken for 30 min. Then, the roots were taken out and the suspension was kept to let the soil particles settle out, after which a range of dilutions was prepared for isolation of rhizosphere microorganisms.

The total numbers of culturable heterotrophic bacteria, actinomycetes, and micromycetes after plant growth were estimated. For isolation and enumeration of the total number of culturable heterotrophs, we used the GRM-agar medium (State Research Center for Applied Biotechnology and Microbiology, Obolensk, Russia) of the following composition: fish meal pancreatic hydrolysate 12 g/L, enzymatic peptone 12 g/L, NaCl 6 g/L, and agar-agar 10–12 g/L. The number of actinomycetes was determined using a starch–ammonium agar medium of the following composition: $(NH_4)_2SO_4$ 1.0 g/L, $MgSO_4 \cdot 7H_2O$ 1.0 g/L, NaCl 1.0 g/L, $CaCO_3$ 3.0 g/L, and agar-agar 20 g/L. The number of microscopic fungi was determined using Martin's medium of the following composition: glucose 10 g/L, KH_2PO_4 5 g/L, $MgSO_4 \cdot 7H_2O$ 0.5 g/L, peptone 5 g/L, agar-agar 20 g/L, and tap water 1 L (pH 5.5). The number of microorganisms resistant to Zn^{2+} and Pb^{2+} ions was determined on LB agar medium [42]. After sterilization, the water-soluble salts of heavy metals $ZnSO_4 \cdot 7H_2O$, $Pb(CH_3COO)_2$, or $CuSO_4$ were added to the culture medium to a final metal concentration of 0.5 mmol/L. The inoculated plates were incubated at 28–30 °C for 5–10 days, after which microbial colonies and colony-forming units (CFU) were counted and the morphological diversity of the microorganisms was measured.

The study of the taxonomic structure of rhizosphere microbial communities was carried out using metagenomic analysis of rhizosphere soil samples for the 16S rRNA gene. The purified DNA preparation was used as a template in the PCR reaction with universal primers (27f/533r) to the variable V4 region of the 16S rRNA gene using GS Junior Technology (Roche 454 Life Sciences, Branford, New Haven, CT, USA). Sequencing was performed on a MiSeq platform (Illumina, San Diego, CA, USA). The obtained data were analyzed using the QIIME v. 1.9.1.

2.8. Statistical Analysis

All experiments were performed in triplicate and the results are presented as mean values ± standard deviations. Basic statistics were completed by using Microsoft Office Excel Microsoft, Redmond, Washington, DC, USA), and all the others were performed with STATISTICA StatSoft (New York, NY, USA). The Student's *t*-test was performed to reveal the differences between the values obtained for background and experimental soils ($p < 0.05$).

3. Results

The main characteristics of the analyzed soils are given in Table 1.

Table 1. Characteristics of urban soils used in the study.

Soil Sample	Background (Yasnaya Polyana)	KMP	Tulachermet	Highway (Lenin Ave.)	Embankment
Soil type	clay loam	clay loam	sandy loam	clay loam	sandy loam
pH	6.20	7.26	7.35	7.29	7.31
Content of essential elements, mg/kg					
$N\text{-}NO_3^-$	20.5 ± 1.2	28.2 ± 1.6	38.6 ± 3.5	41.1 ± 2.4	5.1 ± 1.8
$N\text{-}NH_4^+$	12.9 ± 1.2	6.9 ± 0.7	9.5 ± 0.3	2.3 ± 0.2	5.2 ± 0.4
P_2O_5	51 ± 5	160 ± 4	98 ± 8	218 ± 7	305 ± 7
total carbon, % to air dry soil	4.87 ± 0.05	4.41 ± 0.11	4.84 ± 0.33	3.83 ± 0.06	4.21 ± 0.08
water-soluble carbon, % to the total	0.14 ± 0.05	0.13 ± 0.02	0.14 ± 0.06	0.12 ± 0.03	0.16 ± 0.06
humus, % to air dry soil	8.12 ± 0.16	7.14 ± 0.14	7.62 ± 0.15	6.16 ± 0.12	5.61 ± 0.11
humus carbon, %	4.72 ± 0.09	4.15 ± 0.08	4.43 ± 0.09	3.58 ± 0.07	3.26 ± 0.07
MgO, %	1.06 ± 0.09	0.94 ± 0.08	1.33 ± 0.11	0.62 ± 0.05	1.12 ± 0.09
CaO, %	1.17 ± 0.11	2.83 ± 0.25	2.24 ± 0.23	11.68 ± 1.32	5.48 ± 0.47
Na_2O, %	0.69 ± 0.07	0.55 ± 0.05	0.51 ± 0.05	0.63 ± 0.06	0.59 ± 0.05
K_2O, %	2.48 ± 0.24	2.42 ± 0.21	1.55 ± 0.18	1.21 ± 0.12	2.41 ± 0.22
Trace elements, mg/kg					
Mn	900 ± 87	**6600 ± 260**	950 ± 87	700 ± 58	**1600 ± 156**
Fe	$39{,}250 \pm 1960$	$45{,}750 \pm 2290$	**$116{,}650 \pm 5830$**	$45{,}300 \pm 2260$	**$97{,}700 \pm 8754$**
V	65 ± 3.2	55 ± 4	**145 ± 7.2**	59 ± 2.9	91 ± 5
Cr	**116 ± 8**	137 ± 12	117 ± 4.5	70 ± 2.8	**1260 ± 80**
Ni	41 ± 3	45 ± 4	**42 ± 3.3**	35 ± 2.8	**285 ± 23**
Cu	29 ± 2	52 ± 4	27 ± 0.8	30 ± 0.9	**1188 ± 89**
Zn	23 ± 2	**192 ± 9**	71 ± 3.6	106 ± 5.3	**4579 ± 230**
Pb	28 ± 2	**71 ± 6**	24 ± 0.7	36 ± 1.1	**185 ± 14**
As	5.1 ± 0.4	**9.9 ± 0.5**	4.3 ± 0.2	5.4 ± 0.3	**12.5 ± 0.6**
petroleum products, g/kg	1.5 ± 0.6	2.6 ± 0.4	**4.1 ± 1.0**	2.5 ± 0.7	**9.5 ± 3.2**

* Values exceeding MPC are marked in bold ([43,44]).

The KMP soil was characterized by a high content of Fe (45,750 mg/kg), exceeding the MPC for Mn (by 4.4 times) and Zn (by 41%); Tulachermet soil was characterized by a very high content of Fe (116,650 mg/kg), exceeding the MPC (APC) for V-Mn (by 40–50% for V and 10% for Mn), Ni (by 110–175%), Cu (by 127%), Zn (by 29–192%), As (by 115–220%), and oil products (by 4 times); the soil of Lenin Avenue was characterized by a high content of Fe (45,300 mg/kg), exceeding the permissible concentrations of Mn (by 6%) and Cu (by 186%). The soil of the embankment opposite the arms factory was characterized by the greatest pollution, it exceeded the permissible content of the complex of heavy metals, Mn (by 6%), Cr, Ni (3.6 times), Cu (36 times), Zn (83 times), As (by 25%), and petroleum products (by 9.5 times). In the background soils, the excess of MPC and APC for normalized elements was not noted.

In the performed study, the adaptive characteristics were studied and data on the sowing qualities of seeds and biometric parameters of *Echinochloa frumentacea* on soils contaminated with heavy metals were obtained. The results obtained are presented in Table 2. The soils of sanitary protection zones of metallurgical companies did not affect significantly the decrease in the germination of *Echinochloa frumentacea*. However, the soils of the SPZ of the highways and the soils of the embankment most contaminated with toxic elements caused a decrease in seed germination by 21–23% (Table 2).

Table 2. Effect of soil pollution with heavy meals on germinating ability and biometric parameters of *Echinochloa frumentacea*.

Characteristics	Background (Yasnaya Polyana)	TCh	KMP	Highway Lenin Ave.	Embankment
Germinating ability, %	87 ± 9	80 ± 7	80 ± 8	$67 * \pm 6$	69 ± 7
Height of shoots, cm	15.1 ± 1.3	$9.8 * \pm 1.2$	10.1 ± 1.3	$8.6 * \pm 0.9$	$4.6 * \pm 0.6$
root length, cm	7.6 ± 0.8	$4.5 * \pm 0.3$	$4.3 * \pm 0.4$	$3.5 * \pm 0.3$	$4.3 * \pm 0.3$

*—$p < 0.05$ for the difference between experimental and background samples.

The biometric parameters of *Echinochloa frumentacea* grown on the soils of the experimental zones were measured one month after germination. The results of biometric measurements showed that soil contamination with trace elements had a toxic effect on plants, which was expressed in a decrease in their growth parameters. The maximum toxic effect on the experimental plants was exerted by the soils of the embankment, where there were multiple excesses of MPC (MAC) for a complex of toxic components: Cr, Ni (4 times), Cu (36 times), Zn (83 times), and oil products (more than 9 times). The shoot length of plants on the most polluted soils decreased 3.3 times compared to the background. The soils of the SPZ of metallurgical industries caused a decrease in the growth parameters of the shoot up to 35% compared to the background. The soils of the sanitary protection zone of the highway had an increased toxic effect, reducing the shoot length of the experimental plants by 43% compared to the background (Table 2).

The development of the root system, which performs the function of supplying photosynthetic organs with minerals and water, also suffered from the presence of toxic elements in soils. The length of roots on the soils of the sanitary protection zone of metallurgical industries and the arms factory (embankment) was lower than on the background soils by 19–20%. The most polluted soils of the highway had the maximum inhibitory effect on the development of the root system, reducing the length of the roots by 54% compared to the background (Table 2, Figure A1). Such pronounced impact of the soils of the SPZ of the highway on the growth and development of the root system can also be due to its salinization caused by the use of reagents during the winter.

The formation of plant biomass, which is important to take into account when carrying out phytoremediation measures, depends on the leaf surface area formed by plants, the work of photosynthetic pigments, and enzymatic activity. The quantitative characteristics of the photosynthetic apparatus are the most important indicators of plant adaptation to oxidative stress in the condition of soil contamination. The obtained results showed that the soils of the sanitary protection zone of metallurgical plants had a stimulating effect on the pigment apparatus of *Echinochloa frumentacea*.

The content of chlorophyll a in plants grown on the soils of Tulachermet was higher than in plants grown on the background soils by 36% and on the soils of KMP by 19% (Table 3). The content of chlorophyll a in plants grown on the soils of Lenin Avenue was lower than on the background soils by 7%. The greatest depressing effect on the content of pigments was exerted by the soils of the most polluted experimental zone, the embankment near the arms plant. The content of chlorophyll a in the shoots of Japanese millet was lower by 22%. At the same time, the content of chlorophyll b decreased by 34% compared with the control. On the soils of the embankment, the development of chlorosis and apical necrosis of the leaves of *Echinochloa frumentacea* was observed.

Table 3. Effect of the polyelemental pollution of soils on the quantitative characteristics of the photosynthetic pigments of *Echinochloa frumentacea*.

Pigments' Content, mg/g and Their Ratio	Background	TCh	KMP	Lenin Ave.	Embankment
chlorophyll a	1.46 ± 0.12	$1.99 * \pm 0.14$	1.74 ± 0.15	1.37 ± 0.11	$1.14 * \pm 0.07$
chlorophyll b	0.32 ± 0.03	$0.49 * \pm 0.05$	0.35 ± 0.03	0.24 ± 0.03	$0.21 * \pm 0.02$
chlorophyll sum	1.78 ± 0.15	$2.48 * \pm 0.19$	2.09 ± 0.18	1.61 ± 0.14	$1.35 * \pm 0.08$
carotenoids	0.49 ± 0.05	$0.66 * \pm 0.07$	0.57 ± 0.05	0.43 ± 0.04	0.40 ± 0.03
chlorophyll a/b	4.6	4.1	5.1	5.6	5.4
chlorophyll/ carotenoids	3.6	3.8	3.7	3.7	3.4

*—$p < 0.05$ for the difference between experimental and background samples.

The content of chlorophyll b in the seedlings of *Echinochloa frumentacea* grown on the soils of metallurgical industries was higher than on the soils of the background zones by 9 and 53% for the soils of the SPZ KMP and Tulachermet, respectively. However, in plants growing on the soils of the highway, the content of chlorophyll b decreased by 10% compared to the control.

The same trends were observed for the content of carotenoids in the shoots of *Echinochloa frumentacea*: on the soils of the SPZ KMP and Tulachermet, it increased by 16 and 35%, respectively, compared to the background soils (Table 3). On the contrary, on soils with the highest pollution, it decreased significantly (by 19%). Carotenoids are one of the components of the antioxidant system of the plant, as well as a component of the pigment systems. Therefore, a decrease in their content can lead not only to a decrease in plant biomass but also to the development of severe oxidative stress [45].

The study showed that the pigment apparatus of *Echinochloa frumentacea* is well adapted to soils with a low degree of polyelement pollution (SPZ of metallurgical industries); however, on soils with strong polyelement anomalies (Embankment), the photosynthetic apparatus of the plant suffers due to a drastic decrease in the chlorophylls level (up to 2 times).

An important adaptive characteristic of plants grown on soils contaminated with HMs is their ability to produce antioxidant compounds that prevent the development of oxidative stress, which is accompanied by the oxidation of membrane lipids and disruption of transport and homeostasis processes [46]. The plant antioxidant system (AOS) includes low molecular weight antioxidants such as ascorbic acid (AA) and glutathione (GSH) [47,48]. Due to the ability to reversibly oxidize and reduce, ascorbic acid is involved in the most important plastic and energy processes of the plant cell: photosynthesis and respiration [47,49–51]; it is a recognized antioxidant [52–54], participates in growth and development processes [48,55], and forms plants' resistance to many adverse factors: UV radiation, pathogens [56], ozone [55,57], low temperatures, drought [48], soil salinity [48,58], heavy metals [59,60], and petroleum products [61].

The content of ascorbic acid and glutathione as components of AOS that prevent the development of oxidative stress in seedlings of *Echinochloa frumentacea* grown on soils with polyelement pollution was studied. The results are presented in Table 4.

Table 4. Effect of the polyelemental pollution of soils on the content of ascorbic acid and glutathione in *Echinochloa frumentacea* shoots.

Experimental Soils	Content of Ascorbic Acid, mg/g	Content of Glutathione, mg/g
Background	6.86 ± 0.52	1.45 ± 0.12
TCh	7.12 ± 0.64	1.52 ± 0.14
KMP	8.18 * ± 0.79	1.67 * ± 0.16
Highway	8.05 ± 0.74	1.58 ± 0.15
Embankment	6.45 ± 0.58	1.35 ± 0.12

*—$p < 0.05$ for the difference between experimental and background samples.

The content of low-molecular antioxidants ascorbic acid and glutathione in the shoots of *Echinochloa frumentacea* was 2–2.5 times higher than in *Poa pratensis* grown on the same soils and varied within 6.9–8.2 mg/g (AA) and 1.35–1.67 mg/g (GSH) (Table 4). On the soils of the SPZ Tulachermet, KMP, and the highway, it was higher than in the control by 4, 19, and 17%, respectively. On the soils of the embankment, a slight decrease in the content of AA was observed compared to the control (by 6%). The content of glutathione in the shoots of *Echinochloa frumentacea* varied within 1.35–1.67 mg/g and was also higher on the soils of three experimental zones by 5, 15, and 9% for Tulachermet, KMP, and Lenin Avenue, respectively. The content of glutathione slightly decreased in comparison with the control on the soils of the embankment.

Study of the formation of the biomass of *Echinochloa frumentacea* showed that after 1.5 months of growth on the soils of SPZ of metallurgical plants, the plant formed more biomass than on the background soils. The biomass of Japanese millet grown on experimental soils varied in the range of 428–1024 g/m² (Table 5). The increase in shoot biomass relative to plants grown on background soils was 39% for plants on KMP soils and 48% for plants on Tulachermet soils. The biomass of plants grown on the soils of the SPZ of Lenin Ave. and the embankment was 12.6% and 20% less than on the background soils, respectively. The moisture content in the shoots on the background soils and Tulachermet was 82–84%, on the soils of KMP and highway in the experimental plants, the accumulation of dry matter increased and the water content decreased to 78–65%.

Table 5. Formation of *Echinochloa frumentacea* biomass on the soils with polymetal anomalies.

Soil Samples	M of the Shoot, g	Shoots Biomass, kg/ha	Dry Content, %	Moisture Content, %	Biomass of Dry Weight (Shoots), kg/ha
Background	1.34 ± 0.12	5360 ± 487	15.3	84.8	820
TCh	1.98 * ± 0.17	7920 * ± 713	17.6	82.4	1393
KMP	1.87 * ± 0.19	7480 * ± 778	22.4	77.6	1675
Highway (Lenin Ave.)	1.17 ± 0.11	4680 ± 423	35.3	64.7	1652
Embankment	1.07 * ± 0.09	4280 * ± 419	27.8	72.2	1190

*—$p < 0.05$ for the difference between experimental and background samples.

To assess the phytoremediation capacity of plants, it is important to determine the bioaccumulation of toxic elements in plant organs and to calculate the efficiency of removal of the element from contaminated soils.

Since there are no MPCs established for plants, and the data on the composition of elements in different works differ greatly depending on the applied methods of measurement and sample preparation techniques [62–65] (some studies were performed on unwashed plant material), when determining the bioaccumulative characteristics of plants, a comparison was made with the average data obtained for different plants species and published in Markert [66]—Reference Plants (RP).

The accumulative capacity of the shoots and roots of adult plants (4.5 months) grown on contaminated soils was assessed using atomic absorption spectroscopy.

The content of V in the roots of *Echinochloa frumentacea* grown on contaminated soils varied within 8.9–47.8 mg/kg of dry weight, which is 43–78 times higher than in the aboveground parts of the plant (Table 6). On the soil with a high content of V (Tulachermet), the accumulation of elements by the root system of Japanese millet increased significantly, while in the shoots the content of the element was at a low level of 0.61 mg/kg of dry weight.

Table 6. The average content of heavy metals in the shoots and root system of *Echinochloa frumentacea* grown on soils with polyelement anomalies (full vegetation period), mg/kg dry weight (AAS).

Collection Site	Organ	Pb mg/kg	Cd mg/kg	Cu mg/kg	V mg/kg
Background	shoot	0.20 ± 0.006	0.46 ± 0.014	11.02 ± 0.3	0.40 ± 0.02
	root	3.30 ± 0.09	0.27 ± 0.008	27.84 ± 0.8	8.85 ± 0.4
TCh	shoot	0.14 ± 0.004	0.25 ± 0.008	11.47 ± 0.3	0.61 ± 0.02
	root	1.57 ± 0.05	0.18 ± 0.05	16.44 ± 0.5	47.79 * ± 2.4
KMP	shoot	7.42 * ± 0.22	0.25 ± 0.015	16.34 ± 0.5	0.87 ± 0.04
	root	7.38 * ± 0.22	0.51 ± 0.08	41.05 * ± 1.2	19.26 * ± 1.0
Highway	shoot	0.18 ± 0.05	0.28 ± 0.08	27.56 * ± 0.8	0.43 ± 0.02
	root	6.93 * ± 0.02	1.01 ± 0.03	96.60 * ± 2.9	18.47 * ± 0.9
Reference Plant, Markert, 1992		1	0.05	10	0.5

*—$p < 0.05$ for the difference between experimental and background samples.

The root system of *Echinochloa frumentacea* is an active barrier that prevents the transport of V into the photosynthetic organs of the plant. The coefficient of transfer of the element from roots to shoots was 0.01–0.05. A possible way to clean V-contaminated soil using Japanese millet is rhizofiltration.

The content of Pb in the roots of *Echinochloa frumentacea* was 1.6–7.4 mg/kg, and in the shoots, 0.14–7.42 mg/kg dry weight (Table 6). An analysis of the obtained data showed that the root system of *Echinochloa frumentacea* ceases to perform barrier functions with respect to Pb when the threshold concentrations of the element in soils are exceeded. The content of Pb in Japanese millet grown on soils, in which Pb content exceeded the MPC value (71 mg/kg soil) was the highest and the same for shoots and roots, 7.4 mg/kg dry weight (Table 6). At the obtained values of the removal of the element from soils, which exceeded RP by more than 7 times, it should be expected that the content of the element in the soil during phytoremediation measures will come to acceptable values in 4–5 years.

For most plants, the critical level of copper content is 10–20 mg/kg dry weight [67]. Cu accumulation by the shoots and roots of *Echinochloa frumentacea* exceeded the threshold of normal regulation for plants on KMP-contaminated soils, where the content of the copper was quite high (52 mg/kg) and amounted to 16.3 mg/kg in shoots and 41 mg/kg in roots. In general, the root system of Japanese millet had a high affinity for copper. Therefore, one of the ways to clean soils from Cu in the case of polyelement pollution using *Echinochloa frumentacea* is rhizofiltration. However, in the shoots of KMP, highway, and embankment soils, the content of Cu varied from 16.3 to 96.6 mg/kg of dry weight (Tables 6 and 7). These values are higher than those given in Murillo et al. [68] for sorghum of 4.6–11.7 mg/kg; therefore, it can be assumed that Japanese millet can also be used for the phytoextraction of Cu from soils with polyelement contamination.

Table 7. Bioaccumulation of toxic elements by *Echinochloa frumentacea* grown on soils with polyelement anomalies (vegetation period 1.5 months) (average data for ICP-MS and INAA), mg/kg.

Element	TCh	KMP	Highway	Embankment	Background	RP (Markert, 1992)
Sc	0.070 ± 0.002	0.059 ± 0.002	0.076 ± 0.0021	0.047 ± 0.001	0.057 ± 0.002	0.02
Rb	6.4 ± 0.6	4.8 ± 0.4	3.0 ± 0.3	5.7 ± 0.5	38.3 ± 0.2	50
Mo	$2.2 * \pm 0.01$	$12.7 * \pm 0.6$	$3.5 * \pm 0.01$	$4.2 * \pm 0.02$	0.8 ± 0.02	0.5
Cd	0.31 ± 0.009	0.28 ± 0.008	0.47 ± 0.01	0.37 ± 0.01	0.52 ± 0.02	0.05
Sb	0.023 ± 0.001	0.019 ± 0.001	$0.044 * \pm 0.002$	$0.074 * \pm 0.003$	0.010 ± 0.001	0.1
Ba	6.7 ± 0.34	9.6 ± 0.48	12.3 ± 0.62	4.8 ± 0.24	10.8 ± 0.54	40
Pb	0.50 ± 0.02	0.25 ± 0.01	0.44 ± 0.01	$0.88 * \pm 0.03$	0.22 ± 0.02	1
Th	0.074 ± 0.004	0.050 ± 0.003	0.075 ± 0.004	0.045 ± 0.003	0.063 ± 0.003	0.005
U	0.023 ± 0.001	0.014 ± 0.001	$0.037 * \pm 0.001$	$0.036 * \pm 0.001$	0.015 ± 0.001	0.01
V	0.61 ± 0.03	0.64 ± 0.03	0.74 ± 0.04	0.66 ± 0.03	0.50 ± 0.02	0.5
Cr	1.01 ± 0.06	0.93 ± 0.06	1.62 ± 0.1	$4.57* \pm 0.27$	0.93 ± 0.06	1.5
Co	0.19 ± 0.01	0.19 ± 0.01	0.27 ± 0.02	0.10 ± 0.01	0.16 ± 0.01	0.2
Ni	1.4 ± 0.1	2.4 ± 0.2	3.6 ± 0.3	4.7 ± 0.4	3.4 ± 0.3	1.5
Cu	$20.2 * \pm 0.6$	12.3 ± 0.4	$31.3 * \pm 0.9$	$32.1 * \pm 0.9$	21.2 ± 0.6	10
As	$0.29 * \pm 0.01$	0.21 ± 0.01	0.15 ± 0.01	0.20 ± 0.01	0.12 ± 0.01	0.1
Se	0.15 ± 0.01	0.13 ± 0.01	0.06 ± 0.005	0.19 ± 0.01	0.13 ± 0.01	0.02
Al	295 ± 8.8	322 ± 9.7	$390 * \pm 12$	251 ± 7.5	275 ± 8.2	80
Fe	291 ± 15	$379 * \pm 129$	324 ± 16	327 ± 16	282 ± 14	150
Mn	173 ± 8.6	101 ± 5.5	199 ± 10	$572.8 * \pm 30$	195 ± 9.7	200
Zn	$100 * \pm 4$	41 ± 1.6	$109 * \pm 4.3$	$454 * \pm 18$	53	50
Sr	36 ± 2.9	$91 * \pm 7.2$	48 ± 3.8	45 ± 3.6	33 ± 2.6	

$*$—$p < 0.05$ for the difference between experimental and background samples.

However, on soils with a high content of V and Cr, the absorption of the element by the plant decreased, which should be taken into account when carrying out remediation measures. According to the literature data, when the content of copper in plants exceeds the critical level, the signs of toxicity may develop: the concentration of chlorophyll decreases, and the growth of shoots and roots is suppressed [69]. The complex of HM pollutants in the experimental soils suppressed the growth parameters of Japanese millet; however, on the soils of the sanitary protection zone of metallurgical industries, a compensatory mechanism was observed and the content of photosynthetic pigments increased (Table 3).

Accumulation of Mo by shoots of *Echinochloa frumentacea* on the soils of Tulachermet up to 12.7 mg/kg dry weight, which is 3–6 times more than on the soils of other experimental zones and 12 times more than on the background soils (Table 7). High Mo-uptake can be explained by the slightly alkaline pH of analyzed soils [70].

The accumulation of Cd did not exceed 0.28–0.52 μg of dry weight in most of the experimental variants. These values are 5.5–10.4 times higher than the average data for RP. The Cd enrichment factor for *Echinochloa frumentacea* is 5.6–19 and is at its maximum on the soils of the sanitary protection zone of the highway. Low Cd uptake can be associated with the presence of zinc in soil, which inhibits Cd uptake and translocation under cadmium/zinc combined stress [71].

The accumulation of Cr by the shoots of *Echinochloa frumentacea* on soils with the highest polyelement pollution, where the content of the element was many times higher than APC (1260 mg/kg), was higher than for other soils (Cr content up to 117 mg/kg) and amounted to 4.7 mg/kg of dry weight (Table 7). These values were 3 times higher than RP; however, with the soils' contamination with Cr when its content exceeds TEC 4 times, this level of bioaccumulation of the element is not enough for effective biological treatment in terms less than 10 years. On the soils of other sampling sites, Cr accumulated in the shoots of *Echinochloa frumentacea* was up to three times higher than the upper limit of normal values (0.1–0.5 mg/kg, [62]): 0.93–1.62 mg/kg.

The critical level of Co in most plants starts from 0.4 mg/kg. Cereals are most sensitive to their excess [72] The Co content in Japanese millet ranged from 0.10 to 0.27 mg/kg dry weight and lies within the average values characteristic for vegetation, 0.02–1 mg/kg [62].

The content of Ni in *Echinochloa frumentacea* plants varied in the range of 1.4–4.7 mg/kg dry weight and its maximum accumulation was attained on the most polluted soils. These values are 2–3 times higher than the average values for vegetation grown on the soils of Tulachermet, the highway, and embankment.

The content of As in the shoots of *Echinochloa frumentacea* was 0.15–0.29 mg/kg of dry weight. Its content on the soils of the SPZ of metallurgical industries and the embankment was 2–3 times higher than RP values. However, based on the data on the bioaccumulation and removal of As, Japanese millet is not recommended for the phytoremediation of soils contaminated with arsenic.

The critical concentrations of manganese in plants vary from 220 to 5300 mg/kg dry weight [67]. Japanese millet is not an Mn bioaccumulator. The content of the element in the shoots of *Echinochloa frumentacea* falls within the limits of average values and amounts to 57–199 mg/kg of dry weight and lies below the limits of critical values (Table 7).

Visible symptoms of zinc toxicity appear with its concentrations in plants above 300 mg/kg but are sometimes possible at lower concentrations (100 mg/kg) [67]. The content of Zn in the shoots of *Echinochloa frumentacea* was 53 mg/kg on the background soils and varied in the range of 41–457 mg/kg dry weight on the soils of the experimental zones. The content of the element exceeded the critical concentrations for plants on the most polluted soils of the embankment, which could cause the observed toxic effects. The factor *Echinochloa frumentacea* enrichment with Zn was two on the soils of the KMP and Lenin Avenue and eight on the soils of the SPZ of the arms plant (embankment). These are a good indicator for the phytoextraction of the element from soils, taking into account the accumulation of biomass by plants.

The content of Fe in the shoots of *Echinochloa frumentacea* varied within 282–389 mg/kg of dry weight and was below the critical concentration limits for plants (500 mg/kg [62,67]) but higher than the average data for RP 2 or more times. Apparently, for Fe, which is a soil pollutant in the region, there is no barrier at the level of the root system of the plant.

Culture-Based Assessment of Microbial Communities in the Rhizosphere of E. frumentacea

At the end of the experiment, the total number of heterotrophic bacteria, and the number of actinomycetes and micromycetes, were determined in the rhizosphere of plants. The results of the analysis of the main groups of cultivated heterotrophic soil microorganisms in the rhizosphere of *E. frumentacea* are shown in Figure A2.

The abundance of rhizospheric microorganisms depended on the kind of soil in which the plants were grown. The number of bacteria in the plant rhizosphere varied from 0.8 to 3.3×10^7 CFU/g of soil sampled from the Embankment and Background, respectively. It was found that the soil of the Embankment near the Tula Arms Plant was the most polluted (Table 1). The high content of almost all metals analyzed clearly caused the toxicity of this soil, which led to the inhibition of the number of all groups of microorganisms studied, bacteria, actinomycetes, and micromycetes. In addition, this soil was characterized by a minimum content of humus carbon (Table 1), which is important for the development of soil microflora.

The number of cultivated heterotrophic bacteria in the rhizosphere soil from the Embankment (Figure 1a) was two times lower than in the background soil (Yasnaya Polyana). The almost two-fold excess of the number of bacteria in the Tulachermet soil may be due to the presence of an additional carbon source in this soil (oil products' content was 4.1 g/kg, and humus carbon was higher than in the Embankment soil). However, an even higher concentration of oil products (9.5 g/kg) did not support the number of bacteria in the soil of the Embankment, which was associated with the toxic effect of heavy metals on oxidative enzymes involved in the hydrocarbon degradation. The soil from Lenin Ave. also stimulated the rhizospheric bacteria; however, there was no obvious explanation of this effect from the data of analyses of this soil (Table 1). We can only assume that in terms of background pollution, the Lenin Ave. soil, just like the KMP soil, was the closest to the background one.

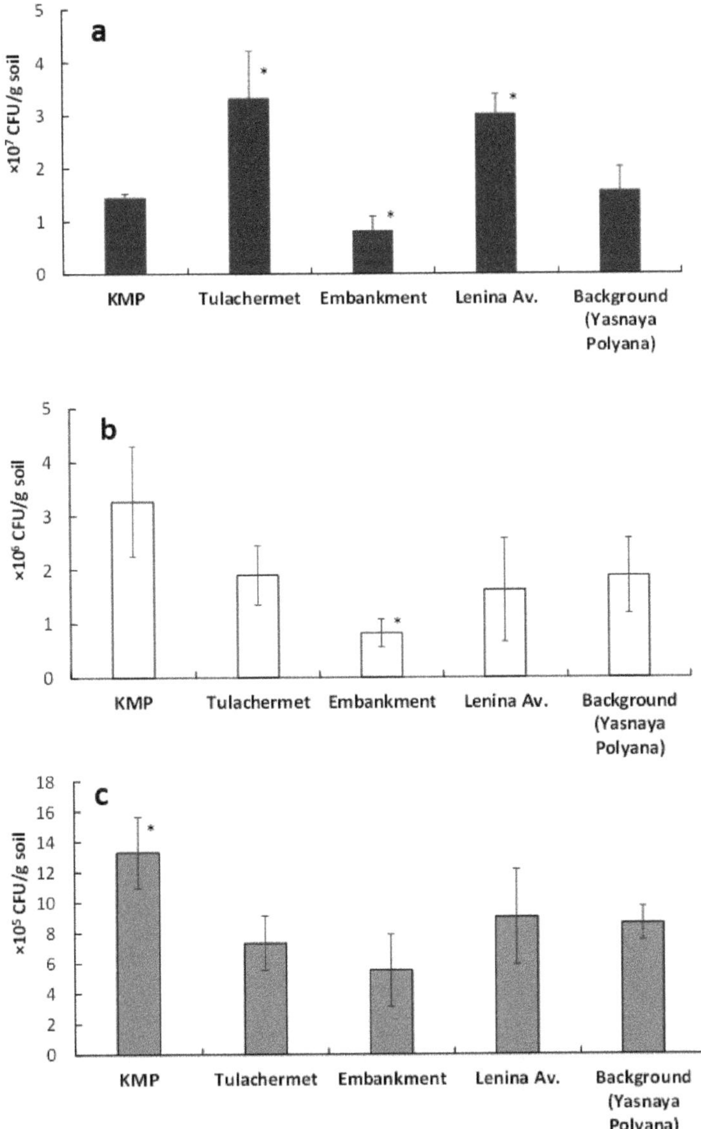

Figure 1. The number of cultivable heterotrophic bacteria, actinomycetes, and micromycetes in the rhizosphere of *E. frumentacea*. (**a**)—bacteria; (**b**)—actinomycetes; (**c**)—micromycetes. Error bars mean standard errors/*—$p < 0.05$ for the difference between experimental and background samples.

The number of actinomycetes in the rhizosphere of *E. frumentacea* varied from 0.8 to 3.3×10^6 CFU/g of the soil in the Embankment and KMP samples, respectively. The influence of urban soils on the actinomycete population in the rhizosphere was most noticeable in the soils of the Embankment and KMP (Figure 1b). In the first case, in relation to bacteria, the inhibitory effect of the contaminated urban soil was visible: the number of actinomycetes in the rhizosphere decreased by 2.3 times compared to the control soil. However, in the KMP soil, stimulation of actinomycetes (1.7 times) was observed. Probably, this could be associated with an increase in the actinomycete taxa resistance to heavy metals and the ability to participate in remediation of the environment [73]. This explanation was also confirmed by our data on the taxonomic structure of the rhizosphere community of *E. frumentacea* grown in the KMP soil. There were no significant differences between the abundance of actinomycetes in the Tulachermet and Lenin Ave. soils and the control soil of Yasnaya Polyana.

The number of micromycetes in the rhizosphere of *E. frumentacea* reached 5.5–13.3 $\times 10^5$ CFU/g of the soil of the Embankment and KMP samples, respectively (Figure 1c). The changes in the abundance of micromycetes depending on the soil type were similar to those described for actinomycetes. The minimal number of micromycetes was observed for the Embankment soil (a 1.6-fold decrease from the control), and the maximal number was observed for the KMP soil (a 1.5-fold increase in comparison with the control), which could be associated with specific stimulation by the plant of micromycetes taxa resistant to heavy metals.

The sequencing of 16s rRNA gene from rhizosphere samples resulted in 133.776 reads. After data denoising and chimera screening, a total of 92.527 sequences were used for further identification. The average number of nucleotide sequences in the library per sample was 15.728.

To characterize biodiversity and carry out a comparative analysis of the communities, the parameters of α-diversity were calculated, the taxonomic composition of the community in all samples was determined and compared, and taxa that reliably decreased or increased their number in the rhizospheres of the plants studied were identified.

The calculation of the sampling effort indicated that, on average, 73.7% of the true diversity of the rhizosphere communities of *E. frumentacea* were covered (Table 8). The α-diversity was assessed using species richness indices (observed OTUs) and the Shannon and the Simpson indices in the rhizosphere of *E. frumentacea* grown in the control soil, and the average number of OTUs was minimal (2013 on average) in comparison with other soils. The species richness index of plants grown in Tulachermet was maximal (3816 observed OTUs) (Table 8).

Table 8. The α-diversity indices for rhizospheric microbial communities of *E. frumentacea* grown on different soils.

Soil Samples	Observed OTUs	PD Whole Tree	Chao1	Shannon	Simpson	Sampling Effort
KMP	3437	94.08	4820.03	10.39	0.99829	71.3
Tulachermet	3816	102.17	5321.48	10.64	0.99842	71.7
Embankment	3059	98.08	4056.96	10.07	0.99738	75.4
Lenin Av.	3403	87.14	4620.18	10.38	0.99811	73.7
Background	2013	66.99	2635.74	8.94	0.99331	76.4

The Shannon and Simpson indices indicated a high taxonomic diversity of bacterial communities in the rhizosphere of plants grown in contaminated urban soils, while the background soil (Yasnaya Polyana) was characterized by the lowest taxonomic diversity (Table 8).

Figure A2 shows the bacterial types that dominate in the rhizosphere of *E. frumentacea*. The Proteobacteria, Actinobacteria, Planctomycetes, Acidobacteria, as well as Bacteroides and Chloroflexi phyla occupied the dominant positions in the taxonomic structure of the rhizosphere community of *E. frumentacea*. However, the ratio of these types was different for plants grown on different soils. Thus, the share of Proteobacteria in the total taxonomic structure of the microbial community in the background soil (Yasnaya Polyana) reached 52%, but it significantly decreased when plants were grown on urban soils. The minimal amount of Proteobacteria was noted in the soils of KMP (27%) and Tulachermet (28%), followed by the soils of Lenin Ave. (30%) and the Embankment (36%). At the same time, the share of another dominant phylum, Actinobacteria, increased from 18% (in control) to 22, 26, 29, and 31% in the soils of the Embankment, Tulachermet, KMP, and Lenin Ave., respectively. In the rhizospheric communities of plants grown in the polluted urban soils, the share of other phyla also clearly increased. Thus, the share of Acidobacteria in the control soil was only 4% but increased to 5, 8, 9, and 9.2 in the soils of Lenin Ave., the Embankment, Tulachermet, and KMP, respectively. The Planctomycetes phylum in the rhizospheric community of *E. frumentacea* grown in the control soil was only 4%, while in contaminated urban soils it increased to 6, 9, and 10% (in the soils of the Embankment, Tulachermet, Lenin Ave., and KMP, respectively). In contaminated soils, an increase in Chloroflexi share from 3% in the control to 6–7% in urban soils was noted. In addition to Proteobacteria, under the influence of pollution, the share of another phylum, Firmicutes, decreased (albeit less significantly), which was more than 4% in clean soil and decreased to 2–3% in polluted soil. Changes in the share of other phyla of the microbial community in the rhizosphere of *E. frumentacea* under the influence of pollution were unsignificant. Thus, a comparative analysis of the taxonomic structure of the rhizosphere microbial community at the phylum level revealed a distinct influence of urban soils characterized by technogenic polyelement anomalies.

The taxonomic analysis of the dominant phylum Actinobacteria made it possible to identify 39 families, among which 16 families occupied a fundamental position, presented in Table A1 and constituting from 60% to 77% of all found families of Actinobacteria in the rhizosphere of *E. frumentacea*. The Gaiellaceae family had the maximal share in the population of rhizospheric actinomycetes in almost all soil samples; only in the soil of the Embankment was the abundance of Gaiellaceae minimal. At the same time, this soil was characterized by an increased share of the Micrococcaceae family (Table A1). Nocardioidaceae, Micromonosporaceae (0.9–3.4%), as well as representatives of the Solirubrobacterales and 0319-7L14 orders were other notable actinomycetes in the rhizosphere of *E. frumentacea*.

As part of another dominant phylum, Proteobacteria, 66 families of bacteria were identified. Thirteen families accounting for 72–86% of all OTUs assigned to Proteobacteria made the most significant contribution to the rhizosphere microbiome structure of *E. frumentacea*, (Table A1). Alphaproteobacteria is the most numerous class of the Proteobacteria phylum, accounting for 41 to 53% of all Proteobacteria. Among Alphaproteobacteria, the dominant position was occupied by the Sphingomonadaceae (2.6–12.7%) and Hyphomicrobiaceae (2.0–5.6%) families. The Betaproteobacteria class accounted for 15 to 32% of all OTUs assigned to the Proteobacteria phylum, and the Comamonadaceae (0.8–5.1%) and Oxalobacteraceae (1.3–3.5%) families were its dominant representatives. The Gammaproteobacteria class accounted for 16 to 30% of all OTUs assigned to the Proteobacteria phylum, and the Xanthomonadaceae family, whose share reached 1.6 to 13.5%, was its dominant representative. It can be assumed that, along with Sphingomonadaceae, the Xanthomonadaceae family can also form the basis of the rhizobiome of *E. frumentacea*. The share of these families decreased when plants were grown in all urban soils, except for Lenin Ave., where it distinctly increased, as did the share of the Hyphomicrobiaceae and Oxalobacteraceae families. Among other bacterial taxa the dominant position in the soil of the Embankment was occupied by representatives of Comamonadaceae, and in the soil of KMP, by Sphingomonadaceae, Hyphomicrobiaceae, and Xanthomonadaceae. The share of Oxalobacteraceae increased in all urban soils compared to the control soil. In the rhizosphere of plants grown in Tulachermet soil, it reached a maximum (3.5%).

In the course of this study, 22 morphologically different strains of bacteria were isolated from the rhizosphere of *E. frumentacea* plants. All isolates were characterized by their resistance to heavy metals in the environment and plant growth-promoting potential. The strains combining these properties were transferred to the collection of rhizospheric microorganisms of the Institute of Biopharmaceutics, Russian Academy of Sciences (http://collection.ibppm.ru, accessed on 14 March 2022).

4. Discussion

Summarizing the data on the quantitative content of the studied plant antioxidant system components of *Echinochloa frumentacea* showed that the plant is characterized by a good adaptive potential on soils with polyelement anomalies caused by technogenic pollution of metallurgical industries: the content of carotenoids increases from 16 to 34% on the soils of the SPZ of metallurgical enterprises. The amount of AA and GSH in plants grown on experimental soils are 4–19% higher compared to the soils of the background zone (Table 9).

Table 9. Components of the *Echinochloa frumentacea* antioxidant system on soils with polyelement anomalies.

Sample of Soils	Carotenoids	AA	GSH
TCh	34%↑	4%↑	5%↑
KMP	16%↑	19%↑	15%↑
Lenin Av.	No significant differences	17%↑	9%↑
Embankment	19%↓	6%↓	7%↓

↑ higher or ↓ lower compared to background soil.

However, intense pollution of the soils of the Embankment caused inhibition of the synthesis of carotenoids and consumption of the synthesized low molecular weight antioxidants (ascorbic acid and glutathione) necessary for the urgent adaptive response of the plant to prevent the effects of oxidative stress [74]. In bean and rice HM-stressed plants, other authors also observed a significant increase in the content of these antioxidants compared to the control [75,76].

Echinochloa frumentacea on soils with multielement contamination accumulated similar amounts of Mn compared to the literature data for *Echinochloa colona* (L.) Link (213 mg/kg) [15]. The accumulation of Zn by shoots of Japanese millet was 2 times lower than for *Echinochloa colona*; however, on soils with the highest Zn content, its accumulation increased by 1.7 times.

The calculation of the removal of elements from soils by *Echinochloa frumentacea* during the growing season of 1.5 months is presented in Table 10.

Table 10. Removal of elements from soils *Echinochloa frumentacea* in the case of polyelement pollution, g/ha.

Element	KMP	TCh	Lenin Av.	Embankment	Background
Mo	3.69 *	17.7 *	5.78 *	4.70 *	0.66
Cd	0.52	0.39	0.78 *	0.41	0.43
Sb	0.04	0.03	0.07 *	0.08 *	0.01
Ba	11.2	13.4 *	20.3 *	5.37	8.86
Pb	0.84 *	0.35	0.73 *	0.98 *	0.18
Th	0.12 *	0.07	0.12 *	0.05	0.05
U	0.04	0.02	0.06 *	0.04	0.01
V	1.02 *	0.89	1.22 *	0.74	0.41
Cr	1.69	1.30	2.68 *	5.11 *	0.76
Co	0.32 *	0.26	0.45 *	0.11	0.13
Ni	2.35	3.35	5.95 *	5.26	2.79
Cu	33.8	17.1	51.7*	35.9 *	17.4
As	0.49 *	0.29 *	0.25	0.22	0.10
Se	0.25 *	0.18	0.10	0.21 *	0.11
Al	494 *	449	644 *	281	226
Fe	488 *	528 *	535 *	366	231
Mn	290 *	141	329 *	64 *	160
Zn	168 *	57.1	180 *	508 *	43.5
Sr	60.3 *	127 *	79.3 *	50.4	27.1

*—$p < 0.05$ for the difference between experimental and background samples.

The calculation of the removal of elements on the dry biomass of plants showed that on experimental soils with polyelement anomalies, the analyzed plant is not suitable for

Pb and As phytoremediation. The removal of these elements from the soil by shoots was 0.35–084 g/ha and 0.22–0.49 g/ha, respectively. Bioaccumulation of Pb was at its maximum in the root system and constituted 1.57–7.38 mg/kg dry weight. The content of Pb in the shoots was lower and varied within 0.14–0.88 mg/kg dry weight. At the same time, in plants grown on soils with a multiple excess of the MPC for Pb by the end of the growing season, up to 7.42 mg/kg of dry weight of Pb was accumulated in the aboveground mass. These figures are lower compared to those available in the literature for other cereals, e.g., sorghum 107–378 g/ha [77].

The content of V in the root system of *E. frumentacea* on contaminated soils reached 9–48 mg/kg dry weight, depending on the soil. At the same time, the root system performs a barrier function and the transfer of V into the shoots was very low, 0.01–0.05. The content of V in the shoots was 0.43–0.87 mg/kg dry weight. The removal of V from contaminated soils by shoots was 0.89–1.12 g/ha.

The amount of Ni and Cu removed from the polluted soils increased with an increase in pollution level and was 5.26–5.95 g/ha for Ni and 34–52 g/ha for Cu. The content of Cu in the shoots was 11.5–27.6 mg/kg of dry weight, and in the root system, 16.5–96.6 mg/kg of dry weight. The plant can be used for long-term phytoextraction of the indicated elements when the MPC is exceeded up to 2 times.

The removal of manganese from soils, in which its content exceeds the MPC more than 4 times, amounted to 290 g/ha. These values are not sufficient for remediation of the soil when MPC is multifold exceeded because the remediation period can be more than 20 years.

The removal of Zn from contaminated soils was 168–508 g/ha and it was maximal on the most polluted soils. The content of zinc in the shoots of *E. frumentacea* grown on soils, where the MPC was exceeded by less than twice, was 109 mg/kg; while at multiple excess of MPC (4.4 times) this increased to 454 mg/kg of dry weight. Zinc is a mobile element and is well accumulated by the plant. The removal of the element for *Sorghum bicolor* and *Zea mais* due to the formation of the maximum biomass, according to the literature data, was 1223–1998 g/ha [77]. An increase in the accumulation of zinc and the removal of the element from contaminated soils makes it possible to recommend this culture for phytoremediation in grass mixtures with other crops, for example, with a representative of cruciferous plants, *Brassica napus*.

Studies on the rhizosphere microflora of *Echinochloa* plants are rare [78–80], and we did not find any data on the impact of technogenic soil pollution on the structure of microbial communities in the root zone of *E. frumentacea*. At the same time, the study of the root microflora of remediating plants and the isolation of rhizospheric microbial strains that are resistant to the presence of pollutants (metals) in the environment and have a plant growth stimulating potential will make it possible to find a suitable microbial inoculant as a tool to increase the effectiveness of the remediation capabilities of *E. frumentacea* used for the restoration of technogenically disturbed soil ecosystems. Previously, a positive effect on the growth of *E. frumentacea* when inoculated with PGPR (plant growth-promoting rhizobacteria) strains was shown both in pure soil conditions [81,82] and in the presence of heavy metals (Cd, Ni, Pb, Cu) and As [83].

It is known that heavy metals have a great effect on bacterial communities of soil, contributing to an increase or decrease in bacterial abundance, species diversity, and alterations in dominant and subordinate species [84]. However, a negative effect of heavy metals on the composition and abundance of soil actinomycetes and fungal communities is also noted [85–87].

In this study, we demonstrated that the abundance of the main groups of microorganisms (bacteria, actinomycetes, and micromycetes) in the rhizosphere of *E. frumentacea* depended on the type of soil in which the plants were grown and, probably, was determined by its characteristics. The minimal number of all groups of microorganisms studied (on average, two times lower than in the uncontaminated control soil) was noted in the Embankment soil, which contains the highest concentrations of both inorganic (heavy metals) and organic (oil products) pollutants. A two-fold increase in the number of bacteria was observed in the rhizosphere of plants grown in the soils of Tulachermet and Lenin Ave., a 1.7-fold increase in the number of actinomycetes, and 1.5-fold increase in the number of micromycetes were observed in the soil of KMP, which could be due to specific plant stimulation of microbial taxa resistant to heavy metals.

The rhizosphere populations of *E. frumentacea* grown in the background soil of Yasnaya Polyana were characterized by the lowest taxonomic diversity compared to the rhizomes of plants grown in contaminated urban soils. The species richness index was maximal for rhizosphere communities of plants grown on Tulachermet soils. The taxonomic structure of the rhizospheric microbiomes of *E. frumentacea* was represented by the dominant bacterial phyla Proteobacteria and Actinobacteria, as well as the phyla Planctomycetes, Chloroflexi, Acidobacteria, and Bacteroides. Compared to the control soil, the cultivation of *E. frumentacea* plants in urban soils led to a change in the ratio of the main phyla. The share of Proteobacteria (~52% in control, 27–36% in urban soils) and Firmicutes (>4% in control, <2–4% in urban soils) decreased, but the shares of Actinobacteria (~18% in control, 22–31% in urban soils), Planctomycetes (~4% in control, 6–10% in urban soils), Chloroflexi (~3% in control, 6–7% in urban soils), Acidobacteria (~4% in control, 5–9% in urban soils), and Bacteroides (~5% in control, 4–8% in urban soils) increased (Table A1).

The data of metagenomic analysis characterizing the rhizospheric microbiome of *E. frumentacea* plants grown on different soils make it possible to assume that the dominant families are Gaiellaceae and Nocardioidaceae (Actinobacteria), Sphingomonadaceae, Hyphomicrobiaceae, Comamonadaceae, Oxalobacteraceae, and Xanthomonadaceae (Proteobacteria), and the families Pirellulaceae (Planctomycetes), Cytophagaceae, and Chitinophagaceae (Bacteroides) can participate in the formation of the rhizobiome of *E. frumentacea* (Table A1). Sun et al. (2018) suggested a potential ecological role for Gaiellaceae in metal-contaminated soils, finding their increased numbers in sequencing libraries and a close significant correlation with various metals' or metalloids' content. In our study, in four out of five soil samples, the Gaiellaceae family was dominant, reaching a maximum in the Tulachermet soil. It can be assumed that representatives of the Gaiellaceae family can form the so-called "core" rhizobiome of *E. frumentacea* plants, being present in its rhizosphere under various conditions. Representatives of the Planctomycetes and Bacteroides phyla also seem to contribute to the formation of the "core" rhizobiome of *E. frumentacea*. Thus, Pirellulaceae (Planctomycetes), Cytophagaceae, and Chitinophagaceae (Bacteroides) families make up a significant share in the rhizospheric microbial populations of plants grown in various soils.

Thus, the qualitative and quantitative changes in the rhizospheric microbial populations of *E. frumentacea* under the influence of technogenic soil pollution have been found.

5. Conclusions

The study of the adaptive characteristics of *Echinochloa frumentacea* revealed that soils with a high degree of contamination with a complex of HMs and petroleum products have an inhibitory effect on the germination and growth processes of *Echinochloa frumentacea*: on the most polluted soils of the Embankment, a reduction in seed germination by 21–23% and a decrease in growth parameters up to 3 times were observed.

A low level of contamination with the complex of HM in the sanitary protection zones of metallurgical industries did not have a toxic effect on seed germination and stimulated the quantitative characteristics of photosynthetic pigments and AOS of *Echinochloa frumentacea*: the content of chlorophylls and low molecular antioxidants such as ascorbic acid and glutathione (GSH) increased by 9–53%, 17–19%, and 5–15% respectively. The data obtained indicate good adaptive characteristics of *Echinochloa frumentacea* to technogenic pollution of soils by metallurgical enterprises.

The accumulation of *Echinochloa frumentacea* biomass on contaminated soils varied within 4280–7920 kg/ha. The increase in shoot biomass in relation to plants on background soils was shown for plants grown on metallurgical enterprises soils. The polyelement pollution of soils of the SPZ and of the highway contributed to the accumulation of the dry mass of plants compared to background soils.

Echinochloa frumentacea is not suitable for Mn, Co, As, and Cd accumulation from soils with polyelement contamination. The bioaccumulation of Pb was maximum in the root system and constituted 1.57–7.38 mg/kg dry weight. V from polluted soils mainly accumulated in the roots. The main method of V accumulation from soils is rhizofiltration. The accumulation of Cu by the shoots and root system of plants makes it possible to consider *Echinochloa frumentacea* as a Cu phytoremediant. The removal of Zn from contaminated soils during a short vegetation period (1.5 months) was 168–508 g/ha. The obtained values make it possible to recommend *Echinochloa frumentacea* for Zn removal from soils with polyelement anomalies in combination with other species accumulating the element, in intercrop mixtures.

The minimal number of all groups of microorganisms studied (on average, two times lower than in the uncontaminated background soil) was noted in the Embankment soil, which contains the highest concentrations of both inorganic (heavy metals) and organic (petroleum products) pollutants. A two-fold increase in the number of bacteria was observed in the rhizosphere of plants grown in the soils of Tulachermet and Lenin Ave., and a 1.7-fold increase in the number of actinomycetes and a 1.5-fold increase in the number of micromycetes were observed in the soil of KMP, which could be due to specific plant stimulation of microbial taxa resistant to heavy metals.

The rhizosphere populations of *Echinochloa frumentacea* grown in the background soil of Yasnaya Polyana were characterized by the lowest taxonomic diversity compared to the rhizobiomes of plants grown in contaminated urban soils. The species richness index was maximal for rhizosphere communities of plants grown on Tulachermet soils.

The data of metagenomic analysis characterizing the rhizospheric microbiome of *E. frumentacea* plants grown on different soils make it possible to assume that the dominant families are Gaiellaceae and Nocardioidaceae (Actinobacteria), Sphingomonadaceae, Hyphomicrobiaceae, Comamonadaceae, Oxalobacteraceae, and Xanthomonadaceae (Proteobacteria), and the families Pirellulaceae (Planctomycetes), Cytophagaceae, and Chitinophagaceae (Bacteroides) can participate in the formation of the rhizobiome of *E. frumentacea*.

Further detailed study of the properties of the selected microorganisms will make it possible to select a promising inoculant to improve plant growth on contaminated soil to increase the efficiency of its phytoremediation.

Author Contributions: Conceptualization S.V.G., A.K. and A.Y.M.; methodology S.V.G., A.Y.M., I.Z. and O.I.O.; software A.Y.M., I.Z. and O.I.O.; validation A.Y.M., S.V.G., I.Z. and O.I.O.; formal analysis S.V.G., A.Y.M., I.Z. and O.I.O.; investigation S.V.G., A.Y.M., I.Z. and O.I.O.; resources RFBR; writing—S.V.G., A.K., A.Y.M. and I.Z.; visualization, S.V.G., A.Y.M. and I.Z.; supervision S.V.G.; project administration S.V.G.; funding acquisition S.V.G. All authors have read and agreed to the published version of the manuscript.

Funding: This research was funded by the Russian Foundation for Basic Research project No 19-29-05257 "Technogenic soil pollution with toxic elements and possible methods for its elimination".

Acknowledgments: The study was carried out with financial support by the Russian Foundation for Basic Research project No 19-29-05257 "Technogenic soil pollution with toxic elements and possible methods for its elimination".

Conflicts of Interest: The authors declare no conflict of interest.

Appendix A

Figure A1. *Echinochloa frumentacea* shoots (1.5 monthly) grown on the soils of the experimental zones: from left to right: background, Tulachermet, KMP, Embankment, highway (Lenin Ave.).

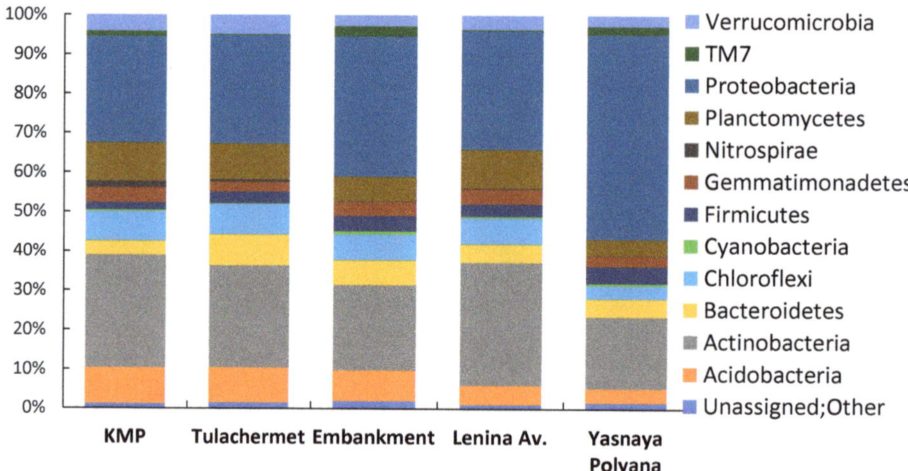

Figure A2. The relative abundances at the phylum level of the microbial communities in the rhizosphere of *E. frumentacea* grown on different soils.

Table A1. Taxonomic structure of rhizosphere microbial communities of *E. frumentacea* (%).

Taxonomic Associations of OTUs	Soil Samples				
	1	2	3	4	5
Unassigned; Other; Other; Other; Other	27.30	28.40	33.10	20.50	23.20
k__Bacteria;p__Acidobacteria;c__Acidobacteria-6;o__iii1-15;f__	4.10	4.10	2.50	0.30	2.40
k__Bacteria;p__Acidobacteria;c__Acidobacteriia;o__Acidobacteriales;f__Acidobacteriaceae	0.00	0.00	0.00	1.50	0.00
k__Bacteria;p__Acidobacteria;c__Acidobacteriia;o__Acidobacteriales;f__Koribacteraceae	0.00	0.00	0.20	1.00	0.00
k__Bacteria;p__Acidobacteria;c__Sva0725;o__Sva0725;f__	0.50	0.10	1.00	0.00	0.20
k__Bacteria;p__Actinobacteria;c__Acidimicrobiia;o__Acidimicrobiales;f__	2.20	1.40	1.70	0.30	1.20
k__Bacteria;p__Actinobacteria;c__Acidimicrobiia;o__Acidimicrobiales;f__C111	1.70	1.90	0.40	0.00	1.00
k__Bacteria;p__Actinobacteria;c__Actinobacteria;o__Actinomycetales;f__	0.20	0.50	0.30	2.40	0.20
k__Bacteria;p__Actinobacteria;c__Actinobacteria;o__Actinomycetales;f__Cellulomonadaceae	0.10	0.30	1.60	0.00	0.10
k__Bacteria;p__Actinobacteria;c__Actinobacteria;o__Actinomycetales;f__Intrasporangiaceae	0.80	1.50	0.60	0.70	1.00
k__Bacteria;p__Actinobacteria;c__Actinobacteria;o__Actinomycetales;f__Micrococcaceae	0.90	1.50	3.60	1.10	1.60

Table A1. *Cont.*

Taxonomic Associations of OTUs	Soil Samples				
	1	2	3	4	5
k__Bacteria;p__Actinobacteria;c__Actinobacteria;o__Actinomycetales;f__Micromonosporaceae	1.00	0.80	0.40	1.30	1.50
k__Bacteria;p__Actinobacteria;c__Actinobacteria;o__Actinomycetales;f__Mycobacteriaceae	0.90	0.10	1.30	0.20	1.40
k__Bacteria;p__Actinobacteria;c__Actinobacteria;o__Actinomycetales;f__Nocardiaceae	0.50	0.00	0.10	0.20	1.60
k__Bacteria;p__Actinobacteria;c__Actinobacteria;o__Actinomycetales;f__Nocardioidaceae	3.10	2.00	1.60	1.80	3.70
k__Bacteria;p__Actinobacteria;c__Actinobacteria;o__Actinomycetales;f__Pseudonocardiaceae	0.80	0.70	0.10	0.40	1.30
k__Bacteria;p__Actinobacteria;c__Actinobacteria;o__Actinomycetales;f__Streptomycetaceae	1.20	0.10	0.30	0.80	1.80
k__Bacteria;p__Actinobacteria;c__MB-A2-108;o__0319-7L14;f__	3.20	2.30	0.10	0.30	1.00
k__Bacteria;p__Actinobacteria;c__Thermoleophilia;o__Gaiellales;f__Gaiellaceae	3.80	6.10	1.00	4.00	4.00
k__Bacteria;p__Actinobacteria;c__Thermoleophilia;o__Solirubrobacterales;f__	3.10	1.80	1.40	1.50	3.80
k__Bacteria;p__Actinobacteria;c__Thermoleophilia;o__Solirubrobacterales;f__Solirubrobacteraceae	0.80	0.40	0.00	0.10	1.90
k__Bacteria;p__Bacteroidetes;c__Cytophagia;o__Cytophagales;f__Cytophagaceae	1.40	3.00	2.00	0.40	2.20
k__Bacteria;p__Bacteroidetes;c__[Saprospirae];o__[Saprospirales];f__Chitinophagaceae	1.00	3.00	1.50	3.50	1.10
k__Bacteria;p__Chloroflexi;c__Ellin6529;o__;f__	2.30	2.80	2.30	0.30	2.10
k__Bacteria;p__Chloroflexi;c__Gitt-GS-136;o__;f__	1.40	1.20	0.80	0.10	0.90
k__Bacteria;p__Chloroflexi;c__Thermomicrobia;o__JG30-KF-CM45;f__	1.00	1.60	0.90	0.10	1.60
k__Bacteria;p__Firmicutes;c__Bacilli;o__Bacillales;f__Bacillaceae	0.80	1.20	1.40	2.80	0.90
k__Bacteria;p__Firmicutes;c__Clostridia;o__Clostridiales;f__Veillonellaceae	0.20	0.50	1.10	0.00	0.00
k__Bacteria;p__Gemmatimonadetes;c__Gemm-1;o__;f__	1.60	0.60	1.50	0.10	1.10
k__Bacteria;p__Gemmatimonadetes;c__Gemmatimonadetes;o__;f__	0.40	0.50	1.00	1.50	0.90
k__Bacteria;p__Nitrospirae;c__Nitrospira;o__Nitrospirales;f__0319-6A21	1.40	0.30	0.00	0.00	0.10
k__Bacteria;p__Nitrospirae;c__Nitrospira;o__Nitrospirales;f__Nitrospiraceae	0.30	0.20	0.10	0.10	0.20
k__Bacteria;p__Planctomycetes;c__Phycisphaerae;o__WD2101;f__	3.50	3.90	1.40	2.90	2.90
k__Bacteria;p__Planctomycetes;c__Planctomycetia;o__Gemmatales;f__Gemmataceae	1.10	1.30	0.80	0.20	1.00
k__Bacteria;p__Planctomycetes;c__Planctomycetia;o__Pirellulales;f__Pirellulaceae	3.20	2.40	3.00	0.50	3.80
k__Bacteria;p__Planctomycetes;c__Planctomycetia;o__Planctomycetales;f__Planctomycetaceae	0.50	0.60	0.70	0.10	1.20
k__Bacteria;p__Proteobacteria;c__Alphaproteobacteria;o__Caulobacterales;f__Caulobacteraceae	0.80	0.70	1.70	2.00	0.90
k__Bacteria;p__Proteobacteria;c__Alphaproteobacteria;o__Rhizobiales;f__Bradyrhizobiaceae	0.40	1.10	1.40	1.70	1.00
k__Bacteria;p__Proteobacteria;c__Alphaproteobacteria;o__Rhizobiales;f__Hyphomicrobiaceae	3.50	2.00	2.70	5.60	2.20
k__Bacteria;p__Proteobacteria;c__Alphaproteobacteria;o__Rhizobiales;f__Phyllobacteriaceae	0.40	0.20	0.40	0.50	1.20
k__Bacteria;p__Proteobacteria;c__Alphaproteobacteria;o__Rhizobiales;f__Rhizobiaceae	0.10	0.10	1.30	0.40	0.20
k__Bacteria;p__Proteobacteria;c__Alphaproteobacteria;o__Rhodospirillales;f__Rhodospirillaceae	1.20	1.10	0.80	1.90	0.90
k__Bacteria;p__Proteobacteria;c__Alphaproteobacteria;o__Sphingomonadales;f__Sphingomonadaceae	4.10	2.60	3.60	12.70	5.40
k__Bacteria;p__Proteobacteria;c__Betaproteobacteria;o__Burkholderiales;f__Comamonadaceae	0.80	1.60	5.10	0.90	1.10
k__Bacteria;p__Proteobacteria;c__Betaproteobacteria;o__Burkholderiales;f__Oxalobacteraceae	2.70	3.50	2.50	3.30	1.30
k__Bacteria;p__Proteobacteria;c__Deltaproteobacteria;o__Myxococcales;f__	0.80	2.90	1.80	0.30	1.00
k__Bacteria;p__Proteobacteria;c__Gammaproteobacteria;o__Pseudomonadales;f__Pseudomonadaceae	0.10	0.10	2.40	1.00	0.40
k__Bacteria;p__Proteobacteria;c__Gammaproteobacteria;o__Xanthomonadales;f__Sinobacteraceae	2.10	1.30	0.70	0.50	1.30
k__Bacteria;p__Proteobacteria;c__Gammaproteobacteria;o__Xanthomonadales;f__Xanthomonadaceae	3.20	2.30	1.60	13.50	6.60
k__Bacteria;p__TM7;c__TM7-1;o__;f__	0.50	0.20	2.10	1.30	0.20
k__Bacteria;p__Verrucomicrobia;c__[Spartobacteria];o__[Chthoniobacterales];f__[Chthoniobacteraceae]	1.40	1.50	0.70	2.20	1.70

Notes: Soil samples: 1—KMP; 2—Tulachermet; 3—Embankment; 4—Lenin Ave.; 5—Background (Control).

References

1. FAO. *The Future of Food and Agriculture*; The Food and Agriculture Organization of the United Nations: Rome, Italy, 2017; pp. 1–52.
2. Kidd, P.; Mench, M.; Álvarez-López, V.; Bert, V.; Dimitriou, I.; Friesl-Hanl, W.; Herzig, R.; Olga Janssen, J.; Kolbas, A.; Müller, I.; et al. Agronomic Practices for Improving Gentle Remediation of Trace Element-Contaminated Soils. *Int. J. Phytoremed.* **2015**, *17*, 1005–1037. [CrossRef] [PubMed]
3. Kolbas, A.; Mench, M.; Herzig, R.; Nehnevajova, E.; Bes, C.M. Copper phytoextraction in tandem with oilseed production using commercial cultivars and mutant lines of sunflower. *Int. J. Phytoremed.* **2011**, *13*, 55–76. [CrossRef] [PubMed]
4. Mench, M.J.; Dellise, M.; Bes, C.M.; Marchand, L.; Kolbas, A.; Coustumer, P.L.; Oustrière, N. Phytomanagement and remediation of cu-contaminated soils by high yielding crops at a former wood preservation site: Sunflower biomass and ionome. *Front. Ecol. Evol.* **2018**, *6*, 123. [CrossRef]
5. Witters, N.; Mendelsohn, R.O.; Van Slycken, S.; Weyens, N.; Schreurs, E.; Meers, E.; Tack, F.; Carleer, R.; Vangronsveld, J. Phytoremediation, a sustainable remediation technology? Conclusions from a case study. I: Energy production and carbon dioxide abatement. *Biomass Bioenergy* **2012**, *39*, 454–469. [CrossRef]
6. Čechmánková, J.; Vácha, R.; Skála, J.; Havelková, M. Heavy metals phytoextraction from heavily and moderately contaminated soil by field crops grown in monoculture and crop rotation. *Soil Water Res.* **2011**, *6*, 120–130. [CrossRef]
7. Mench, M.; Lepp, N.; Bert, V.; Schwitzguébel, J.P.; Gawronski, S.W.; Schröder, P.; Vangronsveld, J. Successes and limitations of phytotechnologies at field scale: Outcomes, assessment and outlook from COST Action 859. *J. Soils Sediments* **2010**, *10*, 1039–1070. [CrossRef]

8. Meers, E.; Van Slycken, S.; Adriaensen, K.; Ruttens, A.; Vangronsveld, J.; Du Laing, G.; Witters, N.; Thewys, T.; Tack, F.M.G. The use of bio-energy crops (*Zea mays*) for "phytoattenuation" of heavy metals on moderately contaminated soils: A field experiment. *Chemosphere* **2010**, *78*, 35–41. [CrossRef]

9. Vangronsveld, J.; Herzig, R.; Weyens, N.; Boulet, J.; Adriaensen, K.; Ruttens, A.; Thewys, T.; Vassilev, A.; Meers, E.; Nehnevajova, E.; et al. Phytoremediation of contaminated soils and groundwater: Lessons from the field. *Environ. Sci. Pollut. Res.* **2009**, *16*, 765–794. [CrossRef]

10. Nowack, B.; Schulin, R.; Robinson, B.H. Critical assessment of chelant-enhanced metal phytoextraction. *Environ. Sci. Technol.* **2006**, *40*, 5225–5232. [CrossRef]

11. Vamerali, T.; Bandiera, M.; Mosca, G. Field crops for phytoremediation of metal-contaminated land. A review. *Environ. Chem. Lett.* **2010**, *8*, 1–17. [CrossRef]

12. Kolbas, A.; Kidd, P.; Guinberteau, J.; Jaunatre, R.; Herzig, R.; Mench, M. Endophytic bacteria take the challenge to improve Cu phytoextraction by sunflower. *Environ. Sci. Pollut. Res.* **2015**, *22*, 5370–5382. [CrossRef] [PubMed]

13. Muratova, A.Y.; Zelenova, N.A.; Sungurtseva, I.Y.; Gorelova, S.V.; Kolbas, A.P.; Pleshakova, Y.V. Comparative Study of the Rhizospheric Microflora of Sunflower Cultivars of Helianthus annuus (*Asteraceae, Magnoliópsida*) Grown on Soils with Anthropogenic Polyelemental Anomalies. *Biol. Bull.* **2021**, *48*, 1904–1911. [CrossRef]

14. Prelac, M.; Bilandžija, N.; Zgorelec, Ž. Potencijal fitoremedijacije teških metala iz tla pomoću Poaceae kultura za proizvodnju energije: Pregledni rad. *J. Cent. Eur. Agric.* **2016**, *17*, 901–916. [CrossRef]

15. Sonowal, S.; Prasad, M.N.V.; Sarma, H. C3 and C4 plants as potential phytoremediation and bioenergy crops for stabilization of crude oil and heavy metal co-contaminated soils-response of antioxidative enzymes. *Trop. Plant Res.* **2018**, *5*, 306–314. [CrossRef]

16. Angelova, V.R.; Ivanova, R.V.; Delibaltova, V.A.; Ivanov, K.I. Use of Sorghum Crops for in Situ Phytoremediation of Polluted Soils. *J. Agric. Sci. Technol.* **2011**, *1*, 693–702.

17. Gorelova, S.V.; Gorbunov, A.V.; Frontasyeva, M.V.; Sylina, A.K. Toxic elements in the soils of urban ecosystems and technogenic sources of pollution. *WSEAS Trans. Environ. Dev.* **2020**, *16*, 608–618. [CrossRef]

18. Nsanganwimana, F.; Pourrut, B.; Mench, M.; Douay, F. Suitability of Miscanthus species for managing inorganic and organic contaminated land and restoring ecosystem services. A review. *J. Environ. Manag.* **2014**, *143*, 123–134. [CrossRef]

19. Sabeen, M.; Mahmood, Q.; Irshad, M.; Fareed, I.; Khan, A.; Ullah, F.; Hussain, J.; Hayat, Y.; Tabassum, S. Cadmium phytoremediation by Arundo donax l. from contaminated soil and water. *BioMed Res. Int.* **2013**, *2013*, 324830. [CrossRef]

20. Park, J.; Kim, K.; Kim, J.; Lee, B.; Lee, J.; Bae, B. Enhanced Phytoremediation by Echinochloa frumentacea Using PSM and EDTA in an Arsenic Contaminated Soil. In Proceedings of the International Symposia on Geoscience Resources and Environments of Asian Terranes; Chulalongkorn University: Bangkok, Thailand, 2008; pp. 487–488.

21. Abe, T.; Fukami, M.; Ogasawara, M. Effect of hymexazole (3-hydroxy-5-methylisoxazole) on cadmium stress and accumulation in Japanese millet (*Echinochloa frumentacea* Link). *J. Pestic. Sci.* **2011**, *36*, 48–52. [CrossRef]

22. Almasi, A.; Mohammadi, M.; Kazemitabar, Z.; Hemati, L. Phytoremediation potential of Echinochloa crus galli and Hibiscus cannabinus in the stabilization of municipal wastewater sludge. *Int. J. Environ. Sci. Technol.* **2021**, *18*, 2137–2144. [CrossRef]

23. Wu, L.; Li, Z.; Han, C.; Liu, L.; Teng, Y.; Sun, X.; Pan, C.; Huang, Y.; Luo, Y.; Christie, P. Phytoremediation of soil contaminated with cadmium, copper and polychlorinated biphenyls. *Int. J. Phytoremed.* **2012**, *14*, 570–584. [CrossRef] [PubMed]

24. Ehsan, N.; Nawaz, R.; Ahmad, S.; Khan, M.M.; Hayat, J. Phytoremediation of Chromium-Contaminated Soil by an Ornamental Plant, Vinca (*Vinca rosea* L.). *J. Environ. Agric. Sci.* **2016**, *7*, 29–34.

25. dos Santos, G.C.G.; Rodella, A.A.; de Abreu, C.A.; Coscione, A.R. Vegetable species for phytoextraction of boron, copper, lead, manganese and zinc from contaminated soil. *Sci. Agric.* **2010**, *67*, 713–719. [CrossRef]

26. Bhatia, R.K.; Sakhuja, D.; Mundhe, S.; Walia, A. Renewable energy products through bioremediation of wastewater. *Sustainability* **2020**, *12*, 7501. [CrossRef]

27. Kurz, H.; Schulz, R.; Römheld, V. Selection of cultivars to reduce the concentration of cadmium and thallium in food and fodder plants. *J. Plant Nutr. Soil Sci.* **1999**, *162*, 323–328. [CrossRef]

28. Schröder, P.; Mench, M.; Povilaitis, V.; Rineau, F.; Rutkowska, B.; Schloter, M.; Szulc, W.; Žydelis, R.; Loit, E. Relaunch cropping on marginal soils by incorporating amendments and beneficial trace elements in an interdisciplinary approach. *Sci. Total Environ.* **2022**, *803*, 149844. [CrossRef] [PubMed]

29. Angelova, V.; Ivanova, R.; Ivanov, K. Heavy metal accumulation and distribution in oil crops. *Commun. Soil Sci. Plant Anal.* **2004**, *35*, 2551–2566. [CrossRef]

30. Sheoran, V.; Sheoran, A.S.; Poonia, P. Phytomining: A review. *Miner. Eng.* **2009**, *22*, 1007–1019. [CrossRef]

31. Russian National Standard 53123-2008. Soil Quality. Sampling. Part 5. Guidance on the Procedure for the Investigation of Urban and Industrial Sites with Regard to Soil Contamination. Available online: https://docs.cntd.ru/document/1200074384 (accessed on 1 March 2022).

32. Russian National Standard 26483-85. Soils. Preparation of Salt Extract and Determination of Its pH by CINAO Method. Available online: https://docs.cntd.ru/document/1200023490 (accessed on 13 March 2022).

33. Malahova, S.G. Interim Guidelines on Soil Pollution Control. Available online: https://files.stroyinf.ru/Data2/1/4294847/4294847412.htm (accessed on 1 March 2022).

34. Russian National Standard 26207-91. Soils. Determination of Mobile Compounds of Phosphorus and Potassium by Kirsanov Method Modified by CINAO. Available online: https://docs.cntd.ru/document/1200023451 (accessed on 1 March 2022).

35. Russina National Standard 26489-85. Soils. Determination of Exchangeable Ammonium by CINAO Method. Available online: https://docs.cntd.ru/document/1200023496 (accessed on 1 March 2022).

36. Russian National Standard 26488-85. Soils. Determination of Nitrates by CINAO Method. Available online: https://docs.cntd.ru/document/1200023495 (accessed on 1 March 2022).

37. PND F 16.1:2:2.2:2.3:3.64-10. Quantitative Chemical Analysis of Soils. Method for Measuring the Mass Fraction of Oil Products in Samples of Soils, Soils, Bottom Sediments, Silts, Sewage Sludge, Production and Consumption Wastes by the Gravimetric Method. Available online: https://files.stroyinf.ru/Data2/1/4293807/4293807051.htm (accessed on 1 March 2022).

38. Lichtenthaler, H.K.; Wellburn, A.R. Determinations of total carotenoids and chlorophylls a and b of leaf extracts in different solvents. *Biochem. Soc. Trans.* **1983**, *11*, 591–592. [CrossRef]

39. Gavrilenko, V.F.Z.T.V. Great Workshop on Photosynthesis. Available online: https://obuchalka.org/2015052484950/bolshoi-praktikum-po-fotosintezu-gavrilenko-v-f-jigalova-t-v-2003.html (accessed on 1 March 2022).

40. Zinicovscaia, I.; Hramco, C.; Chaligava, O.; Yushin, N.; Grozdov, D.; Vergel, K.; Duca, G. Accumulation of potentially toxic elements in mosses collected in the Republic of Moldova. *Plants* **2021**, *10*, 471. [CrossRef]

41. Borisova, G.G.; Maleva, M.G.; Nekrasova, G.V.; Chukina, N.V. Methods for Assessing the Antioxidant Status of Plants: A Teaching Aid for Students Enrolled in the Undergraduate Program in the Field of Study 020400 "Biology". 2012. Available online: http://lib.urfu.ru/mns-urfu/publications/82678/?aid=402 (accessed on 30 January 2022).

42. Bertani, G. Studies on lysogenesis. I. The mode of phage liberation by lysogenic *Escherichia coli*. *J. Bacteriol.* **1951**, *62*, 293–300. [CrossRef] [PubMed]

43. GN 2.1.7.2041-06; Predel'no Dopustimye Koncentracii (PDK) Himicheskih Veshchestv v Pochve: Gigienicheskie Normativy.—M.: Federal'nyj Centr Gigieny i Epidemiologii Rospotrebnadzora. Rospotrebnadzor: Moscow, Russia, 2006. Available online: https://files.stroyinf.ru/Data2/1/4293850/4293850511.pdf(accessed on 14 March 2022).

44. GN 2.1.7.2511-09; Orientirovochno Dopustimye Koncentracii (ODK) Himicheskih Veshchestv v Pochve: GIGIENICHESKIE Norma- 852 tivy—M.: Federal'nyj Centr Gigieny i Epidemiologii Rospotrebnadzora. Rospotrebnadzor: Moscow, Russia, 2009. Available online: https://cyberleninka.ru/article/n/tyazhelye-metally-v-pochvah-svalki-promyshlennyh-othodov-kurskaya-oblast(accessed on 30 January 2022).

45. Karnaukhov, V.N. Carotenoids: Recent progress, problems and prospects. *Comp. Biochem. Physiol.-Part B Biochem.* **1990**, *95*, 1–20. [CrossRef]

46. Merzlyak, M.N. *Activated Oxygen and Oxidative Processes in Plant Cell Membranes*; Multidisciplinary Digital Publishing Institute (MDPI): Basel, Switzerland, 1989; Volume 6.

47. Chupakhina, G.N.; Romanchuk, A.Y.; Platunova, E.V. Ascorbic acid as an anti-stress factor of plants. Introduction, acclimatization and cultivation of plants. *Kaliningr. Izd-Vo KSU* **1998**, *91*, 144. [CrossRef]

48. Noctor, G.; Foyer, C.H. Ascorbate and Glutathione: Keeping Active Oxygen Under Control. *Annu. Rev. Plant Physiol. Plant Mol. Biol.* **1998**, *49*, 249–279. [CrossRef]

49. Bartoli, C.G.; Pastori, G.M.; Foyer, C.H. Ascorbate biosynthesis in mitochondria is linked to the electron transport chain between complexes III and IV. *Plant Physiol.* **2000**, *123*, 335–343. [CrossRef]

50. Yu, C.M.C.; Brand, J.J. Role of divalent cations and ascorbate in photochemical activities of Anacystis membranes. *Biochim. Biophys. Acta-Bioenerg.* **1980**, *591*, 483–487. [CrossRef]

51. Millar, A.H.; Mittova, V.; Kiddle, G.; Heazlewood, J.L.; Bartoli, C.G.; Theodoulou, F.L.; Foyer, C.H. Control of Ascorbate Synthesis by Respiration and Its Implications for Stress Responses. *Plant Physiol.* **2003**, *133*, 443–447. [CrossRef]

52. Bohinski, R. *Modern Views in Biochemistry*; Elsevier: Amsterdam, The Netherlands, 1987.

53. Ronald, J.; Su, C.; Wang, L.; Davis, S.J. Cellular Localization of Arabidopsis EARLY FLOWERING3 Is Responsive to Light Quality. *Plant Physiol.* **2022**, kiac072. [CrossRef]

54. Khan, T.; Mazid, M.; Mohammad, F. A review of ascorbic acid potentialities against oxidative stress induced in plants. *J. Agrobiol.* **2012**, *28*, 97–111. [CrossRef]

55. Smirnoff, N. The function and metabolism of ascorbic acid in plants. *Ann. Bot.* **1996**, *78*, 661–669. [CrossRef]

56. Blokhina, O.; Virolainen, E.; Fagerstedt, K.V. Antioxidants, oxidative damage and oxygen deprivation stress: A review. *Ann. Bot.* **2003**, *91*, 179–194. [CrossRef] [PubMed]

57. Chen, Z.; Gallie, D.R. Increasing tolerance to ozone by elevating foliar ascorbic acid confers greater protection against ozone than increasing avoidance. *Plant Physiol.* **2005**, *138*, 1673–1689. [CrossRef] [PubMed]

58. Shakhov, A.A. *Salt Tolerance of Plants*; Frontiers Media S.A.: Lausanne, Switzerland, 1983.

59. Garifzyanov, A.R.; Gorelova, S.V.; Ivanischev, V.V.; Muzafarov, M.E. Comparative analysis of the activity of the components of the antioxidant system of woody plants in the conditions of anthropogenic stress. *News TSU. Nat. Sci.* **2009**, *2*, 166–179. [CrossRef]

60. Gorelova, S.V.; Garifzyanov, G.A. Ascorbic acid as a factor in the anti-oxidant protection of woody plants at technogenic stress. *Tula Ecol. Bull.* **2012**, 177–183. [CrossRef]

61. Chupakhina, G.N.; Maslennikov, P.V. Plant adaptation to oil stress. *Russ. J. Ecol.* **2004**, *35*, 290–295. [CrossRef]

62. Kabata-Pendias, A. *Trace Elements in Soils and Plants*; CRC Press: Boca Raton, FL, USA, 2000; ISBN 9780429191121.

63. Bargagli, R. Trace Elements in Terrestrial Plants: An Ecophysiological Approach to Biomonitoring and Biorecovery. Available online: http://www.amazon.com/Trace-Elements-Terrestrial-Plants-Ecophysiological/dp/3540645519/ref=sr_1_1?ie=UTF8&qid=1361277936&sr=8-1&keywords=Trace+Elements+in+Terrestrial+Plants.+An+Ecophysiological+Approach+to+Biomonitoring+and+Biorecovery (accessed on 25 February 2022).

64. Kulagin, A.A.; Shagieva, J.A. *Woody Plants and Biological Preservation of Industrial Pollutants*; SciPress Ltd.: Wollerau, Switzerland, 2005.

65. Chernenkova, T.V. *The Response of Forest Vegetation on Industrial Pollution*; Nauka: Moscow, Russia, 2021.

66. Markert, B. Establishing of "Reference Plant" for inorganic characterization of different plant species by chemical fingerprinting. *Water Air Soil Pollut.* **1992**, *64*, 533–538. [CrossRef]

67. Bityutsky, N. *Trace Elements of Higher Plants*; Frontiers Media S.A.: Lausanne, Switzerland, 2011.

68. Murillo, J.M.; Marañón, T.; Cabrera, F.; López, R. Accumulation of heavy metals in sunflower and sorghum plants affected by the Guadiamar spill. *Sci. Total Environ.* **1999**, *242*, 281–292. [CrossRef]

69. Fargašová, A. Phytotoxic effects of Cd, Zn, Pb, Cu and Fe on Sinapis alba L. seedlings and their accumulation in roots and shoots. *Biol. Plant.* **2001**, *44*, 471–473. [CrossRef]

70. Rutkowska, B.; Szulc, W.; Spychaj-Fabisiak, E.; Pior, N. Prediction of molybdenum availability to plants in differentiated soil conditions. *Plant Soil Environ.* **2017**, *63*, 491–497. [CrossRef]

71. Du, J.; Zeng, J.; Ming, X.; He, Q.; Tao, Q.; Jiang, M.; Gao, S.; Li, X.; Lei, T.; Pan, Y.; et al. The presence of zinc reduced cadmium uptake and translocation in Cosmos bipinnatus seedlings under cadmium/zinc combined stress. *Plant Physiol. Biochem.* **2020**, *151*, 223–232. [CrossRef] [PubMed]

72. Il'in, V.B.; Baidina, N.L.; Konarbaeva, G.A.; Cherevko, A.S. Heavy Metals in the Soils and Plants of Novosibirsk. *Eurasian Soil Sci.* **2000**, *33*, 66–73.

73. Sun, W.; Xiao, E.; Krumins, V.; Häggblom, M.M.; Dong, Y.; Pu, Z.; Li, B.; Wang, Q.; Xiao, T.; Li, F. Rhizosphere microbial response to multiple metal(loid)s in different contaminated arable soils indicates crop-specific metalmicrobe interactions. *Appl. Environ. Microbiol.* **2018**, *84*, 701–719. [CrossRef] [PubMed]

74. Mittler, R.; Vanderauwera, S.; Gollery, M.; Van Breusegem, F. Reactive oxygen gene network of plants. *Trends Plant Sci.* **2004**, *9*, 490–498. [CrossRef] [PubMed]

75. Thounaojam, T.C.; Panda, P.; Mazumdar, P.; Kumar, D.; Sharma, G.D.; Sahoo, L.; Panda, S.K. Excess copper induced oxidative stress and response of antioxidants in rice. *Plant Physiol. Biochem.* **2012**, *53*, 33–39. [CrossRef] [PubMed]

76. Cuypers, A.; Vangronsveld, J.; Clijsters, H. Biphasic effect of copper on the ascorbate-glutathione pathway in primary leaves of Phaseolus vulgaris seedlings during the early stages of metal assimilation. *Physiol. Plant.* **2000**, *110*, 512–517. [CrossRef]

77. Marchiol, L.; Sacco, P.; Assolari, S.; Zerbi, G. Reclamation of polluted soil: Phytoremediation potential of crop-related BRASSICA species. *Water. Air. Soil Pollut.* **2004**, *158*, 345–356. [CrossRef]

78. Zhang, L.; Song, L.; Shao, H.; Shao, C.; Li, M.; Liu, M.; Brestic, M.; Xu, G. Spatio-temporal variation of rhizosphere soil microbial abundance and enzyme activities under different vegetation types in the coastal zone, Shandong, China. *Plant Biosyst.* **2014**, *148*, 403–409. [CrossRef]

79. Srivastava, V.B. Investigations into rhizosphere microflora. IV. Fungal association in different root regions of some rainy-season crops. *Acta Soc. Bot. Pol.* **1973**, *42*, 409–422. [CrossRef]

80. Sun, B.; Liu, J.; Meng, B.; Bao, K. Structural Variability and Co-Occurrence Pattern Differentiation in Rhizosphere Microbiomes of the Native Invasive Plant Echinochloa caudate in Momoge National Nature Reserve, China. *Wetlands* **2020**, *40*, 587–597. [CrossRef]

81. Malviya, M.K.; Sharma, A.; Pandey, A.; Rinu, K.; Sati, P.; Palni, L.M.S. Bacillus subtilis NRRL B-30408: A potential inoculant for crops grown under rainfed conditions in the mountains. *J. Soil Sci. Plant Nutr.* **2012**, *12*, 811–824. [CrossRef]

82. Poorniammal, R.; Senthilkumar, M.; Prabhu, S.; Anandhi, K. Effect of Methylobacterium on seed germination, growth and yield of Barnyard Millet (Echinochloa frumentacea Var. COKV 2) under Rainfed Condition. *Phytopathology* **2020**, *9*, 1675–1677. [CrossRef]

83. Lee, A.-R.; Bae, B.-H. Improved Germination and Seedling Growth of Echinochloa crus-galli var. frumentacea in Heavy Metal Contaminated Medium by Inoculation of a multiple-Plant Growth Promoting Rhizobacterium (m-PGPR). *J. Soil Groundw. Environ.* **2011**, *16*, 9–17. [CrossRef]

84. Shirokikh, I.G.; Solov'eva, E.S.; Ashikhmina, T.Y. Actinomycete complexes in soils of industrial and residential zones in the city of Kirov. *Eurasian Soil Sci.* **2014**, *47*, 89–95. [CrossRef]

85. Terekhova, V.A.; Shitikov, V.K.; Ivanova, A.E.; Kydralieva, K.A. Assessment of the ecological risk of technogenic soil pollution on the basis of the statistical distribution of the occurrence of micromycete species. *Russ. J. Ecol.* **2017**, *48*, 417–424. [CrossRef]

86. Joynt, J.; Bischoff, M.; Turco, R.; Konopka, A.; Nakatsu, C.H. Microbial community analysis of soils contaminated with lead, chromium and petroleum hydrocarbons. *Microb. Ecol.* **2006**, *51*, 209–219. [CrossRef] [PubMed]

87. Hemida, S.K.; Omar, S.A.; Abdel-Mallek, A.Y. Microbial populations and enzyme activity in soil treated with heavy metals. *Water Air Soil Pollut.* **1997**, *95*, 13–22. [CrossRef]

www.ingramcontent.com/pod-product-compliance
Lightning Source LLC
LaVergne TN
LVHW070928170726
843515LV00005B/898

MDPI

St. Alban-Anlage 66

4052 Basel

Switzerland

Tel. +41 61 683 77 34

Fax +41 61 302 89 18

www.mdpi.com

Soil Systems Editorial Office

E-mail: soilsystems@mdpi.com

www.mdpi.com/journal/soilsystems